FROM CURRENT ALGEBRA TO QUANTUM CHROMODYNAMICS

A Case for Structural Realism

The advent of quantum chromodynamics (QCD) in the early 1970s was one of the most important events in twentieth-century science. This book examines the conceptual steps that were crucial to the rise of QCD, placing them in historical context against the background of debates that were ongoing between the bootstrap approach and composite modeling, and between mathematical and realistic conceptions of quarks. It explains the origins of QCD in current algebra and its development through high energy experiments, model-building, mathematical analysis, and conceptual synthesis. Addressing a range of complex physical, philosophical, and historiographical issues in detail, this book will interest graduate students and researchers in physics and in the history and philosophy of science.

TIAN YU CAO is the author of *Conceptual Developments of 20th Century Field Theories* (1997) and the editor of *Conceptual Foundations of Quantum Field Theory* (1999), also published by Cambridge University Press.

FROM CURRENT ALGEBRA TO QUANTUM CHROMODYNAMICS

A Case for Structural Realism

TIAN YU CAO

Boston University, USA

CAMBRIDGE UNIVERSITY PRESS
Cambridge, New York, Melbourne, Madrid, Cape Town,
Singapore, São Paulo, Delhi, Mexico City

Cambridge University Press
The Edinburgh Building, Cambridge CB2 8RU, UK

Published in the United States of America by Cambridge University Press, New York

www.cambridge.org
Information on this title: www.cambridge.org/9781107411395

First published 2010
First paperback edition 2012

A catalogue record for this publication is available from the British Library

Library of Congress Cataloguing Publication Data
Cao, Tian Yu, 1941–
From current algebra to quantum chromodynamics : a case for structural realism / Tian Yu Cao.
p. cm.
Includes bibliographical references and index.
ISBN 978-0-521-88933-9
1. Quantum chromodynamics–Mathematical models. 2. Algebra.
3. Quantum chromodynamics–History. I. Title.
QC793.3.Q35C36 2010
539.7'548–dc22

2010016805

ISBN 978-0-521-88933-9 Hardback
ISBN 978-1-107-41139-5 Paperback

In memory of
Dia-dia and M-ma

Contents

Preface

This volume is the first part of a large project which has its origin in conversations with Cecilia Jarlskog and Anders Barany in December 1999, in which the difficulties and confusions in understanding various issues related to the discovery of QCD were highly appreciated.

While the forthcoming part will be a comprehensive historical study of the evolution of various conceptions about the strong interactions from the late 1940s to the late 1970s, covering the meson theory, Pauli's non-abelian gauge theory, S-matrix theory (from dispersion relation, Regge trajectories to bootstrap program), current algebra, dual resonance model and string theory for the strong interactions, QCD, lattice gauge theory, and also briefly the supersymmetry approach, the D-brane approach, and the string–gauge theory duality approach, titled *The Making of QCD*, this volume is a brief treatment, from a structural realist perspective, of the conceptual development from 1962 to 1972, covering the major advances in the current algebraic approach to QCD, and philosophical analysis of the historical movement.

The division of labor between the two parts of the project is as follows. This volume is more philosophically oriented and deals mainly with those conceptual developments within the scope of current algebra and QCD that are philosophically interesting from the perspective of structural realism; while the whole history and all the historical complexity in the making of QCD will be properly dealt with in the longer historical treatise. They will be mutually supportive but have minimal overlap and no repetition.

After the project was conceived, I have visited Santa Barbara, Princeton, Santa Fe, DESY in Hamburg, the Max Planck Institute of Munich, University of Bern, and CERN, and talked to theorists Stephen Adler, Luis Alvarez-Gaume, James Bjorken, Curtis Callan, Sidney Coleman, Richard Dalitz, Freeman Dyson, Harold Fritzsch, Murray Gell-Mann, Sheldon Glashow,

Peter Goddard, David Gross, Roman Jackiw, Heinrich Leutwyler, Juan Maldacena, Peter Minkowski, David Olive, Nathan Seiberg, Arthur Wightman, and Edward Witten, on various issues related to the project. I have learnt numerous conceptual and technical details and many deep insights from them and also from other theorists who were active in the early 1970s through email exchanges. I also had long conversations with two experimenters Jerome Friedman of MIT and Gunter Wolf of DESY, and learnt from them various details of crucial experiments which led to the discovery of scaling and three-jet events. The SLAC archives was very helpful and provided me with the whole set of the original documents without which I would have no way to know how the deep inelastic scattering experiments were actually conceived, planned, performed, and interpreted. Critical exchanges with two historians of science, Charles Gillispie of Princeton and Paul Forman of the Smithsonian Institution greatly helped me in the general conception of the project.

Over years, parts of the research were presented at Harvard, Princeton, DESY, MPI of Munich, University of Bern and other places. Most memorable was a whole-day small workshop (around 20 participants) at Princeton, on April 20, 2005, chaired by Stephen Adler, at which I reported my preliminary researches on the history of QCD. Murray Gell-Mann, Arthur Wightman, Charles Gillispie and Paul Benacerraf, together with those from the Institute, took an active part, examined various issues raised by those preliminary results, made helpful comments, provided much background information, and had interesting exchanges of judgments.

To all those institutions and scholars I owe my most sincere gratitude. My research was interrupted several times by emotional turbulences caused by my mother's death and several deaths of close relatives and friends in the last few years. In the difficult times unfailing support from my wife Lin Chun and sister Nanwei helped me recover from depression and carry the project ahead. I am deeply grateful to them.

1
Introduction

In the 1950s, all hadrons, namely particles that are involved in strong interactions, including the proton and neutron (or nucleons) and other baryons, together with pions and kaons and other mesons, were regarded as elementary particles. Attempts were made to take some particles, such as the proton, neutron and lambda particle, as more fundamental than others, so that all other hadrons could be derived from the fundamental ones (Fermi and Yang, 1949; Sakata, 1956). But the prevailing understanding was that all elementary particles were equally elementary, none was more fundamental than others. This general consensus was summarized in the notion of "nuclear democracy" or "hadronic egalitarianism" (Chew and Frautschi, 1961a, b; Gell-Mann, 1987).

As to the dynamics that governs hadrons' behavior in the processes of strong interactions, early attempts to model on the successful theory of quantum electrodynamics (or QED, a special version of quantum field theory, or QFT, in the case of electromagnetism), namely the meson theory, failed, and failed without redemption (cf. Cao, 1997, Section 8.2). More general oppositions to the use of QFT for understanding strong interactions were raised by Landau and his collaborators, on the basis of serious dynamical considerations (Landau, Abrikosov, and Khalatnikov, 1954a, b, c, d; Landau, 1955). The resulting situation since the mid 1950s was characterized by a general retreat from fundamental investigations to phenomenological ones in hadron physics. The prevailing enquiry was phenomenological because no detailed understanding of what is going on in strong interactions was assumed or even aspired to, although some general principles (such as those of crossing, analyticity, unitarity, and symmetry) abstracted from some model dynamical theories were appealed to for reasoning from inputs to outputs; thereby the enquiry enjoyed some explanatory and predictive power.

At the end of the 1970s, however, none of the hadrons was regarded as elementary any more. The unanimous consensus in the physics community and, through popularization, in the general public became as follows. First, all hadrons were composed of quarks that were held together by gluons; and second, the dynamics of quark–gluon interactions was properly understood and mathematically formulated in quantum chromodynamics (or QCD). As to the strong interaction among hadrons, it could be understood as the uncancelled residual of the quark–gluon super-strong interaction, a kind of Van der Waals force of the hadrons.

Such a radical change in our conception of the fundamental ontology of the physical world and its dynamics was one of the greatest achievements in the history of science. The intellectual journey through which the conception was remolded is much richer and more complicated than a purely conceptual one in which some ideas were replaced by others. The journey was fascinating and full of implications, and thus deserves comprehensive historical investigation. However, even the conceptual part of the story is illuminative enough to make some historical and philosophical points.

While a full-scale historical treatment of the episode is in preparation, (Cao, forthcoming) the present enquiry, as part of the more comprehensive project, has a more modest goal to achieve. That is, it aims to give a concise outline of crucial conceptual developments in the making of QCD. More precisely, its attention is restricted to the journey from the proposal of current algebra in 1962 to the conceptual and mathematical formulation of QCD in 1972–73.

As a brief conceptual history, its intention is twofold. For the general readers, it aims to help them grasp the major steps in the reconceptualization of the fundamental ontology of the physical world and its dynamics without being troubled by technical details. However, it is not intended to be a popular exposition. For experts who are familiar with the details (original texts and technical subtleties), it promises to offer a decent history, in which distorted records will be straightened, the historical meaning of each step in the development clarified, and significance properly judged, on the basis of present understanding of the relevant physics and its historical development, that is, helped by hindsight and present perspective.

The preliminary investigations pursued so far have already revealed something of deep interest, and thus provided a firm ground for making some claims about the objectivity and progress of scientific knowledge, the central topics in contemporary debate about the nature of scientific knowledge and its historical changes.

Pivotal to the debate is the status of unobservable theoretical entities such as quarks and gluons. Do they really exist in the physical world as objective entities, independently of human will, or exist merely as human constructions for their utility in organizing our experiences and predicting future events? If the former is the case, then a related question is whether we can have true knowledge of them, and how? Thus the notion of unobservable entity is central to metaphysics, epistemology and methodology of theoretical sciences.

In the debate there are, roughly speaking, two camps. One is the realist camp, and the other antirealist. Realists take the objective existence of unobservable entities for granted if these entities can consistently give us successful explanation and predictions. They may differ in how to know the entities, but all of them are optimistic in human ability to know them. As a corollary, historical changes of scientific knowledge, according to realists, are progressive in nature. That is, the change means the accumulation of true knowledge of the objective world, consisting of observable as well as unobservable entities structured in certain ways. The necessity of the unobservable entity comes from the hypothetic-deductive methodology, which, in turn, has its deep roots in human desire for explanation.

For antirealists, the status of the unobservable entity is dubious at best. Antirealists find no justification to take it as more than a fictitious device for convenience. They refute the realist argument for its objective existence, mainly the success it has brought in explanation and prediction, as being too naïve, and deploy their own more "sophisticated" arguments, one logical, the other historical, to remove the notion of unobservable entities from our basic understanding of theoretical sciences.

The logical argument is based on the notion of underdetermination. The underdetermination thesis suggested by Pierre Duhem (1906) and W. V. O. Quine (1951) claims that in general no theoretical terms, and unobservable entities in particular, can be uniquely determined by empirical data. That is, given a set of evidence, we can always construct more than one theory, each of them based on some unobservable entities as its basic ontology for explanation and predictions; while all of these theories are compatible with the evidence, the hypothetical entities assumed by these theories may have conflicting features, and thus cannot be all true to the reality.[1] Once the logical ground for inferring the reality of unobservable entities from evidence is removed, the existential status of unobservable entities can never be settled, that is, their status can only be taken as conventional rather than objective.

It has been noticed that the convincing power of the Duhem–Quine thesis rests entirely on taking unstructured empirical data (or more precisely, structured in its existent form) as the sole criterion for determining the acceptability of a hypothetical entity. Once this kind of data is deprived of such a privileged status, the simplistic view of scientific theory as consisting only of empirical, logico-mathematical, and conventional components is to be replaced by a more sophisticated one, in which a metaphysical component (e.g. one that is responsible for the intelligibility and plausibility of a conceptual framework, which is the result of, and also a foundation for, a particular way of structuring data) is also included and plays an important role in selecting acceptable unobservable entities. This would surely put scientific theories in a wider network of entrenched presuppositions of the times and in a pervasive cultural climate, and thus invites cultural and sociological studies of science to join force with history and philosophy of science in our effort to understand scientific enterprise. If this is the case, the Duhem–Quine thesis alone is not powerful enough to discredit the realist interpretation of unobservable entities.

In the last four decades, however, the antirealist has relied more heavily on its historical argument, which is based on the notion of scientific revolution, a notion that was made popular mainly by Thomas Kuhn. If the Duhem–Quine thesis accepts the existence of a multiplicity of conflicting theoretical ontologies, and thus nullifies the debate on which ontology should be taken as the true one, Kuhn rejects the reality of any theoretical ontology: if whatever ontology posited by a scientific theory, no matter how successful it was in explanation and prediction, is always replaced, through a scientific revolution, by another different and often incompatible one posited by a later theory, as the history of science seems to have shown us, and there is no coherent direction of ontological development in the history of science, how can we take any theoretical ontology as the true ontology of the world (Kuhn, 1970)? If there is no reason to believe that there will be an end of scientific revolution in the future, then, by induction, the privileged status of the unobservable entities discovered or constructed by our current successful theories has to be deprived (Putnam, 1978).

Thus the rejection of the reality of unobservable entities is reinforced by the claim of discontinuity in history of science, which takes the pessimistic induction argument just mentioned as its most combative form. A corollary is that, according to antirealists, no claim to progress could be made in terms of accumulation of true knowledge of the objective world. The true role of unobservable entities, in which our knowledge is encapsulated, is not to describe and explain what actually exists and happens in the world. Rather, they are constructed for our convenience in making successful predictions.

A difficult question for the antirealist is: why some constructions are successful and others are not? The realist argues that if the success of science is not a miracle, then the successful theory and its hypothetical, unobservable entities must have something to do with reality. If simply taking every unobservable entity in a successful theory as what actually exists in the world, for the reasons raised by the antirealist, is too naïve an attitude, then at least one can argue that the relational and structural aspects of a successful theory must be real in the sense that some similar aspects exist in the world. If this is the case, then the connection between theory and evidence can be recovered and the continuity of scientific development can be properly argued for. This is the so-called structural realist position, a more sophisticated approach to realism indeed.

Structural realism was first conceived by Henri Poincare (1902), and then deliberated by Bertrand Russell (1927), Ernst Cassirer (1936, 1944) and others. In recent decades, it has been intensely pursued by Grover Maxwell (1970a, b), John Worrall (1989), Elie Zahar (1996, 2001), Steven French (2003a, b), Tian Yu Cao (1997, 2003a, b, c), and others. In its current incarnation, structural realism takes different forms. Common to all these forms is a recognition that a structure posited or discovered by a successful theory, as a system of stable relations among a set of elements or a self-regulating whole under transformations, in contrast with unobservable entities[2] underlying the structure, is epistemically accessible, thus its reality can be checked with evidence (up to isomorphism, of course, due to its relational nature), and the objectivity of our knowledge about it is determinable.

Apparently, structural realism smacks of phenomenalism. But it can be otherwise. A crucial point here is that a structure, while describing a recognized pattern in phenomena, such as those patterns recorded and suggested by global symmetry schemes for hadron spectroscopy, may also point to deep reality, such as quarks and gluons, both in terms of a deep structure underlying the patterns in phenomena, such as the one suggested by the constituent quark model of hadrons, and also in terms of hidden structuring agents that hold components together to be a coherent whole, such as permanently confined color gauge bosons. The conceptual development that will be recounted in the following chapters will illuminate this crucial point in a convincing way.

A vexing question for structuralism in all areas, perhaps with the exception of certain branches in mathematics, is that a structure in a scientific theory has relevance to the real world only when it is interpreted, usually by specifying the nature and properties of its underlying elements. Since a structure can be interpreted in different ways, we are facing underdetermination again. In addressing this vexing question, three different positions have emerged from structural realism.

The first position, known as epistemic structural realism, takes an agnostic attitude toward underlying unobservable entity, and restricts reliable scientific knowledge only to structural aspects of reality, which is usually encapsulated in mathematical structures, without involving the nature and content of underlying entities whose relations define the structure (Worrall, 1989). With such a realistic understanding of structural knowledge, this position has somewhat addressed the pessimistic induction argument: the history of science is nothing less than a process in which true structural knowledge is accumulated, and thus is continuous and progressive in nature. However, as far as unobservable entity is concerned, this position is not different from the antirealist one. For this reason, Kuhn's original claim, that there is no coherent direction of ontological development in the history of science, is evaded rather than properly addressed, if by ontology we mean the fundamental entities in a domain of scientific investigations, from which all other entities and phenomena in the domain can be deduced.

The second position, known as ontic structural realism, is extremely radical in fundamental metaphysics and semantics (French and Ladyman, 2003a, b). It claims that only structures are real, no objects actually exist; and that the phenomenological existence of objects and their properties has to be reconceptualized purely in structural terms. For example, electric charge has to be understood as self-subsistent and a permanent relation, and elementary particles have to be understood in terms of group structures and representations. By taking structures as the only ontology in the world, Kuhn's ontological discontinuity claim is addressed, and the continuity and progress in the historical development of science can be defended. But the price for these gains is that the very notion of unobservable entity is dissolved and eliminated altogether from scientific discourse.

The third version, which may be called constructive structural realism, is much more complicated (Cao, 1997, 2003a, b; 2006). More discussion on this position will be given in Chapter 9, when the conceptual development from current algebra to QCD is clarified and analyzed. For the present purpose, it suffices to list two of its basic assumptions: (i) the physical world consists of entities that are all structured and/or involved in larger structures; and (ii) entities of any kind can be approached through their internal and external structural properties and relations that are epistemically accessible to us. Its core idea that differentiates it from other versions of structural realism is that the reality of unobservable entity can be inferred from the reality of structure. Methodologically, this suggests a structural approach to unobservable entity, as will be illustrated in the following chapters, and further elaborated in Chapter 9.

On the basis of a structural understanding of unobservable entity and of a dialectical understanding of the relationship between a structure and its components,[3] the third position can also address Kuhn's claim of ontological discontinuity by a notion of ontological synthesis, which underlies a dialectic understanding of scientific development (Cao, 1997, Section 1.4; 2003c). More generally, if a scientific revolution can be understood as a reconstruction of fundamental ontology in the domain under investigation by having reconfigured the expanded set of structural knowledge of the world, then the ontological continuity and progress in scientific development may be understood in terms of reconstructing fundamental ontology in the domain through reconfiguring the expanded set of structural knowledge in a way that is different from the ways in previous theories, and in which empirical laws can be better unified. More discussion on this point will be given in Chapter 9.

Metaphysically, the constructive version differs from the ontic version in having retained a fundamental status for entity ontology, while stressing that this fundamental ontology is historically constructed from available structural knowledge of reality. For this reason, the fundamental ontology of the world has an open texture and thus is revisable with the progress of science. This point and the more general relationship between physics and metaphysics will be examined in Chapter 10.

In addition to exemplifying how successful the structural approach is for discovering unobservable entities, such as quarks and gluons, this enquiry will also shed new light on what has been achieved in the formulation of QCD: it is more than merely a discovery of new entities and forces, but rather a discovery of a deeper level of reality, a new kind of entity, a new category of existence.[4] The enquiry would further help historians of science to understand how such a discovery was actually made through a structural approach. Essentially it takes four steps.

But before elaborating the four steps, let me comment on the structuralist understanding of current algebra. First, without a physical interpretation, a purely mathematical structure, here a Lie algebra, would have no empirical content. Second, if we interpret the Lie algebra in terms of physical structures, taking electromagnetic and weak currents as its representations, then we have physical content, but only at the phenomenological level. In order to understand the physical structures (the currents) properly, we have to move deeper onto the level of their constituents (hadrons or quarks) and their dynamics so that we can have a dynamic understanding of the behavior of the currents, and thus of many features of current algebra and of reasons why current algebra is so successful.

Driven by the recognition of this necessity, most physicists took the idea of quark realistically and tried to conceive it as a new natural kind through the structural knowledge of it.[5] The result was fruitful: there emerged a detailed picture of the microscopic world with quarks as important ingredients.

Now let me turn to the four steps I have just mentioned.

First, the notion of unobservable entities (quarks and gluons) was hypothetically constructed under the constraints of acquired structural knowledge about hadronic phenomena, such as various symmetry properties in hadron spectroscopy and hadronic weak and electromagnetic interactions, which were summarized in the achievements of the current algebra approach to hadron physics. The approach itself was based on the flavor SU(3) symmetry and a hidden assumption, implied by the infinite momentum framework adopted in current algebra calculations, that certain types of interactions among quarks during high energy processes should be ruled out.[6]

Second, the reality of some of the defining structural features of these entities, which were expressed in the current algebra results, was established by checking with experiments, such as the experiments of deep inelastic electron–proton scatterings performed at Stanford Linear Accelerator Center (SLAC). The most important features of quarks and gluons established by the observed scaling in the deep inelastic scattering experiments were their point-like nature and their lack of interactions at short distances.

Third, a coherent conceptual framework, such as QCD, is constructed to accommodate various experimental and theoretical constraints, such as the observed scaling and pion-two gamma decay rate in the former, and infrared singularity and scale anomaly in the latter.

And, finally, the distinctive implications (predictions) of the theory (such as the logarithmic violation of scaling and the three-jet structure in the electron–positron annihilation process) were checked with experiments to establish the full reality of the unobservable entities, quarks and gluons. Although these particles may not be understood as Wigner particles with well-defined spin and mass and new superselection rules based on the liberated color charge, the physical reality of these particles, according to the criteria we will elaborate in Chapter 9, is beyond doubt.

It is clear that the reality of discovered unobservable entities, here quarks and gluons, is highly theory dependent. If the theory, QCD, stands firmly with observations and experiments, the reality of quarks and gluons is confirmed. What if QCD turned out to be wrong tomorrow or in the next decade? More discussions on this interesting question will be given in Chapter 9.

The following chapters will also show that structural realism can help historians of science to make proper judgments on what steps taken were original, consequential, crucial and historically effective in the process of discovery. More elaboration on the light shed by this enquiry on the methodology of historiography will be given in the concluding chapter, but a few remarks on presentism seems to be in order before we embark on a very selective treatment of such an important episode in the history of science.

As a historical enquiry, the aim and real content of historiography of science is to clarify the historical significance of scientific endeavor rather than to popularize its cognitive meaning. Since the perceived meaning and effects of scientific explorations change dramatically with the change of perspectives over time, their long-term significance for science, metaphysics and culture in general that was discerned and understood by contributing scientists when the history was in the making is usually quite different from what critical historians understand because of the difference in perspectives. Thus scientists' judgments, even those concerning their own contributions, cannot be unreflectively taken for granted, but have to be critically assessed and properly interpreted by historians of science before they can be adopted in a historical account.

An elementary but crucial point here worth noticing is that data for historical enquiry are almost always too many and too few. Too many so that we have to select relevant, interesting and informative ones from numerous noises; too few for a meaningful picture of what actually or even plausibly happened in the past so that we have to fill the gap with our reconstructive efforts. How can a historian select events from what are available to him, reconstruct a narrative of what happened in the past, and interpret their historical significance without some guiding hypotheses, that is, without some presuppositions about what had meaningfully happened in the past, how and why an event evolved to the next, and what the overall direction is in the evolution? Thus there is simply no way for a historian to escape from taking working hypotheses and imposing a narrative structure in general, and an overall narrative direction in particular, onto a set of selected events in the past under investigation. It is historians' efforts of this kind that have fixed the meaning structure of a chosen set of past events, from which the past events are interpreted and turned out to be a history through historians' narrative.

But then the important question is where these hypothetic moves of historians come from and what the nature of these moves is. Since the intelligibility and significance of scientific events in the past lie in the message they deliver, the lesson they teach and the authority and confidence they give to current engagements in science and/or culture, any specific hypothetic move must come from a historian's response to imperatives that are immanent in the current praxis in science and/or culture and shaping the prevailing knowledge of the past and expectations for the future. Such a response in fact has defined a historian's intellectual horizon. In this sense, presentism is inescapable in any historical enquiry, and what Croce once declared is right that all history is contemporary history. Similarly, we may say that all historiographical work in science is dictated by a contemporary perspective in science and/or culture, which is particularly chosen by a historian from many investigations, some of which may be in conflict with each other.

Does the acknowledgment of inevitable presentism entail an endorsement of Whig history, a practice in historiography that involves selecting only those data that seem to point in the direction leading to the present without taking proper account of their historical context? Of course, a Whig history is not a real history, but only a distorted retrospection guided by a teleological view of history in terms of a unilinear trajectory. While a Whig history is a form of presentism, the practice of presentism may take other forms, in which the selection of events and the interpretation of their meaning are guided by views of history other than teleology.

A crucial notion whose meaning has to be clarified in this context is that of direction. It is difficult to conceive a narrative without some sense of direction. However, a progressive history or a unilinear direction of events in the past, is too speculative and too apriorist to be acceptable. Even the very notion of an internal direction of events is dubious and thus unacceptable because the direction of events, under the pressure of the contingent circumstances and idiosyncratic strategic considerations of the agents, frequently and unpredictably changes. That is, it is impossible for events in the past to have any coherent direction, and thus nothing is predictable for future developments.

Still, we can legitimately talk about a direction in historical enquiry, that is, a direction in the narrative in which events move toward the end of the narrative (such as the discovery of QCD). However, this direction only reflects the selection of the narrator, who knows the significance of each event in the past by hindsight, and thus has nothing to do with the direction of the events themselves.

It should be noted that the narrator's direction is realized only in frequent changes of the direction of events under investigation. More specifically, as will be illustrated in the following chapters, the direction of scientists' activities changes under the pressure of each stage of scientific exploration: each stage has its own major concerns and means and ways of addressing these concerns; however, sooner or later, the explorations would bring some experimental and/or theoretical breakthroughs, which would create a new context for further explorations, most likely with a different direction, and thus trigger a transition from one stage to another, ultimately led to the end of the narrative.

If the direction of the narrative is realized only in the description of changes of direction of scientists' activities, as is the case of this enquiry, and thus its status as an externally imposed artifact is duly recognized, then nothing is farther from Whig history than this form of presentism.

In this enquiry, the narrative will begin with the rise of current algebra, which was proposed by Gell-Mann in 1962, in Chapter 2. The experimental and theoretical context in which the approach was proposed will be briefly reviewed; Gell-Mann's motivation and strategy will be analyzed; the content of the proposal, including its explicit and implicit assumptions, as well as its intended and unintended implications, will be examined carefully; and its real nature will be clarified.

Chapter 3 is devoted to the major achievement of current algebra, namely Stephen Adler's sum rule for neutrino processes proposed in 1966. Conceptually, Adler's rule had provided an opportunity, the first time in history indeed, of peeking into the inside of hadrons through lepton–hadron collisions, and thus laid down the foundation for further theoretical developments, including the scaling hypothesis and the parton idea in the analysis of the SLAC experiments.

Metaphysical implications of current algebra will be clarified, through an examination of Geoffrey Chew's challenge, from the perspective of bootstrap philosophy, to Adler's and Bjorken's work, in Chapter 4. Since Chew's challenge exemplified a competition between two research programs in high energy physics at its deepest level, a level concerning the nature of scientific knowledge: descriptive versus explanatory, conceptual issues involved in this regard will also be discussed.

Chapter 5 will describe, briefly, the interesting interplay between James Bjorken's scaling hypothesis and the recognition and interpretations of the scaling phenomena in the electron–proton deep inelastic scattering experiments performed by the MIT-SLAC experimentalists at SLAC.

Chapter 6 is devoted to a much more complicated course of theory constructions. The heroic and imaginative attempts, made by Richard Feynman and Bjorken, Adler and Kenneth Wilson, Curtis Callan and Kurt Symanzik, Harald Fristzsch and Gell-Mann, at exploring ways of theorizing over the constraints posed by the new conceptual situation created by the SLAC experiments and other evidence that emerged immediately afterwards, will be clearly recounted and analyzed. The symbiosis between theoretical constructions and nature's responses to them, which is displayed in the materials of Chapters 5 and 6, will be further analyzed in Chapter 9 and 10.

The final breakthrough or, more precisely, the first clear articulation of QCD by Fritzsch and Gell-Mann in 1972 will be critically assessed in Chapter 7. Various conceptual and historiographical considerations for justifying such an assessment will also be spelled out.

Three kinds of early justification for QCD that were given soon after its first articulation will be examined in Chapter 8. The first kind addresses individual issues, most important among them are asymptotic freedom and confinement, that are crucial for satisfying the constraints posed by observations. The second kind deals with the clarification of the overall conceptual framework. The explanatory and predictive successes of charmonium and charmed particle spectrum, of the logarithmic violation of scaling, and of three-jet events in the electron–positron annihilation processes, belong to the last category, which had provided QCD with much needed experimental justifications and finally consolidated the status of QCD as an accepted theory in the high energy physics community, leaving a number of important questions to be answered in the future. Some of the early theoretical explorations into the richness of the theoretical structure of QCD, such as those related with the axial U(1) anomaly, instantons, the theta vacuum state, will also be briefly reviewed in this chapter.

Chapter 9 will use conceptual developments from the rise of current algebra to the genesis of QCD to make a case for structural realism; then try to justify what is claimed to be achieved by QCD with some arguments derived from structural realism. Lessons learnt from this enquiry on general issues in science studies will be given as concluding remarks in Chapter 10.

Notes

1 This is only one form of the underdetermination thesis, the contrastive form. It has another form, the holist form, which claims that no evidence can determine the falsity of a hypothesis in isolation, which means no meaningful test of a hypothesis would be possible. The reason for this holistic nature of testing is that, according to the holist underdetermination thesis, what confronts the evidence is not a single hypothesis, but rather a network of hypotheses, auxiliary hypotheses, and background beliefs. Thus a discordant piece of evidence typically leaves open the possibility of revising one of these background elements to save the hypothesis under testing.

2 The term "entity" in its broadest sense can mean anything real. Here and in the later discussions in this work, the term is used in its narrower sense, meaning particular things with certain identity, which may or may not be an individual. An entity in this narrow sense is in an ontological category different from those to which a property, a relation, an event, a state of affairs, or a set belongs respectively. An electron, for example, has no individuality

in conventional formulation of quantum physics. But it still belongs to the ontological category of things with certain identity (an electron is not a muon), rather than those of properties or relations. More on this in Chapter 9.

3 More on this dialectical relationship in Chapter 9.

4 Gell-Mann's mental struggle in accepting the reality of the quark shows that such an understanding is far from being trivial. More discussions on this issue will be given in Chapter 10.

5 Murray Gell-Mann discouraged others to take constituent quarks realistically for various reasons we will discuss in Chapters 2, 9, and 10; but he did take current quarks seriously and, after the formulation of QCD, realistically.

6 For more on this, see Chapter 3.

2

The rise of current algebra

Murray Gell-Mann's proposal of current algebra in 1962 (Gell-Mann, 1962) was an important step in the conceptual development of particle physics. In terms of insights, imagination, and consequences, this strategic move ranks among the highest scientific creativity in the history of fundamental physics. When the proposal first appeared, however, because of its mathematical complication and conceptual sophistication, it was perceived by most particle physicists to be too esoteric to understand or see its relevance and applicability. Even today, more than four decades later, with the advantage of hindsight, it remains difficult to properly assess its role in the history and understand why and how it played the role it did.

The difficulty lies not so much in Gell-Mann's idiosyncratic ingenuity but in, first, the puzzling situation of particle physics to which Gell-Mann's proposal was a response, and, second, in the bewildering trajectory of the subsequent development, which was shaped mainly by the competition between various speculative ideas and, ultimately, by the interplays between experiments and theorizing. The aim of this chapter is to address the first aspect of the difficulty, postponing the second to subsequent chapters.

2.1 Context

When Gell-Mann proposed his current algebra to deal with hadron physics, there was no accepted theory of strong interactions. Worse, there was no theory of *how* hadrons interacted with each other at all.

Meson theory

Ironically, a theory of strong interactions was proposed as early as 1935 by Hideki Yukawa (Yukawa, 1935). According to Yukawa, nuclear interactions, the only strong interactions known to him and his contemporaries then, should be mediated by the exchange of mesons, just as the electromagnetic interactions are mediated by the exchange of photons. Great efforts were made to find the Yukawa mesons, and after confusion related to muons, which have nothing to do with strong interactions, were clarified, they were finally found in 1947 and named π mesons (later called pions) (Lattes, Muirhead, Occhialini, and Powell, 1947; Lattes, Occhialini, and Powell, 1947). In the same year the renormalization theory won its first victory in the calculation of the Lamb shift (Bethe, 1947). Within two years, the renormalization procedure was integrated into the general framework of QED and infinite results, which frequently appeared in QED calculations, could be properly handled.[1] As a result, the previously disreputable QED, as a field-theoretical framework, was rehabilitated and conceived to be a consistent and acceptable conceptual and mathematical scheme for dealing with microscopic phenomena.

Thus it was tempting or even natural to apply the same scheme to the nuclear interactions in the form of pion–nucleon Yukawa couplings in parallel with the photon–electron electromagnetic couplings, that is, to understand the nuclear forces in terms of interacting fields. Formally, a meson theory with a pseudo-scalar Yukawa coupling should be renormalizable and thus was promising. However, an insurmountable difficulty had immediately revealed itself. To see this we only have to note that renormalization is practically possible only in a perturbation framework. This is so because the renormalization procedure involves delicate subtractions of one "infinite" term from another at each step of perturbative calculations. Yet the perturbative approach is possible only when the coupling constant is relatively small. In the case of meson theory, however, it was immediately obvious that perturbative techniques cannot work simply because the pion–nucleon coupling constant was too big ($g^2/hc \approx 15$). That is, the success of renormalization theory in QED was no help to meson theory.

Non-abelian gauge theory

A more interesting, but also failed, attempt at developing a non-abelian gauge theory of nuclear interactions, based on the observed charge-independence or isospin invariance in nuclear interactions, was made by Wolfgang Pauli in 1953. In response to Abraham Pais's suggestion of using isospin symmetry,[2] which was treated globally, and baryon number conservation to deal with nuclear interactions, Pauli asked, during the discussion following Pais's talk,

Regarding the interaction between mesons and nucleons . . . I would like to ask . . . whether the transformation group with constant phase can be amplified in a way analogous to the gauge group for electromagnetic potentials in such a way that the meson-nucleon interaction is connected with the amplified group. . .

(Quoted from Pais, 1986, p. 583)

In his subsequent letters to Pais at The Institute for Advanced Study, Princeton (July 25, 1953 and December 6, 1953; Pauli to Pais, letters [1614] and [1682] in Pauli, 1999), and his two seminar lectures given at the ETH (Swiss Federal Institute of Technology) in Zurich in the fall of 1953, Pauli generalized the Kaluza–Klein theory to a six-dimensional space and through dimensional reduction obtained the essentials of an SU(2) gauge theory.[3]

Pauli's letter of July 25, 1953 to Pais[4] and the note therein, titled "Written down July 22–25 1953, in order to see how it looks. Meson-nucleon interaction and differential geometry", was "written in order to drive you [Pais] further into the real virgin-country." Certainly he had more than Pais in mind, because in the letter he also mentioned his then favorite assistant and Pais's close friend, Res Jost, who was at the Institute then. In fact, the letter was circulated among the most talented theoretical physicists in the Institute. In addition to Pais and Jost, at least two more of their close friends also saw the letter; Freeman Dyson and Chen Ning Yang.[5]

Pauli's aim in developing the mathematical structure of a non-abelian gauge theory was to *derive* rather than *postulate* arbitrarily the nuclear forces of the meson theory therefrom. He succeeded in the derivation but was frustrated by the observation that

If one tries to formulate field equations . . . one will always obtain vector mesons with rest-mass zero. One could try to get other meson fields – pseudoscalars with positive rest mass . . . But I feel that is too artificial.

(Pauli to Pais, December 6, 1953, letter [1682] in Pauli, 1999)

Thus when Yang reported his work with Robert Mills on the same subject at a seminar in the Princeton Institute in late February 1954, Pauli interrupted Yang's talk repeatedly with a question which obviously had obsessed him: "What is the mass of this [gauge] field?" (Yang, 1983, p.20). In a letter to Yang shortly afterwards Pauli stressed again, "I was and still am disgusted and discouraged of the vector field corresponding to particles with zero rest-mass" (Letter [1727] in Pauli, 1999) because it is simply incompatible with the observed short-range behavior of the nuclear forces.

For Pauli, this zero-mass problem decisively discouraged him to take his non-abelian gauge theory seriously as a framework for nuclear forces as his initial exploration hoped for.[6] For others who were attracted by the mathematical and conceptual elegance of the idea, attempts were made to reconcile two seemingly irreconcilable sets of ideas: gauge invariance, zero mass of the gauge bosons and renormalizability on the one hand, and massive gauge bosons for short range nuclear forces on the other, without any success (see, for example, Schwinger, 1957; Bludman, 1958; Salam and Ward, 1959; Glashow, 1959; Salam, 1960; Kamefuchi, 1960; Komar and Salam, 1960; Umezawa and Kamefuchi, 1961). Still, the impact of Pauli's idea on particle physicists' thinking was deep though sometimes subtle. It set an agenda for others to pursue even when the ground for the pursuit was not properly prepared.[7] Thus it was not until the early 1970s that what was pursued, regardless of its fecundity, was understandably transmuted into something else, as we will see in the ensuing chapters. However, the paths to the transmuted would not be broken in the first place without Pauli's original agenda.

Deep conceptual difficulties of QFT

In addition to the failures in formulating a renormalizable field theory of strong interactions, even deeper arguments against quantum field theory (QFT) in general were made by Landau and his collaborators in the mid 1950s (see Landau *et al.*, 1954a, b, c, d; 1955; 1956; Landau, 1955). They claimed to have proven that QED had to have zero renormalized charge, and argued more generally that no local field theory could have any solution except for non-interacting particles. They pointed out that renormalizable perturbation theory worked only when infinite terms were removed from power series, and in general there was no nontrivial, nonperturbative and unitary solution for any QFT in four dimensions; while formal power-series solutions were generated and satisfied unitarity to each order of the expansion, the series itself could be shown not to converge. According to Landau, these conceptual difficulties that haunted QFT had their deep roots in its *foundational concepts*, such as those of *quantized local field operators, micro-space-time, micro-causality and the Lagrangian approach*, which were too speculative to be justifiable; and thus indicated inadequacy of QFT as a research framework for dealing with microscopic phenomena. The arguments Landau developed against QFT had a strong impact on some leading figures in theoretical particle physics in the late 1950s and 1960s.

Another conceptual difficulty in the field-theoretical approach to strong interactions was related to the so-called atomistic thinking, one of QFT's basic assumptions: fields described by a Lagrangian were associated with "elementary" particles. Already in the early 1930s, it was realized that despite the existence of beta decay, the neutron could not be considered as a composite of a proton and an electron; and that because it was impossible to tell whether the proton or the neutron was more elementary, each had to be regarded as equally elementary. The same idea seemed to be applicable to a profusion of new particles discovered in the 1950s and early 1960s. It was intuitively clear that these entities could not all be elementary. QFT was then forced to make a sharp distinction between elementary and composite particles. Not until the early 1960s did research on this topic appear in the literature, but no satisfactory criterion for such a distinction was given.[8] Thus, within the scope of strong interactions, it seemed pointless to explore QFT with a few particles being treated as elementary merely because they happened to be discovered first.

The S-matrix theory

Against this backdrop, a new trend manifested in the developments of dispersion relations, Regge poles, and bootstrap were orchestrated by Geoffrey Chew and his collaborators into a highly abstract scheme, the analytic S-matrix theory (SMT), which, in the late 1950s and 1960s, became a major research program in the domain of strong interactions.

The first version of SMT was proposed by Werner Heisenberg in the early 1940s (Heisenberg, 1943a, b; 1944), as an autonomous research program to replace QFT that was plagued by meaningless infinite quantities. S-matrix was defined in terms of asymptotic states and hence was a quantity immediately given by experiments, involving no infinite quantities at all. One of the purposes of Heisenberg's work on SMT was to extract from the foundations of QFT those concepts which would be universally applicable and hence contained in a future true theory, such as Lorentz invariance, unitarity and analyticity.[9] Questions posed by this program, such as "what is the connection between causality and analyticity of S-matrix elements?" and "how to determine the interaction potential of Schrodinger theory from the scattering data?" (see Jost, 1947; Toll, 1952, 1956) led to the dispersion theory of Gell-Mann and Marvin Goldberger in the early 1950s (see Gell-Mann and Goldberger, 1954; Gell-Mann, Goldberger and Thirring, 1954. For a historical review, see Cushing, 1990).

The original motivation for the dispersion theory was to extract exact results from field theory. The analyticity of the scattering amplitudes and the dispersion relations they obeyed were derived from the assumption of microcausality for observable field operators. Another principle, namely crossing symmetry, was regarded as a general property satisfied by the

Feynman diagrams representing the perturbation expansion of a field theory.[10] The results obtained in the dispersion theory, however, were quite independent of the details of field theory. Unitarity relates the imaginary part of any scattering amplitude to a total cross section that involves squares of scattering amplitudes; and causality (via analyticity) relates the real and imaginary part of the scattering amplitude to each other. In this way a closed, self-consistent set of nonlinear equations was generated that would enable physicists to obtain the entire S-matrix from fundamental principles, and free them from the conceptual and mathematical difficulties of QFT.[11]

However, as a dynamical scheme, dispersion theory lacked a crucial ingredient: a clear concept of what in it was the counterpart of the notion of force. The calculational scheme in this theory was carried out on the mass shell, and hence involved only asymptotic states, in which particles were outside each other's regions of interaction. Thus the crucial question for enabling the scheme to describe the interaction is to define an analytic function representing the scattering amplitude. The original dispersion relations formulated by Gell-Mann, Goldberger, and Thirring explicitly exhibited particle poles, which were mathematically necessary for defining an analytic function, since otherwise, by Liouville's theorem, the latter would be an uninteresting constant. Initially, however, no emphasis was given to this feature of the theory, let alone a proper understanding of the role played by the singularities of the S-matrix, namely a role in providing a concept of "force" in the new program which was different from that in QFT.

Progress in this direction was made in three steps. First, by employing analytic functions and adopting Goldberger's relativistic dispersion relations, the Chew–Low theory of pion–nucleon scattering allowed more direct contact with experimental data. The poles of the analytic functions were the key to such contact. The position of the pole was associated with the particle mass, and the residue at the pole was connected with "force strength" (coupling constant). Making this semi-relativistic model fully relativistic led to a new concept of force: ***Force resided in the singularities of an analytic S-matrix***. Furthermore, the dispersion relations themselves were to be understood as Cauchy–Riemann formulae expressing an analytic S-matrix element in terms of its singularities (Chew, 1953a, b; Low, 1954; Chew and Low, 1956; Goldberger, 1955a, b; Chew, Goldberger, Low and Nambu, 1957a, b).

The second step involved Mandelstam's double dispersion relations, which extended the analytic continuation in the energy variable to the scattering angle, or using Mandelstam's variables, in s (the energy variable in s-channel) to t (the invariant energy in the t-channel or, by crossing, the momentum transfer variable in the s-channel). An application of these relations to the crossing symmetry, which was understood by its discoverers, Gell-Mann and Goldberger, as a relation by which incoming particles became outgoing antiparticles, converted it into a new dynamical scheme in which the ***force in a given channel resides in a pole of the crossed channel*** (Gell-Mann and Goldberger, 1954; Mandelstam, 1958).

This new scheme led to the bootstrap understanding of hadrons. The double dispersion relations enabled Chew and Mandelstam to analyze pion–pion scattering in addition to pion–nucleon scattering. They found that a spin-1 bound state of two pions, later named the ρ meson, constituted a force which, by crossing, was the agent for making the same bound state; that is, the ρ meson as a "force" generated the ρ meson as a particle. The bootstrap concept was introduced as a result of this calculation. It did not take long for Chew and Frautschi to suggest that, similar to the ρ meson case, all hadrons were bound states of other hadrons sustained by "forces" represented by hadron poles in the crossed channel. The self-generation of all hadrons by such a bootstrap mechanism could be used, hopefully, to self-consistently and uniquely determine all of their properties (Chew and Mandelstam, 1960; Chew and Frautschi, 1961a, b).

The Mandelstam representation was proven to be equivalent to the Schrödinger equation in potential scattering theory (Blankenbecler, Cook and Goldberger, 1962), and the Schrödinger equation could be obtained as an approximation to dispersion relations at low energies (Charap and Fubini, 1959). This lent support to the conjecture that an autonomous program that had a well-established nonrelativistic potential scattering theory as its limiting case could replace QFT. In this program, *the dynamics were not specified by a detailed model of interaction in spacetime*, but were determined by *the singularity structure of the scattering amplitudes*, subject to the requirement of maximal analyticity. The latter required that no singularities other than those demanded by unitarity and crossing be present in the amplitude (Chew and Frautschi, 1961a, b). A further difficulty remaining in these bootstrap calculations was posed by the problem of asymptotic conditions. For potential scattering, however, the difficulty was settled by Tullio Regge.

This leads us to the third step. In proving a double dispersion relation in potential theory, Regge was able to show that it was possible to continue the S-matrix into the complex energy (s) plane and the complex angular momentum (j) plane simultaneously. Consequently, the position of a particular pole in the j-plane was an analytic function of s, and a fixed pole ($\alpha(s) = \text{constant}$) was not allowed. If at some energy ($s > 0$) the value of Re $\alpha(s)$ passed through a positive integer or zero, one had at this point a physical resonance or bound state for spin equal to this integer. So, in general, the trajectory of a single pole in the j-plane as s varied corresponded to a family of particles of different js and different masses. When $s < 0$,[12] however, the Regge trajectory $\alpha(s)$ was shown by Regge to control *the asymptotic behavior of the elastic-scattering amplitude*, which was proportional to $t^{\alpha(s)}$, with $\alpha(s)$ negative when s was sufficiently negative. Therefore, only in those partial waves with $l < \alpha(s)$ could there be bound states. This implied that the trajectory for $s < 0$ could be detected in the asymptotic behavior of the t-channel, and, most importantly, that the divergence difficulty that plagued QFT could at last be avoided (Regge, 1958a, b; 1959; 1960).

These results were noted by Chew and Frautschi and were assumed to be true also in the relativistic case. The resulting extension of the maximal-analyticity principle from linear ("first kind") to angular momentum ("second kind") implied that the ***force in a channel resides in the moving singularity of the crossed channel***, and made it necessary to restate the bootstrap hypothesis as ***all hadrons, as the poles of the S-matrix, lie on Regge trajectories***. (Chew and Frautschi, 1960, 1961c).

The bootstrap hypothesis provided a basis for a dynamic model of hadrons. In this model, all hadrons could be regarded as either composite particles or as constituents, or binding forces, depending on the specific process in which they are involved. The basic concern was not with the particles, but with their interaction processes, and *the question of the structure of hadrons was reformulated in terms of the structure of hadron-reaction amplitudes*.[13]

As a fundamental framework opposed to QFT, SMT's central ideas and principles were mainly provided by Landau and Chew. Landau stressed the failures of QFT in dealing with any interactions, strong interactions in particular, and argued vehemently against QFT as a framework for dealing with microscopic phenomena. He was the first to recognize the generality of the correspondence between hadrons and the poles in the analytic S-matrix (cf. Chew, 1989 and Capra, 1985). In addition, Landau also formulated the graphical representation and rules for S-matrix singularities.[14] The Landau graph technique was distinct from Feynman's, and was by no means equivalent to perturbation theory. Contributions to a physical process from all relevant particles were included in the singularities,[15] and lines in the graph corresponded to physical hadrons representing asymptotic states, so no renormalization had to be considered. The Landau graphs representing physical processes thus became a new subject of study and an integral part of the foundation of SMT.[16] In fact, Landau's paper (1959) on graph techniques became the point of departure for an axiomatic study of analytic SMT developed in the 1960s (see Stapp, 1962a, b; 1965; 1968; Chandler, 1968;

Chandler and Stapp, 1969; Coster and Stapp, 1969; 1970a, b; Iagolnitzer and Stapp, 1969; and Olive, 1964).

However, the unifying force and real leader of SMT was Chew. Chew passionately argued against the atomistic thinking adopted by QFT, and resolutely rejected the idea of arbitrarily designating some particles as elementary ones. His anti-field-theoretical position had two components: bootstrap and maximal analyticity. According to bootstrap, all hadrons are composite, and all physical meaningful quantities can be self-consistently and uniquely determined by the unitarity equations and the dispersion relations; no arbitrarily assigned fundamental quantities should be allowed. The hypothesis of maximal analyticity entails that the S-matrix possesses no singularities other than the minimum set required by the Landau rules, or that all of the hadrons lie on Regge trajectories. Since the latter property was generally understood to be characteristic for composite particles, Chew argued that the Regge hypothesis provided the only way to realize bootstrap. Since low-mass hadrons, particularly the nucleons, were found to lie on Regge trajectories, Chew was encouraged to break with the traditional belief that the proton was elementary, and to claim that the whole of hadron physics would flow from analyticity, unitarity, and other SMT principles.

For Chew, QFT was unacceptable mainly because it assumed the existence of elementary particles which inevitably led to a conflict with bootstrap and analyticity. In addition to his own arguments, Chew's rejection of elementary particles was also supported by previous works in which the difference between a composite particle and an elementary particle was rejected so far as scattering theory was concerned: all particles gave rise to the same poles and cuts irrespective of their origin (see Nishijima, 1957, 1958; Zimmermann, 1958; and Haag, 1958). Chew could even appeal to Feynman's principle that a correct theory should not allow a decision as to which particles are elementary.[17] For these reasons, together with some of SMT's successes in hadron phenomenology,[18] Chew exerted great pressure upon field theorists and strongly influenced the hadron physics community from the early 1960s. No physicists working on hadron physics in the 1960s could avoid responding to Chew's challenge to atomistic thinking.

Composite models

The atomistic thinking, however, was quite resilient all the time. Although it was difficult to tell whether the proton or neutron was more elementary, when pions were discovered in 1947, however, their elementarity was soon questioned, and an effort was made in 1949 by Enrico Fermi and Yang to conceive them as nucleon–antinucleon composites (Fermi and Yang, 1949), assuming the nucleons as the basic building blocks of matter. With the taking shape of the strangeness scheme in the 1953–1955 period,[19] enlarged composite models were suggested by Maurice Goldhaber and Shoich Sakata in 1956, with the strangeness-carrying particle as the new fundamental particle.[20] The composite model enjoyed some success in the classification of hadrons, although its success in the case of mesons was tarnished by its failure in the case of baryons.

The Sakata model occupied a prominent place in the history of particle physics for at least two reasons. First, it provided the ontological basis (a "logic of matter") and direct stimulus for the exploration of U(3) symmetry (a "logic of form") among hadrons (Ogawa, 1959; Ikeda, Ogawa and Ohnuki, 1959; see also Klein, 1959; and Ohnuki, 1960), which was the starting point of a series of consequential developments as we will see in the next section. Second, when the fundamentality cannot be maintained at the hadron level in terms of nucleons, lambda particles, and their antiparticles, Sakata pointed out in early 1963 that it could be easily moved to a deeper level, the level of the so-called Ermaterie (primordial matter) on which the existence of three kinds of Urmaterie (urproton, urneutron, and urlambda particle, which correspond to proton, neutron, and lambda particle) was assumed,[21] and that hadrons produced by these urbaryons could be taken as Regge poles; this was in consistency with Chew's vision about the hadron, although its metaphysical underpinning was radically different (Sakata, 1963; more on Chew's vision in Chapter 4).

However, without new insights at the fundamental level of particle dynamics – how the composites (hadrons) are dynamically formed from its ingredients (urmaterie) – the composite model could not go very far. Long-term impact aside, its immediate achievement was mainly to give a heuristic account for a selection rule, namely the Nishijima–Gell-Mann rule for the processes in which the strange particles were involved. Although Sakata aspired "to find out a more profound meaning hidden behind this rule," the supposed ontological basis for explaining the empirically successful rule was too *ad hoc* and too speculative to be taken realistically.

Surely it was a calling for physicists to find out the material basis and underlying mechanism for the observed behavior patterns of hadrons, the Nishijima–Gell-Mann rule included. But the mission was to be accomplished not through direct speculations about what ingredients underlie the patterns, but through a series of steps in the spirit of a sophisticated structural realist methodology, as the present enquiry attempts to show, although the insistence of the existence of a deeper layer, whose properties and dynamical behavior would underlie the observed pattern at the hadron level, was surely one of the necessary conditions for the mission to be accomplished. Speculations were not of much help. But the positivist attitude of exclusively adhering to the experimentally and theoretically accessible level at the time, as the S-matrix theorists insisted, without any attempt to find out ways of penetrating into a deeper level, would also foreclose any substantial progress in knowing the structurally layered reality.

Symmetries

A less "profound" (or less speculative) but empirically more productive way of dealing with the proliferation of elementary particles and the discovery of particles' new behavior patterns was to appeal to the notion of *quantum number* that characterizes the known particles: old quantum numbers include the electronic charge and spin, and the new ones include isospin and strangeness.

The notion of quantum number can serve as a basis for the notion of *multiplet*. The members of a multiplet share the same value of a quantum number but, in the cases of a non-scalar quantum number, differ in the value of its components. For example, in the isospin doublet of nucleons, proton and neutron have the same isospin, 1/2, but a different third component of the isospin: the proton's is 1/2, and the neutron's is $-1/2$. The notion of multiplet in turn serves as a basis for a *classification* scheme, in which different multiplets with different values of the quantum numbers can be arranged in an orderly way.

The notion of quantum number can also be used to express the conservation (or non-conservation) law (and the related symmetry or symmetry-breaking) of any particlar quantum number in interactions among elementary particles. An important observation made by Pais in 1953 was about the existence of a hierarchy of interactions, strong, electromagnetic, and weak interactions, with progressive less symmetry (less quantum numbers being conserved) (Pais, 1953). Although the deep reason and actual mechanisms for symmetry-breaking were unknown to him, such a hierarchy certainly provided a rationale for observation-based selection rules.

Mathematically, a conservation law points to the existence of an invariance or a symmetry, and vice versa. For example, the conservation of electric charge points to a U(1) symmetry of electromagnetic interactions; and the charge-independence or isospin invariance of strong interactions points to the conservation of isospin in strong interactions. As to the strangeness, it is conserved in strong and electromagnetic interactions, but not in weak interactions.

In the symmetry approach to hadron physics, the isospin invariance of strong interactions was a point of departure. As we mentioned earlier, the invariance of strong interactions under the isospin group was treated by Pais globally, and by Pauli locally as a non-abelian gauge invariance for specifying the dynamics. Pauli's idea was attractive, but its applicability was hindered by the zero-mass problem. When the concern was mainly on the hadron spectrum (classification and mass formulae) and selection rules in hadron processes, however, the global symmetry approach carried the day. After Pais's first attempt at enlarging the isospin group to accommodate other quantum numbers in 1953, various attempts within the global symmetry scheme were made for a similar purpose (such as Schwinger, 1957; and Gell-Mann, 1957). Most important among them was the one first conceived by Shuzo Ogawa and Oskar Klein, and then articulated by Mineo Ikeda, Ogawa and Yoshio Ohnuki in 1959 (the OK-IOO scheme for short) (Ogawa, 1959; Klein, 1959; Ikeda, Ogawa and Ohnuki, 1959; see also Ohnuki, 1960).

The basic idea of the OK-IOO scheme was simple: as far as strong interactions were concerned, three fundamental hadrons, proton, neutron and lambda particles behaved as if they were the same particles, that is, there was a full symmetry, a three-dimensional unitary symmetry, among them. This was simply an extension of the isopin SU(2) symmetry of the nucleons to a U(3) symmetry for the Sakata particles. The novelty of the scheme, however, lies mainly in its group theoretical approach to the study of hadron physics.

In the OK-IOO scheme, the mathematical structures of the U(3) group were constructed, and the hadrons and their resonances were classified by using the irreducible representations of the U(3) group. According to this approach, π mesons and K mesons were treated as two body systems of the fundamental triplet and their antiparticles and thus belonged to an eight-dimensional representation of U(3) group, together with an iso-singlet meson having the same spin and parity. The existence of such an iso-singlet meson was unknown in 1959, but it was found in 1961 and named η meson (Pevsner *et al.*, 1961). "The prediction of the η meson was the most brilliant success of this approach" (Maki, Ohnuki and Sakata, 1965). But the case for baryons was much more complicated. In the spirit of the Sakata model, it treated proton, neutron, and lambda particle as the fundamental triplet (a three-dimensional representation) of U(3) group. Then it shared the same difficulties with the Sakata model in finding acceptable ways to represent other baryons as the composites of the fundamental particles.

In terms of physics, the OK-IOO scheme as the group-theoretical version of the Sakata model, or the symmetrical Sakata model, had all the strengths and weaknesses of the original Sakata model. But its group-theoretical approach and its mathematical constructs of the U(3) symmetry were adopted by many others and thus made a direct impact on subsequent developments, most importantly on Gell-Mann's thinking of a new scheme with a higher but broken unitary symmetry.[22]

The eightfold way

As a response to the difficulties encountered by the symmetrical Sakata model, Gell-Mann and Yuval Ne'eman proposed a new group-theoretical scheme, the so-called eightfold way, for the classification of hadrons in 1961 (Gell-mann, 1961; Ne'eman, 1961).

The major novelty of the eightfold way, as compared with the symmetrical Sakata model, lay in its way of assigning irreducible representations to low-mass baryons and mesons. In the symmetrical Sakata model, elementary particles (proton, neutron, and lambda particle) formed a fundamental triplet and belonged to the three-dimensional fundamental representation of U(3) group; all other particles were composites of the elementary particles and their antiparticles, and thus belonged to the presentations formed from the fundamental presentation and its conjugate. In the eightfold way, however, the unitary symmetry was treated in an *abstract* way, namely, had no reference to the elementary baryons. This opened the door for different assignments of hadrons to irreducible representations of the unitary group.

More specifically, the low-mass baryons with spin 1/2 (nucleons, lambda, sigma and xi particles) were treated as having equal status and assigned to the same irreducible representation of the unitary group as an octet.[23] But the more striking difference with the symmetrical Sakata model was in its treatment of mesons. In general, mesons were not treated as the composites of elementary baryons, and their representation assignments were constrained only by their *invariant couplings* with baryons. Since baryons were prescribed to the octet representation 8, the mesons coupled to them must have one of the following representations: $8 \times 8 = 1 + 8 + 8 + 10^* + 10 + 27$ (10^* as the conjugate of 10). In the case of vector mesons, their status as elementary particles was derived from the idea that they were gauge bosons of the unitary symmetry,[24] and thus belonged to an octet representation due to their necessary couplings with the unitary symmetry currents.[25] Also constrained by their invariant couplings to the octet baryons, the known low-mass pseudoscalar mesons (pions, kaons, and a missing one, which Gell-Mann called chi but later was named eta) were assigned an octet representation under unitary symmetry.[26]

Each irreducible representation gives a supermultiplet of particles which, in the limit of unitary symmetry, are degenerated. The symmetry limit was obviously an unrealistic situation, and the symmetry was in fact badly broken. But Gell-Mann's aim was to extract useful physical information, namely structural knowledge about the relevant hadrons, from the broken symmetry. For this purpose, he derived mass formulae, sum rules and selection rules – and then submitted them to experimental examinations – from a hierarchy of symmetry breakings that were hinted at by phenomenological observations, although he had no idea about the mechanisms responsible for symmetry breakings. For example, once the mass splitting was taken into consideration, a unitary supermultiplet would break up into the known isospin multiplets, leaving inviolate only the conservation of isotopic spin (which would be further broken by electromagnetism), of strangeness (which would be further broken by the weak interactions), and of baryons and electric charge. Such a hierarchical view of symmetry-breaking in fundamental interactions of hadrons naturally led Gell-Mann to speculate "about the possibility that each kind of interaction has its own type of gauge and its own set of vector particles and that the algebraic properties of these gauge transformations conflict with one another." How to deprive the gauge bosons of strong interactions from hadron's weak and electromagnetic interactions (currents) would have bothered Gell-Mann until his clear formulation of QCD in 1972.

The eightfold way enjoyed impressive phenomenological successes in hadron physics. For example, to first order in the violation of unitary symmetry, the masses were predicted to obey the sum rule $(m_N + m_\Xi)/2 = (3m_\Lambda + m_\Sigma)/4$, which agrees surprisingly well with observations, the two sides differing by less than 20 MeV. In addition, it also gave rise to the selection rules in the weak interactions of hadrons, $|\Delta I| = 1/2$ and $\Delta S/\Delta Q = 0, +1$, naturally. But the most important contribution of the eightfold way was to break a new path in exploring the physical consequences of a *broken symmetry* in a unifying way, namely to apply it to physical situations where the strong, weak and electromagnetic interactions of hadrons are inseparably intertwined. But before examining this kind of exploration in the form of current algebra, let us have a closer look at the conceptual situation in the domain of weak interactions.

Weak interactions

Substantive progress in understanding the nature and structure of the weak interactions was inspired by the advance in theoretical analysis and experimental confirmation of parity violation in weak interactions.[27] In Fermi's original theory, β-decay was written in vector (V) current form. The observation of parity violation in 1957 required an axial vector (A) coupling. Assuming maximal violation of parity, Robert Marshak and George Sudarshan, and also Feynman and Gell-Mann, proposed in 1958 that the weak currents took a V-A form (Marshak and Sudarshan, 1958; Feynman and Gell-Mann, 1958). A universal Fermi theory, in which the weak interaction Hamiltonian was written as a *product of weak currents*, was also suggested and adopted by the particle physics community as an effective theory, even though the nature and constitution of the hadronic weak currents were only a subject for speculation.

However, once *currents* were assumed to be the *basic entities* responsible for weak processes, a door was open for further investigations of their (external and internal) *structural properties*. This kind of *phenomenological* investigation with a seemingly evasive nature, as we will see, would pave the way for *model-theoretical* researches, which inevitably pointed to the investigations of *underlying mechanisms and their ontological basis*.

CVC and form factors

In accordance with the idea first speculated but discarded by Sergey Gershtein and Yakov Zel'dovich,[28] and with the more sophisticated idea of Schwinger's,[29] Feynman and Gell-Mann assumed that the vector part of the weak current, together with the isotopic vector part of the electromagnetic current, form a triplet of conserved currents under the isospin transformations.

The assumption was known as the hypothesis of conserved vector current (CVC). According to CVC, the physical weak currents of hadrons were part of a representation of the SU(2) isospin symmetry of strong interactions.[30] Here the hadronic currents involved in electromagnetic, weak and strong interactions were supposed to be well-defined Lorentzian four-vectors. They might be constructed out of fundamental hadron fields, but nobody knew how to analyze them field-theoretically in a reliable way. However, these currents were supposed to interact with each other through the exchange of some unknown vector quanta.

The matrix elements of hadron currents to the lowest order in the coupling constant are direct observables, obeying the dispersion relations and being supposed to be dominated by one pole. Putting aside the kinematical factors which are easy to determine, the rest of the matrix element can be expressed, by using symmetry principles, in terms of a few unknown scalar functions of four-momentum transfer, the so-called *form factors*, such as the electromagnetic form factors $F_i(q^2)$ ($i = 1, 2, 3$) defined by

$$\left\langle p' \middle| {}_a j_\mu^{em}(0) \middle| p \right\rangle_b = u(p')_a \left[F_1(q^2)(p'+p)_\mu + F_2(q^2)(p'-p)_\mu + F_3(q^2)\gamma_\mu \right] u(p)_b \quad (2.1)$$

where $q^2 = (p'-p)^2$, and the weak form factors $f_v(q^2)$, $f_m(q^2)$, $f_s(q^2)$, $g_A(q^2)$, $g_P(q^2)$, and $g_E(q^2)$, defined by

$$(2\pi)^3 \langle p' |_{s'p} V_\mu(0) | p \rangle_{sn} = iu(p')_{s'p} \left[\gamma_\mu f_V(q^2) + \sigma_{\mu\nu} q_V f_m(q^2) + i q_\mu f_S(q^2) \right] u(p)_{sn} \quad (2.2)$$

and

$$(2\pi)^3 \langle p' |_{s'p} A_\mu(0) | p \rangle_{sn} = iu(p')_{s'p} \left\{ \gamma_5 \left[\gamma_\mu g_A(q^2) + i q_\mu g_P(q^2) + i(p'+p)_\mu g_E(q^2) \right] \right\} u(p)_{sn} \quad (2.3)$$

where V_μ and A_μ are the weak vector current and weak axial vector current respectively, and $q_\mu = (p'-p)_\mu$.

These form factors were supposed to contain all physical information about the modifications of basic weak (or electromagnetic) interactions by virtual strong interactions. There was no way to predict the quantitative details of form factors, because these were determined by the dynamics of strong interactions which was not available in the 1950s and 1960s. Nevertheless, these form factors can still be studied effectively. First, the symmetry considerations put strong constraints on them, limiting their number and establishing correlations among them. Second, the pole-dominance hypothesis helps to provide concrete suggestions concerning the contributions to the form factors by virtual particles. For example, the existence of a vector meson was first conjectured by Nambu in 1957 to explain the electron–proton scattering form factors,[31] and then, in 1961, confirmed by experimental discovery (Maglic, Alvarez, Rosenfeld and Stevenson, 1961). Third, since the form factors are observables, all the analyses based on symmetry (and/or analyticity) considerations can be tested by experiments.

The original motivation of CVC was to give an explanation of the observation that the vector current coupling constant $G_v (= G f_v(0))$ in nuclear β-decay was very close to the vector current coupling constant G in muon β-decay. The observation implied that the renormalization factor $f_V(0)$[32] for the vector current coupling constant G was very close to one, or that G was not renormalized by strong interactions.[33] As part of the isotopic vector current, the charged vector current V_μ, which raises electric charge in β-decay, can be identified with the isotopic raising current $J_{\mu+}^v = J_{\mu 1}^v + iJ_{\mu 2}^v$, and V_μ^+ with the isotopic lowering current $J_{\mu-}^v = J_{\mu 1}^v - iJ_{\mu 2}^v$, whereas the isovector part (j_μ^v) of the electromagnetic current $j_\mu^{em} (= j_\mu^s + j_\mu^v)$ can be identified with the third component, $J_{\mu 3}^v$, of the same isotopic vector current triplet.

This way, CVC suggestively made it possible to relate the known electromagnetic processes to some weak processes through isotopic rotations. For example, since for a pion at rest

$$\langle \pi^+ | J_{03}^V = j_0^{em}(I=1) | \pi^+ \rangle = 1, \tag{2.4}$$

the relevant S-matrix element for pion β-decay must be

$$< \pi^0 | V_0 | \pi^+ > = 2^{1/2}. \tag{2.5}$$

Thus the scale of the weak vector current can be fixed.

Another example is the relation between the form factors for the vector part of nuclear β-decay and those of electron scattering of neutron and proton:

$$F_i(q^2)^+ = F_i(q^2)^p - F_i(q^2)^n.^{34} \tag{2.6}$$

From these relations, Gell-Mann (1958) made a definite prediction about "weak magnetism" $f_m(0)$:

$$f_m(0) = F_2(0)^P - F_2(0)^N = (\mu_P - \mu_N)/2M = (1.79 + 1.91)/2M = 3.70/2M \tag{2.7}$$

which was subsequently confirmed in 1963 (Lee, Mo and Wu, 1963).

The CVC hypothesis occupies a prominent place in the history of particle physics. Most importantly, it suggested, though not explained, the connections among the weak, electromagnetic and strong interactions, thus revealing their unitary character: The vector part V_μ in the weak currents as representations of the SU(2) isospin symmetry of the strong interactions, were inseparably connected with the total electromagnetic currents under the operations of the same symmetry.

PCAC

Then what about the axial part A_μ in the weak current $J_\mu(= V_\mu - A_\mu)$? The experimental value of the renormalization factor $g_A(0)$[35] for the axial coupling constant deviated, although not by much, from unity ($g_A/f_V \approx 1.25$). Such a renormalization effect, although mild, for the axial coupling, together with several other experimental constraints, made it impossible to conceive the axial current A_μ as a conserved one.

The real impetus to the study of the axial current, however, came from a dispersion-relation treatment of the pion decay by Goldberger and Sam Treiman (1958). In their treatment, the absorptive part of the pion decay constant f_π was regarded as a function of the dispersion variable q^2 (the square of the off-shell pion mass) and involved transitions from off-shell pion to a complete set of intermediate states (which was dominated by the nucleon–antinucleon pair state, and coupled through the axial current to the leptons). The final result was

$$f_\pi = Mg_A/g_{\pi NN}. \tag{2.8}$$

where M is the nucleon mass, g_A the axial vector coupling constant in the nucleon β-decay, $g_{\pi NN}$ the pion–nucleon strong interaction coupling constant.

In addition to its excellent agreement with experiment, the Goldberger–Treiman relation (2.8) also exhibits a connection between the quantities of strong and weak interactions. This remarkable success, which had nothing to do with symmetry or the conservation of current, challenged theoreticians to provide an explanation or derivation of it from persuasive hypotheses, and inspired researches along various directions which finally led to the PCAC (an acronym for partially conserved axial current) hypothesis.

One of the popular approaches to PCAC was initiated by Gell-Mann and his collaborators (1960a, b).[36] Working on field-theoretical models, such as the gradient coupling model, the σ-model, and the nonlinear σ-model, they constructed, following the example of John C. Taylor (1958), axial currents whose divergence was proportional to a pion field:

$$\partial A^i_\mu/\partial x_\mu = \mu^2 f_\pi \varphi^i. \tag{2.9}$$

The crucial hypothesis they made in investigations was that the divergence of the axial current was a gentle operator. This means that its matrix elements satisfy the unsubtracted dispersion relations (with the pion mass variable as its dispersion variable) and vary slowly; at low frequencies they were dominated by contribution from one pion intermediate state, and at high frequencies they were vanishing, which justified the claim that the axial current was almost conserved.

The gentleness hypothesis is conceptually independent of symmetry consideration. Its importance lies in allowing a continuation of the matrix elements of the axial current from physical pion mass shell to off-shell ones. Frequently interesting statements can be made when the off-shell pion mass variable goes to zero. The hypothesis gives a hope that these statements may remain more or less true when the variable moves back to the physical mass shell. This way, the hypothesis enjoyed some predictive power and thus can be verified by experiments.

Practically, the PCAC hypothesis had its applications often intertwined with the idea of current algebra because of its involvement in the low energy pion physics (see below). Yet it did have some independent tests. For example, Stephen Adler showed in 1964 that assuming the validity of CVC and PCAC, the matrix element M of a neutrino-induced weak reaction, $\nu + p \rightarrow l + \beta$ (here β is a hadron system), which can be written as

$$M_{q2\rightarrow 0} \sim G < \beta \,|\, \partial A_{\mu}^{1+i2}/\partial x_{\mu} \,|\, p >, q = P_{\nu} - P_{l}. \quad (2.10)$$

and the matrix element M_{π} of the strong pion reaction $\pi^{+} + \mathrm{p} \rightarrow \beta$, which, by using the reduction formula, can be written as

$$M_{\pi} = \lim_{q2\rightarrow -\mu 2} [(q^{2} + \mu^{2})/(2)^{1/2}] < \beta \,|\, \varphi^{1+i2} \,|\, p > . \quad (2.11)$$

(where q is the pion momentum and $\varphi^{1+i2}/2^{1/2}$ is the π^{+} field operator), were related to each other (through PCAC: by using Eq. (2.9) and continuing M_{π} off mass shell in q^{2} to $q^{2} = 0$, which is supposed to result in an amplitude that is not very different from the physical pion amplitude) in the following way:

$$|M|^{2}_{q2\rightarrow 0} \sim G^{2} f_{\pi}^{\ 2} |M_{\pi}|^{2}. \quad (2.12)$$

The relation entails that the energy, angle, and polarization distributions of the particles in β are identical in the reaction $\nu + p \rightarrow l + \beta$ and in the reaction $\pi^{+} + p \rightarrow \beta$. The implication is experimentally testable and can be used to test PCAC (Adler, 1964).

The above example, similar to the Goldberger–Treiman relation, exhibited another relation between the weak and strong interactions. In fact, PCAC led to a whole class of relations connecting the weak and strong interactions. These relations allow one to predict the weak interaction matrix element $< \beta|\partial_{\mu} J_{\mu}^{A}|\alpha >$ if one knows the strong interaction transition amplitude $T(\pi^{+} + \alpha \rightarrow \beta)$. By using this kind of connection, Adler showed in 1965 that in certain cases where only the Born approximation contributes and $< \beta|\partial_{\mu} J_{\mu}^{A}|\alpha >$ can be expressed in terms of weak and strong coupling constants, the weak coupling constants can be eliminated and a consistency condition involving the strong interactions alone can be obtained. A celebrated consistency condition for pion–nucleon scattering obtained by Adler is a nontrivial relation among the symmetric isospin pion–nucleon scattering amplitude $A^{\pi N(+)}$, the pionic form factor of the nucleon $K^{NN\pi}$, and the renormalized pion–nucleon coupling constant g_{r}:

$$g_{r}^{\ 2}/M = A^{\pi N(+)}(\nu = 0, \nu_{\beta} = 0, k^{2} = 0)/K^{NN\pi}(k^{2} = 0), \quad (2.13)$$

[M is the nucleon mass, $-k^{2}$ is the $(\text{mass})^{2}$ of the initial pion, $\nu = -(p_{1} + p_{2}) \cdot k/2M, \nu_{\beta} = q \cdot k/2M, p_{1}, p_{2}$, and q are the four-momenta of the initial and final nucleon and the final pion] which was shown to agree with experiment to within 10 percent (Adler, 1965c).

2.2 Gell-Mann's proposal of current algebra

In the early 1960s, the notion of currents was widely recognized as a convenient instrument for conceptualizing physical interactions, as we have seen in the summary of the last few pages about the universal Fermi theory, CVC and PCAC. However, whereas leptonic currents could be derived from a Lagrangian field theory, for hadronic currents, there was no acceptable Lagrangian field theory to start with. One difficulty in having a field theory for hadrons was caused by hadrons' "population explosion" in the 1950s. Since no decision could be made as to which hadrons were elementary and which were not, no decision could be made about the field content in the hadronic section of a Lagrangian.[37] As a result, it was not clear as to which hadrons, and how many, would contribute to a hadronic current. Thus the nature, constitution and structure of hadronic currents could only be the subject for conjecture. In addition, the absence of vector bosons that were supposed to be coupled with hadronic currents made the role of the currents quite mysterious.

However, Gell-Mann noticed that the difficulties resulted from the ignorance about hadronic currents could be bypassed or bracketed by taking hadronic currents as primary entities, rather than derivatives from a fundamental field theory. Along this line, model field theories might be used, but the purpose would only be to abstract properties the hadronic currents might have, the most important among them being their symmetrical properties suggested by empirical data. This evasive move or phenomenological approach was somewhat justified by the fact that the matrix elements of the hadronic currents can be treated by dispersion relations and thus, together with the assumption of pole-dominance, were direct observables that can be submitted to further experimental testing and theoretical investigations.

Yet, Gell-Mann pointed out, "homogeneous linear dispersion relations, even without subtractions, do not suffice to fix the scale of these matrix elements; in particular, for the nonconserved currents, the renormalization factors cannot be calculated, and the universality of strength of the weak interactions is undefined. More information than just the dispersion relations must be supplied" (Gell-Mann, 1962). What Gell-Mann proposed to supply the information needed was an esoteric scheme called current algebra.

The underlying idea of current algebra was the notion of an *approximate unitary symmetry*[38] *SU(3)* of strong interactions. The notion was first explored in the eightfold way, as we discussed earlier, but then it was applied mainly to the study of hadron spectroscopy, in terms of approximately degenerate supermultiplets, such as the octets of baryons and mesons. Currents were mentioned, but only briefly in the context of CVC; their role in constituting the approximate symmetry, however, was not recognized, and the consequences of their symmetrical properties were not explored.

Mathematically, a symmetry can be expressed in two ways. First, it can be expressed in terms of conservation laws. Yet the utility of conservation laws for hadron physics was quite limited. The main trouble with this approach lies in the fact that most symmetries involved in hadron physics are approximate ones, and the problem of determining how approximate they are cannot be solved by looking at conservation laws. This problem is related with another physically more precise one, namely the problem of how to fix the scale of matrix elements of the currents, which are representations of some approximate symmetry, and thus are subject to renormalization by the non-symmetrical part of strong interactions. Conceptually, a clarification of the meaning and implications of the approximate symmetries is the prerequisite for properly addressing this problem.

In an attempt to make such a clarification and address relevant physical problems, Gell-Mann adopted another way to express symmetry, namely in terms of a Lie algebra, which is satisfied by the generators of a symmetry group and closed under equal-time canonical commutation (ETC for short) relations.

Historically, Gell-Mann's approach marked a third phase of the use of commutators in modern physics. In the first phase, the canonical, non-relativistic commutator between the position q and the momentum $p \equiv \delta L/\delta q, i[p,q] = (h/2\pi i)\ I$ was introduced by Heisenberg in 1925 as an expression for quantization condition. This commutator is independent of the specific form of the Lagrangian L and can be used to derive useful interaction-independent results such as the Thomas–Kuhn sum rule. The second phase was associated with Thirring's approach to express the microcausality in terms of the commutator of two field operators, which should vanish for spacelike separations of their arguments. In form, Gell-Mann's way of defining symmetries in terms of commutator algebra resembles Thirring's commutator definition for causality, because both of them are adapted to contemporary formulation of QFT. In spirit, however, Gell-Mann was closer to Heisenberg rather than to Thirring because the set of commutator algebra was designed to exhibit interaction-independent relations in relativistic dynamics, which can be used to extract physical information without solving the theory.

Mathematically, if there is a finite set of n linearly independent Hermitian operators R_i, and a commutator of any two R_i is a linear combination of the R_i:

$$[R_i(t), R_j(t)] = ic_{ijk}R_k(t), \tag{2.14}$$

the system is called the Lie algebra of an n-parameter continuous group, which is defined by a set of n infinitesimal unitary operators $1 + i\varepsilon R_i(t)(i = 1, \ldots \mathrm{n})$.

Such an abstract mathematical system is physically interesting only when the generators $\mathrm{R_i}$ are physically interpreted. But this was not a difficult task for Gell-Mann. For example, in the case of CVC, the conservation of the weak vector currents $J^v_{\mu i}$ implies that the associated weak charges $Q^v_i(= -i\int J^v{}_{4i}d^3x)$ are the generators of SU(2) weak isospin symmetry group. Taking the system of the leptonic currents as a model in which the leptonic weak charges can be easily shown, by explicit calculations, to satisfy ETC relations

$$\left[Q^{vl}_i(x_0)\ Q^{vl}_j(x_0)\right] = i\varepsilon^{ijk}Q^{vl}_k(x_0), \quad i,j,k = 1,2,3 \tag{2.15}$$

and the corresponding ones

$$[J^{vl}_{4i}(x,t), J^{vl}_{4j}(x,t)] = -i\varepsilon^{ijk}\ J^{vl}_{4k}(x,t)\delta(x-x'), \tag{2.16}$$

for the densities, Gell-Mann suggested that the hadronic weak charges $Q^v{}_i$ generate a similar SU(2) Lie algebra satisfying similar ETC relations

$$\left[Q^v_i(x_0), Q^v_j(x_0)\right] = i\varepsilon^{ijk}Q^v_k(x_0), \quad i,j,k = 1,2,3 \tag{2.17}$$

and the corresponding ETC relations

$$[J^v_{4i}(x,t), J^v_{4j}(x',t)] = -i\varepsilon^{ijk}J^v_{4k}(x,t)\delta(x-x'), \tag{2.18}$$

for the densities.[39]

The ETC relations (2.17) and (2.18) were in fact the starting point of current algebra. It was generated by the charges of the strangeness-preserving charged weak vector currents and of the isovector part of the electromagnetic current. The extension to the entire charged weak vector current, including strangeness-changing terms, and the entire electromagnetic current, based on the approximate SU(3) symmetry of the strong interactions,[40] or on the more sophisticated Cabibbo model (Cabibbo, 1963), was mathematically a straightforward step, but conceptually a bold conjecture about the symmetrical properties of hadronic currents without the empirical support of conservation laws similar to the conservation of isospin charges in strong interactions.

In the extended system of ETC relations of hadronic currents, all weak and electromagnetic vector currents of hadrons, including the isoscalar electromagnetic current, the triplet of isovector electromagnetic and weak strangeness-conserving currents, and the two doublets of weak strangeness-changing currents, belong to the same octet of currents, and the associated charges $Q^v_\alpha (= -i \int J^v{}_{4\alpha} d^3x)$ generate a larger SU(3) Lie algebra satisfying ETC relations

$$[Q^v_\alpha(x_0), Q^v_\beta(x_0)] = if_{\alpha\beta\gamma} Q^v{}_\gamma(x_0), \quad \alpha, \beta, \gamma = 1, 2, \ldots\ldots 8 \tag{2.19}$$

where $f_{\alpha\beta\gamma}$ are the SU(3) structure constants.

A similar proposal was also made about the weak axial vector currents $J^A{}_{\mu\alpha}$ and the associated charges $Q^A{}_\alpha$, which satisfy the ETC relations

$$[Q^V_\alpha(x_0), Q^A_\beta(x_0)] = if_{\alpha\beta\gamma} Q^A_\gamma(x_0) \tag{2.20}$$

$$[Q^A_\alpha(x_0), Q^A_\beta(x_0)] = if_{\alpha\beta\gamma} Q^V{}_\gamma(x_0).^{41} \tag{2.21}$$

Note that (2.21) not only closed the system, but also made $Q^+{}_\alpha = (Q^V{}_\alpha + Q^A{}_\alpha)/2$ and $Q^-{}_\alpha = (Q^V{}_\alpha - Q^A{}_\alpha)/2$ two commuting charges. The $Q^+{}_\alpha$ and $Q^-{}_\alpha$ can be used to define a chiral $SU(3)_L \times SU(3)_R$ algebra for the chiral symmetry of the strong interactions, which satisfy ETC relations

$$[Q^+_\alpha(x_0), Q^+_\beta(x_0)] = if_{\alpha\beta\gamma} Q^+_\gamma(x_0), \tag{2.22}$$

$$[Q^-_\alpha(x_0), Q^-_\beta(x_0)] = if_{\alpha\beta\gamma} Q^-_\gamma(x_0), \tag{2.23}$$

and

$$[Q^+_\alpha(x_0), Q^-_\beta(x_0)] = 0. \tag{2.24}$$

The corresponding relations for densities can be expressed in a way similar to (2.18):

$$[J^+_{4\alpha}(x, t), J^+_{4_\beta}(x, t)] = -if_{\alpha\beta\gamma} J^+_{4\gamma}(x, t)\delta(x - x'), \tag{2.25}$$

$$[J^-_{4\alpha}(x, t),\ J^-_{4\beta}(x', t)] = -if_{\alpha\beta\gamma} J^-_{4\gamma}(x, t)\delta(x - x'), \tag{2.26}$$

and

$$[J^+_{4\alpha}(x, t),\ J^-_{4\beta}(x', t)] = 0. \tag{2.27}$$

This esoteric scheme of current algebra was what Gell-Mann recommended to supplement the dispersion relations for calculating the matrix elements of the weak and electromagnetic currents.

It is worth noting that the currents in current algebra were characterized by their ETC relations rather than the symmetry itself. A subtle yet crucial difference between these two ways of characterization is this. Formally, the ETC relations depend only on the structure of the currents when the currents were regarded as functions of the canonical field variables. If the symmetry-breaking terms in the Lagrangian involve no derivative couplings, then independent of dynamical details, which nobody knew in the 1960s, the currents will retain the original structure, and their ETC relations will remain unchanged. In this way, the vector and axial vector currents, even when the chiral symmetry of the strong interactions was broken, would still remain as the octet representations of the SU(3) $\times$ SU(3) chiral symmetry, not in the sense that their charges are conserved, in fact they are not, but in the sense that their charges satisfy the ETC relations of the chiral symmetry. Thus a precise though abstract meaning was specified for the notion of approximate symmetries, in terms of ETC relations, which now can be regarded as an exact property of strong interactions.

Gell-Mann's novel understanding of the approximate symmetry allowed exact relations among measurable quantities to be derived from a broken symmetry. These relations can be used to fix the scales of matrix elements of currents and to connect the matrix elements of the axial currents with those of the vector currents. In addition, the hypothesis of universality, concerning the strength of the leptonic and hadronic weak interactions, also acquires a precise meaning: the ETC relations for the total leptonic weak current are the same as those for the total hadronic weak current. With the help of algebraic relations involving the symmetry-breaking term u_0, the deviation of the axial coupling from unity, namely its renormalization, can also be calculated. Thus, a workable research program in hadron physics was available to the physics community after Gell-Mann's classic papers (1962, 1964a,b) were published. The hard core of the program is the notion of the approximate chiral symmetry SU(3) $\times$ SU(3), which Gell-Mann suggested "as the one most likely to *underlie the structure of the system of baryons and mesons*" (1962, my emphasis).

In this program, a central topic was the processes involving pions. This was so mainly because the matrix elements of pions in the chiral limit ($m_\pi \to 0$) can be calculated by a direct application of ETC relations of relevant currents combined with the PCAC notion of pion pole dominance for the divergence of the axial current. For example, the matrix elements of the process ($i \to f + \pi$) are related, by using PCAC and the reduction formula, to

$$< f\,|\partial^\mu A^a_\mu|\, i> = ik^\mu < f|\, A^a_\mu|\, i>, \tag{2.28}$$

and the processes involving two soft pions can be studied by analyzing the matrix element

$$< f\,|T\partial^\mu j^a_\mu(x)\partial^\nu j^b_\nu(0)|i> . \tag{2.29}$$

In pulling out the differential operators from the time-ordered product symbol T, we obtain, in addition to a double divergence of a matrix element of two currents, an equal-time commutator of the two currents, which turns out to be crucial for the calculation of the matrix element.

2.3 Current algebra and quark model

The broken symmetry SU(3) × SU(3) of strong interactions, as the underlying idea of current algebra, was first explored in the eightfold way. The advantage of using such an ***abstract*** approach, in comparison with the more ***concrete*** symmetrical Sakata model, was its flexibility in assigning the low mass baryons to an octet with equal status without any discrimination of the fundamental triplet (proton, neutron, and the lambda particle) from other baryons as composites. But if any version of mathematical positivism or mysticism was to be rejected, a justification of such an abstraction, or "some ***fundamental explanation*** of the situation"[42] would be desirable. It was such a desire that drove Gell-Mann to the idea of quarks.[43]

In Gell-Mann's approach,[44] the algebraic properties of hadron interactions were abstracted "from a formal field theory model," and the latter "must contain some basic set of entities, with non-zero baryon number, out of which the strongly interacting particles can be made." Then a question crucial to the fundamental explanation was what this basic set should be. If this set were a unitary octet, the underlying symmetry group would be SU(8) instead of SU(3), and the theory would be very clumsy. And thus Gell-Mann felt tempted "to try to use unitary triplets as fundamental objects."

A unitary triplet consists of an isotopic singlet s of electric charge z (in units of e) and an isotopic doublet (u, d) with charges z + 1 and z respectively. The anti-triplet has the opposite signs of the charges. Complete symmetry among the members of the triplet gave the exact eightfold way, and this was what was desired from a triplet-based field theory as an explanation of current algebra, while a mass difference between the isotopic doublet and singlet gave the first order violation.

Among various schemes of constructing baryon octets from the basic triplet was one that allowed non-integral values for the charges. It was the most elegant scheme since it could dispense entirely with the basic neutral baryon (which was necessary for integral charges in the construction) if the triplet was assigned the following properties: spin 1/2, z = −1/3, and baryon number 1/3. The members $u^{2/3}$, $d^{-1/3}$, and $s^{-1/3}$ of the triplet have been referred to ever since as "quarks" q, and the members of the anti-triplet as anti-quarks $\bar{q}$. Baryons can be constructed from quarks by using the combinations (qqq), $(qqqq\bar{q})$, etc., while mesons can be made out of $(q\bar{q})$, $(qq\bar{q}\bar{q})$, etc. It was assumed that the lowest baryon configuration (qqq) gave just the representations 1, 8, 10 that had been observed, while the lowest meson configuration ($q\bar{q}$) similarly gave just 1 and 8.

A formal mathematical model based on field theory for the quarks was exactly the same as the one for proton, neutron, and the lambda particle in the old Sakata model, "with all strong interactions ascribed to a neutral vector meson field interacting symmetrically with the three particles." Within this framework, the electromagnetic current (in units of e) and the weak current took the forms

$$i\{2/3\bar{u}\gamma_\alpha u - 1/3\bar{d}\gamma_\alpha d - 1/3\bar{s}\gamma_\alpha s\} \tag{2.30}$$

and

$$i\bar{u}_\alpha(1+\gamma_5)(d\cos\theta + s\sin\theta) \tag{2.31}$$

respectively [θ is the Cabibbo angle (Cabibbo, 1963. See also Gell-Mann and Levy, 1960) characterizing the mixture of strange-conserving and strange-changing components in the weak current], and thus "all the features of Cabibbo's picture of the weak current, namely the rules $|\Delta \boldsymbol{I}| = 1$, $\Delta \boldsymbol{Y} = 0$ and $|\Delta I| = 1/2$, $\Delta Y/\Delta Q = +1$, the conserved $\Delta \boldsymbol{Y} = 0$ current with coefficient $\cos\theta$, the vector current in general as a component of the current of the F-spin, and the axial vector current transforming under SU(3) as the same component of another octet" could be obtained.

Most importantly, "the equal-time communication rules for the fourth components of the [vector and axial vector] currents," which yield the group of SU(3) $\times$ SU(3), can also be obtained. This implies that current algebra can be abstracted from the field theory model of quarks, or the field-theoretical model of quarks had provided the demanded fundamental explanation of the current algebra.

It is worth stressing that when Gell-Mann conceived the idea of quarks, he did not conceive them as real building blocks (or constituents) of hadrons. A major reason, among many others, for inhibiting him from such a conception was that no idea of forces that would hold quarks together to be hadrons was available to him, and without such an idea of gluing forces, the very notion of constituents would make no sense at all. Rather, he conceived quarks as the conceptual foundation for a mathematical model, "which may or may not have anything to do with reality"; the important thing for him was to "find suitable algebraic relations that hold in the model, postulate their validity, and then throw away the model."

Guided by this methodological principle, even though "quarks presumably are not real" but "purely mathematical entities" in the limit of infinite mass and infinite binding energy or infinite potential, Gell-Mann still wanted to "use them in our field theory model" to abstract presumably valid algebraic relations.

It should be noted that such a usage of quarks was structuralist but not purely instrumentalist in nature. To see this clearly, let us turn to Gell-Mann's other conception of quarks: the non-appearance of *mathematical* quarks "could certainly be consistent with the *bootstrap* idea, and also possibly with a theory containing *a fundamental triplet which is hidden*, i.e., has effective infinite mass. Thus, without prejudice to the independent questions of whether the bootstrap idea is right and whether *real triplets* will be discovered, we may use a mathematical field theory model containing a triplet in order to abstract algebraic relations [my emphasis]."

Here the boundary line between ***purely mathematical*** entities and (presumably permanently) ***hidden but real*** physical entities was not clearly drawn. The fuzziness of the dividing line became the source of confusions, and heated debates about the nature and reality status of quarks ensued.[45]

The debate was mainly between the advocates of the so-called constituent quark model on the one side and Gell-Mann and his followers on the other. The former group held a kind of entity realist position and claimed that the constituent quarks were real particles or real constituents of hadrons; while the latter group was sticking to the idea of current quarks, to their mathematical (or at least hidden) nature, and thus was hesitant to give credit to the idea of constituent quarks.[46]

Conceptually, the notion of constituent quark is quite different from that of current quark. Current quark is a notion within the framework of quantum field theory, whereas constituent quark is a nonrelativistic notion. What current quarks describe are elementary excitations of the quark field, or the bare quark degrees of freedom, which can serve as a mathematical basis for current algebra, and are consistent with the bootstrap idea of hadrons; but what constituent quarks describe are phenomenological entities, or the dressed quarks, which can be used additively to build up hadrons: the same phenomenological quarks are used for baryons and mesons, which explained the ratio of 3/2 between nucleon–nucleon and meson–nucleon scattering (Levin and Frankfurt, 1965), and helped in understanding a great portion of hadron spectroscopy (Dalitz, 1966, 1967). While the innate physical reality of current quarks as objects of experience was not clear until the establishment of QCD, the constituent quarks were claimed to be very real rather than a mere mnemonic device. At the risk of simplification, the difference can be grasped as one between phenomenological and the fundamental. The achievement of the constituent quark model was real and impressive, but the path to QCD was broken by the notion of current quarks.

An explicit assumption about current quarks was that they were fractionally charged. The charge conservation law implies that different from the case of integral charges, the fractionally charged quarks cannot be separately found in any known baryons or mesons. Gell-Mann appreciated such a confinement of quarks greatly because it made his idea of quarks compatible with the bootstrap idea of hadrons, which in fact was one of the background assumptions of current algebra.

Gell-Mann believed that the correct dynamical description of hadrons required either the bootstrap theory or the current quark model. "In neither case do the familiar neutron and proton play any basic role." Thus in a sense, the difference was irrelevant and no prejudice in favor of one and against the other should be taken seriously. But what Gell-Mann did not realize when he proposed the notion of quark was that ontologically the notion pointed downward to the inside of hadrons, and that such an implication was not acceptable to the bootstrap theorist, as we will see in Chapter 4.

Historically, the theoretical primacy was shifted, significantly, from current algebra to current quarks. When the quark model was proposed, it was only a means for justifying current algebra. But with the unfolding of the ontological implications of current quarks, current algebra gradually became a means for structurally probing current quarks: first through the external structural relations of hadrons described by current algebra, and then through internal ones revealed by current algebra analysis, as we shall see in later chapters.

2.4 Remarks

A clear understanding of current algebra, as of any other esoteric scheme in mathematical physics, requires a firm grasp of its characteristic features. Thus it is desirable to highlight these features in separate remarks.

Empirical roots

Logically speaking, the point of departure for current algebra was the approximate symmetry $SU(3) \times SU(3)$. However, the very idea was conjectured on the basis of indirectly observed patterns manifested in hadron spectroscopy and mass formulae, which motivated the initial conjecture of the eightfold way.

The idea of further exploiting symmetrical properties of a dynamic system without recourse to its dynamical details, of which nobody had any knowledge in the late 1950s and 1960s, was suggested by the universal occurrence of physical currents in hadron physics since the late 1950s. Mathematically, the connection between symmetry and conserved current was made well known by Noether's work: a conserved current followed from a symmetry group of a dynamical system and constituted a representation of the symmetry group. Although physical currents in current algebra were taken to be the primary entities for investigations, and regarded as representations of a symmetry group, instead of being derived from a dynamical system, they in fact came first from empirical descriptions of hadrons' weak and electromagnetic interactions. It was not true that the approximate symmetry that had underlain current algebra was chosen a priori, but rather, it was the conservation or partial conservation of physical currents, such as CVC or PCAC, suggested by exact or approximate regularities appearing in the experimental data that justified the selection of the symmetry.

Phenomenological utilities

In current algebra, the matrix elements of currents were supposed to describe observable processes that took place among real particles. The description depended on two phenomenological moves. First, all unknown effects of strong interactions involved in the process were bracketed into empirical parameters, such as form factors. Second, all unknown intermediate states that might be relevant to the process were summed over in sum rules. With the help of such moves, current algebra, combined with dispersion relations, was useful in calculating the matrix elements of hadronic (weak and electromagnetic) currents.

The adherence to directly observable processes (the corresponding S-matrix elements and form factors, etc.) gave current algebra a clear phenomenological character, and helped it to circumvent all the difficulties caused by the introduction of unobservable theoretical structures in local field theories.

Local field theory as its matrix

The previous sections have shown that the whole scheme of current algebra was extracted from local field theory, and can be studied and explored with field-theoretical models.

According to Gell-Mann's notion of current quarks, the current operators were supposed to be constructed out of local field operators of quarks, and could be manipulated in the same way that the field operators were manipulated in local field theory. For example, a current operator can be inserted into two physical states to express its action upon the initial state, which causes a transition to the final state.

Formal manipulations

In current algebra, symmetry properties of a local field theory were taken to transcend dynamic details of the theory and have validity beyond the first few orders of perturbation. This defining feature of current algebra gave manipulations of algebraic relations (current commutators, etc.) a formal character, which not only simplified real situations and made them manipulatable, but also appeared to be universally valid.

A structuralist breed

What current algebra described were abstract *structural* features of a local field theory, which were hoped to be valid in a future true theory. The whole scheme was a mathematical structure among currents, which expressed the structures of physical processes of some physical entities (hadrons or current quarks) instead of being physical entities themselves, and thus can be regarded as a structure of structures. But then what was the physical relevance of such a mathematical structure? Physically, it was only a *conjecture* to be tested. It should be stressed that although the ingredients of structures at both levels (currents themselves and hadrons or current quarks) were uncertain when the scheme was proposed, the scheme was taken to be the underlying structure of the hadron system, and thus to be an expression of the external structural features of hadrons. Further discussion of the last point will be given, in the form of deliberations on its application to the conceptualization of current quarks.

Inherent limitations

The neglect of unobservable entities and processes in general, and of microscopic dynamics in particular, in current algebra had hampered progress in deeply understanding hadron physics. Moreover, without a proper understanding of dynamics, there was no guarantee that purely formal manipulations, which often inclined to oversimplify real situation, were universally valid. In fact, anomalous behaviors of current commutators, or deviations from *formal* manipulations of current algebra, were discovered as early as 1966, only two years after Gell-Mann's proposal (Johnson and Low, 1966; for more detailed discussion, see section 8.7 of Cao, 1997). These findings suggested that one of the basic assumptions of current algebra, namely that current commutators were independent of dynamical details, might be *wrong*. Great efforts had been made to clarify the situation, which soon led to field-theoretical investigations of current algebra (Adler, 1969; Bell and Jackiw, 1969). The result was the profound discovery of *anomalies* in some local field theories, which had numerous implications. Most important among them, in the concerns of this book, was to provide a conceptual basis for properly understanding the renormalization group equations, as we will see in Chapter 6.

Hints of deep reality

When the question of the structure of hadrons was addressed in the 1950s and 1960s, there was a sharp contrast between the phenomenological approach of the S-matrix theory, in which the question was reformulated in terms of the analytical structure of hadron scattering amplitudes, and the symmetry approach, in which a global symmetry was taken to underlie the structure of the hadron system.

In the symmetry approach, more elaborately, the patterns in hadron spectroscopy and hadron scatterings recorded and suggested by global symmetry schemes in fact pointed to deeper reality in two forms. In the first form, the symmetry pointed to deeper entity structures, such as those suggested by the symmetrical Sakata model and constituent quark model, which underlay the patterns in hadron phenomenology.

In the second form, the symmetry underlay a conceptual movement from the exploration of the nature and constitution of hadronic current to the exploration of the constitution of hadrons, which was exemplified by the movement triggered by current algebra. The ignorance of hadronic currents was replaced by the assumption of current algebra that the weak and electromagnetic currents of hadrons are representations of the symmetry of strong interaction, which was suggested and somewhat justified by the phenomenological success of CVC and PCAC.

The introduction of quarks for underlying the symmetry broke a new path for further movements in which quark currents[47] were used to probe into hadrons, as we will see in later chapters. The idea of quarks summarized all knowledge and information about hadron physics accumulated in the previous decade, and thus was empirically solid. But hidden in the idea was an assumption that certain types of interactions among quarks during the high energy processes are absent, which was consequential and was realized only much later. In fact, when the eightfold way of the unitary symmetry was first introduced by the analogy to the case of leptons, which had no strong interactions at all (Gell-Mann, 1961), it was already implicitly smuggled into the scheme. Later, as we will see in Chapters 3, 4, and 6, when the infinite momentum frame and the idea of partons were introduced so that in the phenomenological applications of current algebra, low energy complexity caused by quark interactions could be avoided, quarks were effectively free from certain types of interactions or, in the case of partons, even from any interaction at all. Conceptual difficulties of various kinds related to this assumption and the struggles for properly dealing with it finally led to the discovery of structuring agents at the same level of quarks that hold quarks together to be hadrons, as we will see in later chapters.

Notes

1 For the details, see Cao, 1997, section 7.6.
2 It was made at the Lorentz-Kamerlingh Onnes centenary Conference in Leiden, June 22–27, 1953. For the details, see Pais, 1986, 583–587. I am grateful to Professor Peter Minkowski who made additional relevant materials available to me in October 2005, and then helped me to interview Professor Norbert Straumann, a disciple of Pauli's former assistant Res Jost, in July 2006; to Professor Norbert Straumann for interviews and relevant materials; and also to Professor Freeman Dyson for email exchanges and materials. See also Paolo Gulmanelli, 1954; Straumann, 2000; Dyson's letter to Res Jost (October 10, 1983). A fuller discussion on this subject will be given in Cao, forthcoming.
3 The major ideas and technical details were given in the note attached to his first letter to Pais, additional details were reported in Gulmanelli, 1954.
4 This was preceded by another letter dated July 3, 1953, in which Pauli mentioned a possible answer in terms of a generalized Kaluza–Klein theory. See Pais, 1986, p.584.
5 Yang stated in his (1983): "He had written in German a rough outline of some thoughts, which he had sent to A. Pais. Years later F. J. Dyson translated this outline into English" (p. 20). In his letter to Jost (1983), Dyson stated: "As you can see, on page 11 there are a few words in German. The rest was written by Pauli in English. Thus the statement on page 20 of Frank Yang's Selected Papers that the original was in German is incorrect. I do not have any record of the date of my handwritten copy, but it was probably done in 1953."
6 For this reason, Pauli did not even bother to publish his results. This is another manifestation of Pauli's behavioral pattern. Compare this with Pauli's other famous non-publication of his ingenious idea, namely his idea of neutrino suggested in 1930 (see Brown, 1978).
7 Sakurai, 1960. Note that at the level of hadrons, there is simply no gauge invariance. Only when physicists' exploration penetrated into the next layer, a fertile ground for pursuing Pauli's idea was broken.
8 The earliest research on this topic occurred in 1957–58. See Nishijima, 1957, 1958; Zimmermann, 1958; Haag, 1958. The result was that there was no difference between a composite particle and an elementary particle as far as the theory of scattering was concerned. Later came the conjecture of "$Z = 0$" as a criterion for the compositeness of particles, which was first suggested and verified for the Lee model (Lee, 1954) in Houard and Jouvet, 1960 and Vaughin, Asron, and Amado, 1961.
9 Kramers (1944) suggested the importance of analyticity, Kronig (1946) connected analyticity with causality.
10 At the Rochester Conference held in 1956, Gell-Mann claimed that crossing, analyticity and unitarity would determine the scattering amplitude if suitable boundary conditions

were imposed in momentum space at infinite momenta. He further claimed that this would almost be enough to specify a field theory; but he also suggested that the dispersion program, if treated non-perturbatively, was reminiscent of Heisenberg's hope of writing down the S-matrix directly instead of calculating it from field theory. See Gell-Mann, 1956b.

11 It is such an inner logical connection between the dispersion program and Heisenberg's program that makes Gell-Mann's "casual" mention of Heisenberg's SMT a strong impetus for the further development of analytic SMT in the 1950s and the 1960s. This claim is justified by following facts: (i) Mandelstam's double dispersion relation (1958), which has been generally acknowledged as a decisive step in the making of SMT, started with Gell-Mann's conjecture that dispersion relations might be able to replace field theory; (ii) Chew quoted Gell-Mann's Rochester remarks at the Geneva conference of 1958; (iii) in addition to various contributions to the conceptual development of the S-matrix theory (cf. Section 8.5 of Cao, 1997), Gell-Mann was also among the first physicists to recognize the importance of Regge's work and to endorse the idea of bootstrap. As Chew remarked in 1962, Gell-Mann

> has for many years exerted a major positive influence both on the subject (SMT) and on me; his enthusiasm and sharp observations of the past few months have markedly accelerated the course of events as well as my personal sense of exitement. (1962)

However, Gell-Mann's later position concerning the status of SMT was quite different from Chew's. For Chew, SMT was the only possible framework for understanding strong interactions because QFT was hopeless in this regard. For Gell-Mann, however, the basic framework was QFT, since all of the principles and properties of SMT were actually abstracted from or suggested by QFT, and SMT itself could only be regarded as a way of specifying a field theory, or an on-mass-shell field theory. Gell-Mann's position on the relationship between SMT and QFT was very much in the tradition of Feynman and Dyson, and of Lehmann, Symanzik, and Zimmerman, (1955), in which the S-matrix was thought to be derivable from a field theory, or at least not incompatible with the field-theoretical framework. But for Chew, the conflict between QFT and SMT was irreconcilable because, he maintained, the notion of elementary particles assumed by QFT led to a conflict with the notions of analyticity and bootstrap.

12 In this case, s represents momentum transfer squared in the t-channel, in which energy squared is represented by t.

13 For a discussion about the ontological commitment of the S-matrix theory, see Capra, 1979.

14 Landau, 1959. Cutkoski (1960) gave a further rule for including the unitarity condition.

15 Thus there is a one-many correspondence between Landau graphs and Feynman graphs.

16 This is partly due to the shift of emphasis in SMT from objects (particles) to processes (scatterings or reactions).

17 Feynman expressed this principle in a talk given at the Aix-en Provence conference on elementary particles in 1961. See footnote 2 in Low's paper (1962). This idea of Feynman's was widely quoted by Chew and Frautschi (1961a), Salam (1962), and many other physicists in the early 1960s.

18 The most convincing of these successes was that a large number of baryon and meson resonances were found to lie on nearly linear Regge trajectories. Another impressive demonstration of the bootstrap dynamics were the calculations of the mass and width of the ρ resonance, the results of which were found to be close to the experimental values. See Chew and Mandelstam, 1960, and Zachariasen and Zamech, 1962. Chew's reciprocal bootstrap calculation of the nucleon, pion and (3,3) resonance (1962) provided a more complicated example of this kind.

19 Gell-Mann, 1953, 1956a; Gell-Mann and Pais, 1954; Nakano and Nishijima, 1953; Nishijima, 1955. For a clear recount of how the scheme was formed in order to understand complicated experimental discoveries of associated productions, quick production

and slow decay of "new" particles and resonances (K-mesons, Lambda, Sigma and Xi particles), see Pais, 1986, 517–520.

20 Goldhaber, 1956; Sakata, 1956. Note that Goldhaber identified his "dionic charge" with Gell-Mann's strangeness, and that Sakata explained the mysterious concept of strangeness conservation in an atomistic term as the conservation of the fundamental strangeness-carrying lambda particles.

21 It could be taken as a precursor of Gell-Mann's idea of quark. But as we will show below, Gell-Mann's motivation and underlying rationale were quite different from Sakata's when he introduced the entities at the sub-hadron level.

22 See Gell-Mann's acknowledgment in his footnote 6 of (1961).

23 Mathematically, Gell-Mann in his proposal of eightfold way derived the octet from the direct product of fundamental representation (triplet or 3) and its conjugate 3*. This left a conceptual question unanswered until he proposed the quark model (1964a): what would be the ontological basis for the fundamental representation?

24 Gell-Mann's "solution" to Pauli's dilemma mentioned earlier was this: "We may, however, take the point of view that there are vector mesons associated with a gauge-invariant Lagrangian plus a mass term, which breaks the gauge invariance of the second kind [local gauge invariance] while leaving inviolate the gauge invariance of the first kind [global symmetry] and the conservation law." (1961)

25 "For low mass states we can safely ignore the representations 27, 10, and 10*." As for "the vector meson transforming according to 1," Gell-Mann took it as one "associated with a guage-invariant Lagrangian plus a mass term, which breaks the gauge invariance of the second kind while leaving inviolate the gauge invariance of the first kind and the conservation law" (Gell-Mann, 1961, 22–24).

26 "The representations 1, 10, 10* and 27 may also correspond to mesons, even pseudoscalar ones, but presumably they lie higher in mass, some or all of them perhaps so high as to be physically meaningless" (Gell-Mann, 1961, p. 19).

27 For theoretical analysis, see Lee and Yang, 1956; for experimental confirmation, see Wu, Ambler, Hayward, Hoppes, and Hudson, 1957, and Garwin, Lederman, and Weinrich, 1957.

28 Gershtein and Zel'dovich (1955) first mentioned and discarded the idea that a conserved vector current of isospin (and the corresponding vector bosons) would be somewhat relevant to the β-decay.

29 Schwinger speculated about the existence of an isotopic triplet of vector bosons whose universal couplings would give both the weak and electromagnetic interactions: the massive charged Z particles would mediate the weak interactions in the same way as the massless photon mediates the electromagnetic interactions. This was the beginning of the intermediate vector boson theory of the weak interactions (1957).

30 Sidney Bludman in (1958) made an extension from the isotopic SU(2) symmetry to the chiral SU(2) × SU(2) symmetry, which presupposed the conservation of the axial vector part of the weak currents.

31 Nambu, 1957. Nambu called this meson ρ^0, it was renamed as the ω meson.

32 $f_v(0)$ is defined by $(2\pi)^3 \langle p'|V_\mu(0)|p\rangle = iu(0)\gamma_4 f_v(0)u(0)\delta_{\mu 4}$, which is the limiting case of Eq. (2.2) when $p = p' \to 0$.

33 A simple proof that the CVC hypothesis implies the non-renormalization of $f_v(0)$ by the strong interactions can be found in Coleman, 1967.

34 Define

$$T_+ = \int dr(J^V_{01}(r,0) + iJ^V_{02}(r,0)).$$

From the isotopic commutation relations we have $J^v_{\mu+} = [J^v_{\mu 3}, T_+]$. Thus,

$$< p'|_P J^V_{\mu+}|p >_N = < p'|_P[J_{V\mu 3}, T_+]|p >_N = < p'|_P J^V_{\mu 3}|p > -$$
$$< p'|_N J^V_{\mu 3}|p >_N = < p'|_P j^{em}_\mu|p >_P - < p'|_N j^{em}_\mu|p >_N.$$

Here the property of T_+ as an isotopic raising operator and $j_\mu{}^{em} = j_\mu^s + j_\mu^v \, (= J_{\mu 3}^v)$ are used. Define $F_1(q^2)^+ = f_v(q^2)$, $F_2(q^2)^+ = f_m(q^2)$, then Eq. (2.6) follows from Eq. (2.1) and Eq. (2.2).

35 $g_A(0)$ is defined, in a way similar to $f_V(0)$, in the limiting case of Eq. (2.3).

36 Another influential one was suggested by Nambu, 1960.

37 To include fields other than elementary fields, such as composites or excited states, in a Lagrangian would make the Lagrangian "miserably complicated." See the discussion and debate between Gell-Mann and Eugene Wigner at the 1st International Meeting on the History of Scientific Ideas, which was held in Catalonia, Spain, September 1983; in *Symmetries in Physics (1600–1980)* edited by Doncel, Hermann, Michel and Pais (Bella Terra, 1987), 542–543.

38 Note the difference between the notion of approximate symmetry and that of broken symmetry in general. The approximate symmetry is broken at classical level, explicitly by small non-invariant terms in the Lagrangian responsible for the mass differences within supermultiplets. The mass differences are too big to be explained away by external disturbances, such as effects caused by electromagnetic and weak interactions, and thus the symmetry is intrinsically broken by the strong interaction itself. The broken symmetry in general, however, refers to the symmetries which may be explicitly broken, or spontaneously broken by the non-invariant vacuum state while the Lagrangian is invariant, or anomalously broken at the quantum level. More discussion on this can be found in Cao, 1997, Sections 8.7 and 10.3.

39 When the charge algebra was extended to the ETCs of time-time components of the currents, the formal reasoning was safe. Yet for the space time commutators, the formal rule had to be modified by some gradients of delta-functions, the so-called Schwinger terms. Cf. Goto, 1955 and Schwinger, 1959.

40 See Gell-Mann, 1961 and Ne'eman, 1961. Note that this symmetry was a global symmetry rather than a gauge symmetry (such as the later color SU(3) symmetry), later called the flavor symmetry. It was only a manifestation of the mass degeneracy, first of the three Sakatons [or proton and neutron in the case of SU(2)], and then of quarks. In the massless limit and assuming the conservation of parity in the strong interactions, the SU(3) or SU(2) flavor symmetry can be extended to a $SU(3)_L \times SU(3)_R$ [$SU(2)_L \times SU(2)_R$] chiral symmetry, cf. Bludman, 1958.

41 We shall not give all the complications involved in the definition of this commutation relation. For details, see Gell-Mann, 1962, 1964. But note that relation (2.21) posed a severe constraint on the constitution of the currents. The relation does not follow from symmetry considerations. The currents made from baryon and pseudoscalar meson octets satisfy the relations (2.19) and (2.20), but not (2.21). If the currents are made by lepton or quark fields obeying the equal-time anticommutation relations, then the relation (2.21) will be satisfied. For an interesting discussion on this point, see Pais, 1986, p. 565.

42 Gell-Mann, 1964a, 214.

43 An apparently similar scheme was proposed by George Zweig (1964a, b, c). But Zweig's motivation and conception were very different from Gell-Mann's. Readers may get some sense of the difference from a brief discussion, below, of the debate between the advocates of constituent quark model and those of current algebra (Gell-Mann and his followers).

44 All the ideas and quotations in this section are taken from Gell-Mann, 1964a, b.

45 Gell-Mann often confused "hidden" with "mathematical" and "real" with "singly observable." More discussions on the relationship between reality and observability in the case of confined entities will be given in Chapter 10.

46 The reader can get a glimpse of the debate from Lipkin, 1997.

47 Quark currents here refer to currents consisting of quarks rather than of hadrons (as the hadronic currents were previously conceived before the introduction of quarks).

3

Sum rules

Current algebra, as a hypothesis about hadron physics, offers a set of algebraic relations relating one physically measurable quantity to another, although it cannot be used to calculate any from first principles. Presumably, the assumed validity of a hypothesis should first be tested before it can be accepted and used for further explorations. However, any exploration of the implications or applications of the hypothesis, within the general framework of hypothetic-deductive methodology, if its results can be checked with experiments, functions as a test.

The test of current algebra, however, requires delicate analyses of hadronic processes and effective techniques for identifying relevant measurable quantities in a justifiable way. For this reason, there was no rapid progress in current algebra until Sergio Fubini and Giuseppe Furlan (1965) suggested certain techniques that can be used, beyond its initial applications, to derive various sum rules from current algebra, which can be compared with experimental data. The importance of the sum rules thus derived, however, goes beyond testing the current algebra hypothesis. In fact, they had provided the first conceptual means for probing the constitution and internal structure of hadrons, and thus had created a new situation in hadron physics and opened a new direction for the hadron physics community to move, as we shall see in the ensuing sections and chapters.

3.1 The Fubini–Furlan method and the Adler–Weisberger sum rule

The Fubini–Furlan method consists in three steps. The first step is to insert a complete set of intermediate states into the matrix element of a commutator of two operators from one state to another, which is related to another matrix element of an operator through the current algebra rule, such as those listed in Section 2.2 (Equations (2.19) to (2.27)). The second is to separate out the one-nucleon contribution to the matrix elements. The third is to calculate the matrix elements in the infinite momentum frame, thereby obtain a sum rule relating observable matrix elements of the operators.

In their original paper, Fubini and Furlan applied their method to the commutator of vector current charges and thereby calculated the radiatively induced renormalization of the vector current. Inspired by their ingenious idea of going to the infinite momentum frame for calculation, Stephen Adler (1965a) and, independently, William Weisberger (1965) soon applied the Fubini–Furlan method, together with the assumption of PCAC, to the case of the axial-vector currents. When a complete set of physical intermediate states was introduced in the right-hand side of the matrix elements of the commutation rule of current algebra

$$2I_3 = [Q_a^+, Q_a^-] \tag{3.1}$$

between physical one-proton states

$$\begin{aligned}(2\pi)^3\delta^{(3)}(\mathrm{p}_2-\mathrm{p}_1) &= \langle P(p_2)|[Q_a^+, Q_a^-]|p(p_1)|\rangle \\ &= \left\{\sum_{spin}\int\frac{d^3k}{(2\pi)^3}\langle P(p_2)|Q^+(t)|n(k)\rangle\langle n(k)|Q^-(t)|P(p_1)\rangle\right. \\ &\quad \left.+\sum_{j\neq N}\langle P(p_2)Q^+(t)|j\rangle\langle j|Q^-(t)|P(p_1)\rangle\right\} - (Q^+\leftrightarrow Q^-)\end{aligned} \tag{3.2}$$

(where the charges are defined by $I^i = \int d^3x\, V_0^i$, $Q_a^i = \int d^3x\, A_0^i$, $i = 1, 2, 3$, $Q_a^{\pm} = Q_a^1 \pm iQ_a{}^2$, and V_0^i, A_0^i are the time components of the isovector members of the vector and axial-vector current octets), the contribution of one-neutron states was isolated, and the contributions for higher intermediate states were calculated in the infinite momentum frame, a sum rule expressing the axial form factor g_A in terms of $\sigma^{\pm}{}_0(W)$ (the zero-mass $\pi^{\pm}$-proton scattering total cross sections at center-of-mass energy W) was obtained:

$$1-\frac{1}{g_A^2} = \frac{4M_N^2}{g_r^2 K^{NN\pi}(0)^2}\frac{1}{\pi}\int_{M_N+M_\pi}^{\infty}\frac{WdW}{W^2-M_N^2}\left[\sigma_0^+(W)-\sigma_0^-(W)\right], \tag{3.3}$$

with M_π and M_N the pion and nucleon masses, g_r the renormalized pion–nucleon coupling constant ($g_r{}^2/4\pi \approx 14.6$), and $K^{NN\pi}(0)$ the pionic form factor of the nucleon, normalized as that $K^{NN\pi}(-M_\pi^2) = 1$. This formula yielded a prediction $|g_A| = 1.24$ by Adler, or 1.16 by Weisberger,[1] which agreed reasonably well with the then available experimental data $g_A^{exp} \approx 1.18$, or the later settled value $g_A^{exp} = 1.257 \pm 0.003$.

Adler (1965b) soon converted the sum rule to an exact relation, involving no PCAC assumption, for forward inelastic high energy neutrino reactions. The generalization was based on his pioneering understanding of the connections between inelastic high energy neutrino scattering reactions and the structure and properties of the weak currents. In his earlier study of the weak pion production (1964), Adler had already noticed that at zero squared leptonic four-momentum transfer, the matrix element could be reduced to just the hadronic matrix element of the divergence of the axial-vector current, which by PCAC is proportional to the amplitude for pion–nucleon scattering. This observation was shown to be only a specific case of a general theorem, which says that in a general inelastic high energy neutrino reaction, when the lepton emerges forward and the lepton mass is neglected, the leptonic matrix element is proportional to the four-momentum transfer; hence when the leptonic matrix element is contracted with the hadronic part, the vector current contribution vanishes by CVC, and the axial-vector current contribution reduces by PCAC to the corresponding matrix element for an incident pion. Thus inelastic neutrino reactions with forward leptons can be used to obtain information about hadronic currents and, further, as potential tests of CVC and PCAC.

When the connection between high energy neutrino scattering reactions and the properties of the weak currents was explored without assuming PCAC, the result for the sum rule was that the cross sections at the right-hand side of the sum rule (3.3) would not be those for the pion–proton scattering processes, but rather those for the high energy neutrino–nucleon inelastic reactions:

$$\nu_l + N \to l + \beta, \tag{3.4}$$

with ν_l a neutrino, l a lepton, N a nucleon, and β a system of hadrons with $M_\beta \neq M_N$.

The Adler–Weisberger sum rule was the first most important success of Gell-Mann's current algebra. In its wake, a string of applications of current algebra combined with PCAC appeared in rapid succession, such as those by Steven Weinberg (1966a, b), Curtis Callan and Sam Treiman (1966), and Yukio Tomozawa (1966).

In order to properly appreciate the importance and physical content of the infinite momentum frame, which was adopted in the derivation of the Adler–Weisberger sum rule and many other kinds of sum rules for some years to come, let us look at a parameter associated with an intermediate state $|j\rangle$, the invariant momentum transfer, $q^2 = (P - P_j)^2$, between the one-proton state and the intermediate state (which is also equal to the invariant momentum transfer between the leptons, $(k_\nu - k_l)^2$, in the case of (3.4)). The importance of this variable, which will be seen clearly in later discussions, lies in the fact that the cross sections in the sum rule as Lorentz scalars can depend on P and q only through the scalar products q^2 and another variable $\nu = P \cdot q$. That is, sum rules are characterized by the invariant momentum transfer q^2.

Since translational invariance in forward neutrino reactions implies $\mathbf{P} = \mathbf{P}_j$, $q^2 = [(M_P{}^2 + \mathbf{P}^2)^{1/2} - (M_j{}^2 + \mathbf{P}^2)^{1/2}]^2$, and in general (non-forward) cases implies $\mathrm{P} + \mathrm{q} = \mathrm{P}_j$, $q^2 = -\mathrm{q}^2 + [(M_P{}^2 + \mathbf{P}^2)^{1/2} - (M_j{}^2 + (\mathbf{P} + \mathbf{q})^2)^{1/2}]^2$, (which approaches the spacelike four-momentum transfer $q^2 = -\mathrm{q}^2$ in the infinite momentum frame: this is how $q^2 \neq 0$ sum rules are obtained, see below), it is clear that q^2 depends on $\mathbf{P}$, and for any finite $\mathbf{P}$, it also depends on the mass of the intermediate state, M_j. Two difficulties can be seen clearly from this observation. First, in terms of intermediate states, we have no idea at all as to how many intermediate states, elementary or not, would contribute to the sum rule. Second, since q^2 increases without limit as the mass of intermediate state increases, it is not clear about the rate at which the sum over the intermediate states can be expected to converge, putting the experimental accessibility to the ever increasing q^2 aside. This is a serious problem since in applications one always truncates the sum over states.

In the infinite momentum frame, however, both difficulties can be somewhat bypassed. When $|\mathbf{P}| \to \infty$, which implies $q^2 = [(M_P^2 + \mathbf{P}^2)^{1/2} - (M_j^2 + \mathbf{P}^2)^{1/2}]^2 \to 0$ in the forward scattering case, and $-\mathrm{q}^2$ in the general case, sum rules are independent of M_j. Thus, the messy situation related to the intermediate states is simplified and the difficulty bypassed. As to the rate of convergence, when q^2 is fixed to zero, although nothing rigorous can be said, at least some educated guesses can be made on the basis of our knowledge of high energy scatterings.[2]

Physically, taking $|\mathbf{P}| \to \infty$, which is equivalent to assuming an unsubtracted dispersion relation for a scalar function related with the cross sections (Adler, 1967), can also be seen as effectively suppressing the experimentally inaccessible processes. As a result, the dependence of the sum rule on so-called "Z-diagram"[3] intermediate states which have particle lines not connected to currents (or of the dispersion relation on the subtraction) is removed. But the removal of the Z-diagram contributions is possible only because current conservation or "partial conservation"[4] was appealed to for providing the necessary damping infinite denominator for the contribution. Thus it is clear that current conservation or "partial conservation," which is presumed in current algebra, is essential for the validity of the infinite momentum frame.

3.2 Adler's local current algebra sum rules

Gell-Mann was highly interested in Adler's work (1965b) in which the relation between the current algebra of vector and axial-vector charges and forward high energy neutrino reactions was explored for testing the validity of current algebra, and urged Adler to extend it to the test of local current algebra, that is, to the local equal-time communication relations satisfied by the fourth components of the vector and axial-vector current octets, rather than by their spatial integrations or charges.

Local current algebra is considerably more restrictive than the integrated one. Mathematically, even if derivatives of the delta function were present on the right-hand side of local ETC relations, the integrated relations would still be valid. Physically, what local current algebra explores is the localized rather than integrated property and behavior of hadrons scattering leptons. The shift of attention from the integrated to the local was a critical step and was historically consequential: a step from phenomenology to the probing of the inside of hadrons through the investigations of local hadronic currents.

Mobilized by Gell-Mann, Adler (1965c, see also 1966) tried to generalize his result in which the integrated algebra was tested in forward (with zero invariant momentum transfer q^2 between neutrino and outgoing lepton) high energy inelastic neutrino reactions, to nonforward ($q^2 < 0$) neutrino reactions, and to derive from local current algebra ((2.25), (2.26) and (2.27) or their variations) a sum rule involving the elastic and the inelastic form factors measurable in high energy neutrino reactions, which is valid for each fixed q^2. This way, Adler claimed, the local current algebra could be tested.

To the particle physics community, this line of reasoning was explorative but also highly conjectural. First, concerning the nature of the dynamics of lepton inelastic scattering at high momentum transfer, there was hardly any soundly based qualitative idea as to whether it is a local coupling or it is mediated by intermediate bosons. Second, and more importantly, concerning the very notion of local hadronic currents, the ground was quite shaky: local fields for hadrons were simply too far away from observations to be regarded as reliable descriptive elements; and a local hadronic current without local fields of hadrons as its constituents would not be much more reliable. Although Gell-Mann suggested to use the totality of the matrix elements of hadronic electromagnetic and weak currents between hadron states as operationally defined descriptive elements, upon which a phenomenology with some predictive power could be based, a thus-defined experimentally meaningful notion of local hadronic currents[5] was at best an educated guess, the only ground for which was an analogy to the situation in QED, where physicists did have enough confidence with local current densities, which are measurable and commute at spacelike separations.

Conjectural as it was, assuming the validity of locality principle, in the sense of local couplings between local leptonic and hadronic currents, in neutrino reactions or, more generally, in any semileptonic process (Lee and Yang, 1962; Pais, 1962) Adler still believed that the reasoning could be used to test the hypothetical idea of local current algebra.

At the International Conference on Weak Interactions held at Argonne National Laboratory in late October of 1965, Adler gave a talk titled "High energy semileptonic reactions" (Adler, 1965c) in which he gave the first public presentation of the local current algebra sum rules for the β deep inelastic neutrino structure functions.

Adler examined the high energy neutrino reaction

$$v + N \rightarrow l + \beta,$$

where v is a neutrino, l is a lepton (electron or muon), N is a nucleon, and β is a system of hadrons. In the laboratory frame in which N is at rest, he defined $q(= k_v - k_l)$ as the lepton four-momentum transfer, q^2 as the invariant momentum transfer between the leptons, W as the invariant mass of the system of β $\left[W = \left(2M_N q_0 + M_N^2 - q^2\right)\right]$, and M_N as the nucleon mass. Assuming that the local coupling of the leptons is correct, the differential cross section for the reaction would have the form

$$d\sigma = \sum_j f^j_{KNOWN}(E_\nu, E_1, \Omega_\nu, \Omega_1) F^j_{UNKNOWN}\left(q^2, r_\beta, s_N\right) \tag{3.5}$$

where E_v, E_l, Ω_v, Ω_l are respectively the energies and the angular variables of the neutrino and lepton, r_β the internal variables of β, and S_N the polarization of N. Equation (3.5) is the analog for neutrino reactions of the Rosenbluth formula, with the form factors $F^j{}_{UNKNOWN}$ the analogs of the electromagnetic form factors. Since only a small number of terms are present in the sum, Equation (3.5) can be tested by doing experiments at a large number of different values of the neutrino and lepton energy and angle variables.

Furthermore, when only the neutrino and lepton are observed, Adler derived the general form for the neutrino reaction leptonic differential cross section as

$$\begin{aligned} &d^2\sigma\left(\begin{pmatrix} \nu \\ \bar{\nu} \end{pmatrix} + p \rightarrow \begin{pmatrix} l \\ \bar{l} \end{pmatrix} + \beta(S=0)\right) \Big/ d\Omega_l dE_l \\ &= \frac{G^2 \cos^2\theta_c}{4\pi E_\nu^2} \frac{E_l}{E_\nu} \left[q^2\alpha^{(\pm)} + 2E_\nu E_l \cos^2\left(\frac{1}{2}\phi\right) + \beta^{(\pm)} \mp (E_\nu + E_l) q^2 \gamma^{(\pm)}\right] \end{aligned} \tag{3.6}$$

for strangeness-conserving reactions.[6] A similar equation holds for strangeness-changing reactions. By measuring $d^2\sigma/d\Omega_l dE_l$ for various values of E_ν, E_l, and the lepton neutrino angle Φ, the form factors α, β, and γ can be determined for all $q^2 > 0$ and for all W above threshold; these form factors can be explicitly expressed in terms of matrix elements of the vector and the axial-vector currents.

Equation (3.6) can also be written as

$$d^2\sigma\left(\begin{pmatrix}\nu\\ \bar{\nu}\end{pmatrix} + p \to \begin{pmatrix}l\\ \bar{l}\end{pmatrix} + \beta(S=0)\right) \Big/ d(q^2)dq_0$$
$$= \frac{G^2\cos^2\theta_c}{4\pi E_\nu^2}\left[q^2\alpha^{(\pm)} + \left(2E_\nu^2 - 2E_\nu q_0 - \frac{1}{2}q^2\right)\beta^{(\pm)} \mp (2E_\nu - q_0)q^2\gamma^{(\pm)}\right]. \tag{3.7}$$

Here the form factors α, β, and γ are functions of q^2 and q_0 which depends only on q^2 and the invariant mass W of the final state hadrons $\beta\left[q_0 = \frac{W^2 - M_N^2 + q^2}{2M_N}\right]$. A similar equation can also be written for strangeness-changing reactions. By making measurements at a variety of neutrino E_ν for fixed q^2 and q_0, the form factors α, β, and γ can be separately determined.

From local current algebra, Adler was able to derive the sum rule for β form factors:

$$\begin{aligned} 2 = {} & g_A(q^2)^2 + F_1^V(q^2)^2 + q^2 F_2^V(q^2)^2 \\ & + \int_{M_N+M_\pi}^{\infty} \frac{W}{M_N} dW \left[\beta^{(-)}(q^2, W) - \beta^{(+)}(q^2, W)\right] \end{aligned} \tag{3.8}$$

and two more sum rules for α and γ, which for reasons given below will be ignored here.[7] The β sum rule holds for each fixed q^2 but has a notable feature that the left-hand side of (3.8) is independent of q^2, even though the Born term contributions and the continuum integrand on the right are q^2-dependent.

When the integration variable is changed from W to q_0 and the elastic contribution is not separated off, the sum rule takes the form

$$2 = \int_{(q^2/2M_N)-}^{\infty} dq_0\left(\beta^{(-)} - \beta^{(+)}\right). \tag{3.9}$$

When $q^2 = 0$, the sum rule tests only the charge algebra; when $q^2 < 0$, however, it tests the local current algebra. The reason for this difference is simple, but its implications are profound.

In all kinds of sum rules, q^2 is equal to zero for forward reactions because the lepton three-momentum transfer $\mathbf{q}$ is equal to 0; while for nonforward reactions, q^2 $(=-q^2) < 0$ because $\mathbf{q}$, which is transferred from the lepton to and thus is carried by the hadronic current, is greater than 0. Intuitively this seems to mean that if local coupling in the nonforward reactions is correct, the lepton is locally probing hadrons with its localized nonzero momentum transfer $\mathbf{q}$ or q^2. But more precisely, it seems to mean that it is the hadronic current density that is locally probing hadrons with its own localized nonzero $\mathbf{q}$ or q^2.

Although the first understanding was experimentally intuitive, and thus later became very popular, the second understanding seems to make more physical sense. The fundamental reason for this is that leptons simply cannot have any interaction with hadrons unless it is mediated by intermediate bosons or photons for the weak or the electromagnetic interactions, or through direct local coupling with hadronic currents, which was assumed in Adler's derivation of the local current algebra sum rules. Furthermore, the very idea of local hadronic current density that was assumed in the derivation makes sense only when the corresponding local operators can be defined, which, when acting on a hadron state, create another hadron state, and thus when they are sandwiched between hadron states and coupled to a local lepton current, would give the lowest order S-matrix element for a weak or electromagnetic process. In fact, only with the second understanding can we make physical sense of the intermediate hadron states, namely take them realistically, and give ground for taking the form factors or structure functions as the sums of effects of hadronic processes mediated by the local hadronic current densities.[8] In other words, the experimentally feasible investigations of hadrons' form factors' behaviors predicted by local current algebra sum rules became an effective way of investigating hadronic currents and hadronic processes carried out by these currents in a larger context of semileptonic process.

If this understanding of the local current algebra sum rules is accepted or assumed, then the crucial question would be how to define the hadronic currents with the desired functions mentioned above. According to Gell-Mann, current algebra had to be fundamentally explained by the notion of current quark (Section 2.3), and the hadronic electromagnetic and weak currents should take the form of (2.30) and (2.31), from which all the achievements of the Cabibbo scheme of weak currents and weak interactions can be reproduced. But the nature of hadronic currents was not a settled issue in the mid 1960s when the sum rule industry was launched: the quark model of the hadronic currents was only one among many, and which model would make the local current algebra sum rule valid was only part, even though a very important part, of the big issue of saturation mechanism in the conceptual scheme of local current algebra sum rules, which will be further explored in the next chapter.

In order to derive the sum rule, in addition to local current algebra, Adler also assumed that "a certain amplitude obeys an unsubtracted energy dispersion relation." As we noted before, this is equivalent to adopting the infinite momentum frame. In the published discussion following his talk, in answer to a question by Fubini, Adler noted that the β sum rule had been rederived by Curtis Callan (unpublished) using the infinite momentum frame limiting method, but that the α and γ sum rules could not be derived this way, reinforcing suspicions that "the integral for β is convergent, while the other two relations (for α and γ) really need subtractions." In fact, the two sum rules for the α and γ were shown by Dashen to be divergent and hence useless, and the one for the β deep inelastic amplitude to be a convergent and useful relation (Adler, 2006, 17–18).

In the talk Adler also pointed out that the sum rule on $\beta^{(\pm)}$ "has an interesting consequence for the limit of neutrino cross sections as the neutrino energy goes to infinity." From (3.7), it is easy to have:

$$\begin{aligned}&\lim_{E_\nu\to\infty}\left\{d\sigma(\overline{\nu}+p\to l+\beta(S=0))/d(q^2)-d\sigma(\nu+p\to l+\beta(S=0))/d(q^2)\right\}\\&=\frac{G^2\cos^2\theta_c}{2\pi}\int_{(q^2/2M_N)-}^{\infty}dq_0\left(\beta^{(-)}-\beta^{(+)}\right)=\frac{G^2\cos^2\theta_c}{\pi}.\end{aligned}\tag{3.10}$$

Adding the similar result for the strangeness-changing case, he obtained the result for the total cross section

$$\begin{aligned}&\lim_{E_\nu\to\infty}\left\{d\sigma_T(\overline{\nu}+n)/d(q^2)-d\sigma_T(\nu+n)/d(q^2)\right\}\\&\quad=\left(G^2/\pi\right)\left(-\cos^2\theta_c+\sin^2\theta_c\right)\\&\lim_{E_\nu\to\infty}\left\{d\sigma_T(\overline{\nu}+p)/d(q^2)-d\sigma_T(\nu+p)/d(q^2)\right\}\\&\quad=\left(G^2/\pi\right)\left(\cos^2\theta_c+2\sin^2\theta_c\right)\end{aligned}\tag{3.11}$$

which says that local current algebra entails that in the limit of large neutrino energy, $d\sigma_T(\overline{\nu}+N)/d(q^2)-d\sigma_T(\nu+N)/d(q^2)$ becomes independent of q^2.

Adler's work on the local current algebra sum rules for the deep inelastic region of high energy and momentum transfer gave the first working out of the structure of deep inelastic neutrino scattering. In the work, he explored the consequences of the current-algebra/sum rule approach for neutrino reactions with great thoroughness. Most important among the consequences is that now leptons can be seen as local pairs that are precisely probing the smeared out nucleons, in terms of form factors or structure functions and their shape and behavior, through their local coupling to local hadronic current densities. With this work, the current algebra as a research program had shifted its focus from phenomenology, such as hadron spectroscopy or low energy theorems, to the examination of the probing of the inside of hadrons by deep inelastic scatterings.

At the Third International Symposium on the History of Particle Physics that was held in 1992, James Bjorken commented on Adler's work: "His work provided a most important basis for what was to follow when the ideas were applied to electroproduction" (Bjorken, 1997, p. 590).

3.3 Bjorken's inequality for electroproduction

When Adler spoke at the Argonne conference on local current algebra sum rules, Bjorken from SLAC was in the audience and was intrigued by the β sum rule results. Soon afterwards, Bjorken (1966a, b) converted Adler's results into an inequality for deep inelastic electron scattering, a step which, although conceptually straightforward, had brought the current algebra program to a closer contact with experiments, and thus, opened up vast vistas for its further development, with numerous consequences unexpected from the original perspectives of current algebra for bypassing the ignorance of hadronic currents and for using the approximate symmetry to understand the structure of the system of baryons and mesons.

Bjorken had been highly interested in experiments ever since his undergraduate years at MIT, and maintained close contact with SLAC experimentalists, such as Richard Taylor and Henry Kendall. When Adler talked about using neutrino reactions to test local current algebra, SLAC was in the midst of the construction of the first very large scale linear accelerator. Thus it was natural for Bjorken to concentrate his efforts on the electron scattering opportunities the new machine presented, especially since many of his close friends and colleagues at SLAC were preparing to do those experiments.

From the very beginning, it was clear to theoreticians and experimenters at Stanford that the linear accelerator ("linac") should be an ideal instrument for observing the instantaneous charge distribution inside the proton via inelastic scattering. There was a long tradition at Stanford of using inelastic electron–nucleus scattering and sum rule techniques to study the constituent nucleons (Drell and Schwartz, 1958). The extension of such ideas to search for constituents of the nucleon itself was natural to consider. In fact, Bjorken heard Leonard Schiff, who played an important part in the creation of SLAC, in a colloquium devoted to the first announcement of a SLAC Project, "describing this opportunity – in particular showing that the energy- and momentum-transfer scales were more than adequate for seeing the insides of the proton" (Bjorken, 1997).

Knowledge of electromagnetic form factors of the nucleon was indispensable for understanding the electron–nucleon high energy interaction, and would provide an important test of new theoretical ideas adopted in the derivation of Adler's sum rules. However, form factors as Fourier transformations of spatial distributions of charge and magnetization, together with the idea of ingredients, were *nonrelativistic intuitions*, while applying these ideas to the electron scattering program at SLAC linac essentially required, for the first time in the history of high energy physics, a good control of *highly relativistic kinematics*. Two questions immediately attracted Bjorken's attention, and none of them was easy to address.

First, Bjorken noticed, local current algebra techniques were relativistic and provided quite solid ground for explorations. In particular, Adler's sum rules admitted corollaries for the electroproduction channel. Since in Adler's sum rule, the behavior of matrix elements of currents as the momentum q carried by the currents approaches infinity can be determined in terms of local current algebra, Bjorken found that by isospin rotation, this rule can be converted into a useful inequality for inelastic electron–nucleon scattering, and thus the lower bound for the scattering at high momentum transfer q can be determined on the same theoretical basis. But then, the crucial step in the exploration turned out to be to control the ranges of energy transfer, that was needed for the convergence of the sums, and the values of momentum transfer, that were necessary for the asymptotic sum rules to be relevant experimentally. This, the so-called saturation problem, was not an easy task, and its exploration went beyond the scope of current algebra and required additional theoretical resources and insights, although the task itself was first raised in the research program of current algebra.

Second, when Bjorken started the work, it was not at all clear to him that the nonrelativistic intuitions about the ingredients and spatial distributions of charge and momentum captured by form factors or structure functions could be legitimately retrieved by going to the highly relativistic motion. Still, his conversion of Adler's neutrino sum rule to the electroproduction inequality recounted below did set the train in motion, which soon moved to a new territory of hadron physics.

The kinematic relation (3.7), when q^2 (<0) is fixed and the center-of-mass energy of lepton and hadron goes to infinite, takes the form:

$$\frac{d\sigma\left(\begin{matrix}\nu\\ \bar{\nu}\end{matrix}\right)^{p}}{dq^2 dq_0} \underset{E\to\infty}{\longrightarrow} \frac{G^2}{2\pi}\beta^{(\pm)}. \tag{3.12}$$

The kinematic relation for the electron–nucleon scattering similar to (3.7) under the same conditions takes the form:

$$\frac{d(\sigma^{ep}+\sigma^{en})}{dq^2 dq_0} \underset{E\to\infty}{\longrightarrow} \frac{4\pi\alpha^2}{q^4}\left(\sigma_{1p}+\sigma_{1n}\right). \tag{3.13}$$

Here σ_{1p} and σ_{1n} are the electromagnetic structure functions of the proton and neutron respectively. By isotopic rotation and adapting Adler's sum rule (3.9) to the case in which the contribution of axial current is left out, gives an inequality:

$$\begin{aligned}\int dq_0\left[\left(\sigma_{1p}+\sigma_{1n}\right)\right] &= \int dq_0[(\sigma_{1v}+\sigma_{1s})] \geq \int dq_0\sigma_{1v} \geq \int dq_0\sigma_{1v}^{\left(I=\frac{1}{2}\text{final state only}\right)} \\ &\geq \int dq_0\left[\sigma_{1v}^{I=\frac{1}{2}} - \sigma_{1v}^{I=\frac{3}{2}}\right] = \frac{1}{2}\int dq_0\left(\beta^{-}-\beta^{+}\right) = \frac{1}{2}.\end{aligned} \tag{3.14}$$

Here v in σ_{1v} and s in σ_{1s} indicate the isovector and isoscalar part respectively. From (3.13) and (3.14) it is easy to get

$$\lim_{E\to\infty}\frac{d(\sigma^{ep}+\sigma^{en})}{dq^2} = \int dq_0\frac{d(\sigma^{ep}+\sigma^{en})}{dq^2 dq_0} = \frac{4\pi\alpha^2}{q^4}\int dq_0\left(\sigma_{1p}+\sigma_{1n}\right) \geq \frac{2\pi\alpha^2}{q^4}. \tag{3.15}$$

In comparison to Adler's sum rule, which was hard to test quantitatively because it involves a difference between neutrino and antineutrino scattering cross section, Bjorken's inequality involves no such difference, and thus had the prospect of being tested experimentally relatively quickly. In fact, although Adler's sum rule was historically the first to be proposed, it was the last to be tested experimentally, only long after all the action was over.

Bjorken's inequality and his subsequent works on electroproduction, together with the parallel experiments at SLAC, changed the landscape of high energy physics. From then on, all the conceptual developments initiated from Gell-Mann's local current algebra and Adler's sum rules were almost exclusively centered around the electron–nucleon scattering and its variants. This historical fact has not only revealed the fertility and profundity of current algebra, which had provided conceptual means for exploring hadron physics through electron–nucleon scattering, but also vindicated the importance of Bjorken's work in bringing current algebra to close contact with feasible experiments at SLAC.

3.4 Remarks

Adler's derivation of sum rules for neutrino processes was an important step in the conceptual development of hadron physics. It revealed the potential of current algebra, which first appeared to be descriptive of hadron phenomenology, to be explorative and explanatory, in terms of fundamental constituents and dynamics of hadrons, by applying it to experimentally measurable physical processes. Thus it has been regarded as the cornerstone of many theoretical edifices used to describe inelastic lepton hadron scattering, most important of which, of course, were those initiated by Bjorken about inelastic electron–nucleon scatterings. Lack of direct theoretical knowledge of hadron physics, the information obtained through sum rules from semileptonic processes seemed to be the only source for theorizing over hadron physics. Yet such an indirect approach appeared to have some special features that deserve to be noted.

Inputs–outputs

In the derivation of sum rules from local current algebra, Adler relied on additional assumptions. Most important among them were the validity of the locality principle and of the infinite momentum frame. The latter, as noted earlier, was physically equivalent to the validity of unsubtracted dispersion relations of observable matrix elements (or related cross sections or form factors or structure functions) and conceptually rested on the assumption of quasi-free field behavior[9] for the commutators of the space components of hadronic currents, first conjectured by Richard Feynman, Gell-Mann and George Zweig (1964).

The sum rules put constraints on, and helped to reveal the regularity of, hadrons' dynamical behavior in the deep inelastic scattering. The implications of these constraints and regularities might be interpreted, as we will see in later chapters, as pointing to the existence of point-like constituents for hadrons interacting with leptons electromagnetically or weakly, which is compatible with the input assumption of locality, but could never reveal any details of the strong interactions the hadron states incurred by the action of hadronic currents without additional assumptions. The reason is simple: one can never get more output (about interactions) than what was assumed (quasi-free field assumption).

A related remark is about the field-theoretical underpinning of the sum rules. The very adoption of local operators for hadronic current is compatible with local field theory. In particular, Bjorken in his work on inequality and sum rules introduced a special way of expressing the commutator of currents in terms of the spectral functions related to directly measurable matrix elements (asymptotic behavior of the forward Compton amplitudes),[10] which rendered the commutators field-theoretically calculable. But, again, this calculation was feasible only for the electromagnetic and weak part. As to the strong interaction, the calculation was confined to the free field behavior, revealing no dynamic details at all.

Possible loopholes

Adler's sum rule was derived for the purpose of experimentally testing the validity of local current algebra. If it was confirmed by experiments, local current algebra would become more credible. But what if it failed the experimental test? In fact, Bjorken himself was suspicious about the assumed validity of Adler's sum rule as late as 1972 (Bjorken and Tuan, 1972). What if the integral in the sum rule does not converge or converges quite slowly (which would fail the sum rule at the available experimental energy)? In particular, what if the free field assumption of Feynman, Gel-Mann and Zweig on which the asymptotic sum rules rested was wrong? Logically speaking, even if one of these situations appears and sum rules died, it does not necessarily invalidate local current algebra, since the fault may lie in the additional assumptions introduced in the derivation of the sum rules.

New task in the new context

Once Adler's sum rule and Bjorken's inequality were derived, and thus the relevance of local current algebra for hadron physics outside of the domain of phenomenology was revealed to the hadron physics community, a new task appeared on the agenda for further explorations, which would not have been conceivable before the derivation. The new task mentioned here refers to the search for mechanisms by which the neutrino sum rule and electron scattering inequality could be saturated at high energy and large momentum transfer. As we will see in later chapters, the pursuit of this new agenda will subsequently create another different context, which would raise its own tasks for further advances. These tasks would not have been perceivable when Gell-Mann proposed his current algebra and Adler derived his sum rule.

Notes

1 The difference came from the inclusion by Adler of a pion off-shell correction associated with the Δ (1232) resonance.
2 Based on the Regge pole model, Adler (1967) found that sum rules involving forward scatterings would converge fairly rapidly.
3 These nuisance diagrams plagued the rest-frame sum rules that Dashen and Gell-Mann had been studying around 1965.
4 This means that the divergence of a current is no more singular than the current itself.

5 The notion of local hadronic currents, which are mathematically represented by corresponding local operators, is experimentally meaningful because when the local current operators are sandwiched between hadron states and coupled to a local lepton current, they would give the lowest order S-matrix element for a weak or electromagnetic process.
6 Originally, Adler separated the cross section into strangeness-conserving and strangeness-changing pieces, whereas the later convention is to define them as the sum of both.
7 Take the derivative for the vector part of equation (3.8) at $q^2 = 0$, gives

$$0 = 2F_1^V + \left(F_2^V(0)\right)^2 + \frac{\delta}{\delta q^2} \int_{M_N + M_\pi}^{\infty} \frac{W}{M_N} dW \left[\beta^{(-)}(q^2, W) - \beta^{(+)}(q^2, W)\right],$$

which is the Cabibbo–Radicati sum rule and seems to be in agreement with experiments.
8 Regarding the difference between the local and integrated current algebra, Adler also pointed out in the talk that "if the commutators are changed by adding terms which vanish when integrated over all space, the sum rule for $\beta^{(+)}$ is modified."
9 Free from certain interactions, such as those represented by the Z-diagram contributions mentioned in Section 3.1.
10 In later literature, it was called the BJL method (Bjorken, 1966b; Johnson and Low, 1966).

4

Saturation and closure

Gell-Mann's idea of current algebra as a physical hypothesis was rendered testable by Adler's sum rules. The success of the Adler–Weisberger zero momentum transfer sum rule testing the integrated algebra gave credit to Gell-Mann's general idea. But what about Adler's nonzero momentum transfer sum rules testing local current algebra? When Bjorken raised the issue of how the nonzero momentum transfer sum rule might be satisfied to Adler, at the Varenna summer school in July 1967 when both Adler and Bjorken were lecturers there, it seemed to be a very serious problem to Adler, since all the conceptual development up till then left undetermined the mechanism by which the nonforward neutrino sum rules could be saturated at large q^2. After the Solvay meeting in October, 1967, Adler discussed this issue with his mentor Sam Treiman, and "then put the saturation issue aside, both because of the press of other projects and concerns, and a feeling that both shared that how the sum rule might be saturated 'would be settled by experiment'" (Adler, 2003, 2009, private communications).

The response by Adler and Treiman to Bjorken's question seemed to be natural. How could anything else but experiments be the ultimate arbitrator for a physical hypothesis? However, as we will show, solely relying on experiments missed an opportunity to use the exploration of mechanisms for saturating the sum rule as a way to go beyond the conservative philosophy of the current algebra program, which was to abstract from field theory relations that might hold in general field theories, rather than searching for particular attractive field theory models that might underlie the general relations. In this chapter, we will first look at Adler's analysis of the saturation problem before moving to Bjorken's approach and Chew's challenge to the underlying ideas of saturation and the whole current-algebra/sum-rules research program.

4.1 Saturation: Adler's empirical analysis

In a paper titled "Neutrino or electron energy needed for testing current communication relations,"Adler and Frederick Gilman (1967) gave an empirical analysis of the saturation of local current algebra sum rules for small momentum transfer q^2. They started with a question: "what is *effectively* an infinite incident neutrino or electron energy" that is required for the neutrino reaction sum rules to hold or the inequalities on electron–nucleon scattering to hold, which are independent of the leptonic invariant four-momentum transfer q^2? They argued that although "no general answer to this question can be given . . . the experimental information used in evaluating the sum rules for the axial-vector coupling constant and for the nucleon isovector radius and magnetic moment suffice to" give an answer to the question in special cases. In particular they noted that SLAC would have enough energy to confront the saturation of the small nonzero q^2 sum rules and inequalities in a meaningful way.

At the end of the paper, however, they cautioned the reader that "as q^2 increases, the needed energy will be expected to increase rapidly. This is clear from the experimental fact that the single-nucleon and (3,3) resonance contributions, i.e., the small W contributions, to the sum rules decrease rapidly with increasing q^2."

The caution was related to a fundamental "puzzle," perceived from their empirical perspective, posed by the β sum rule: the left-hand side of the sum rule [see equation (3.8)] is a constant, while the Born terms on the right are squares of nucleon form factors, which vanish rapidly as the momentum transfer q^2 becomes large; the low-lying nucleon resonance contributions on the right were expected to behave like the Δ (1232) contribution, which is form factor dominated and also falls off rapidly with q^2. Thus, they noticed, in order "to maintain a constant sum at large q^2, the high W states, which require a large E to be excited, must make a much more important contribution to the sum rules than they do at $q^2=0$." That is, something new and interesting – a new component in the deep inelastic cross section, that did not fall off with form factor squared behavior, or perhaps a continuum contribution of some type – must happen in the deep inelastic region if the sum rule were to be satisfied for large q^2.

But then, relying exclusively on experiments, "we were cautious (too cautious, as it turned out!) and did not attempt to model the structure of the deep inelastic component needed to saturate the sum rule at large q^2" (Adler, 2006). The result was that "the calculations of this paper shed no light on the important question of how rapidly E increases with q^2, but only serve to indicate at what energies E it may pay to begin the experimental study of the local current algebra." Thus the goals of the Adler–Gilman paper were limited; they did not attempt to give a conceptual analysis of mechanisms by which the sum rules or inequalities can be saturated. Without such an analysis, however, not only could the important question of how rapidly E increases with q^2 not be addressed, but, more importantly, the door for further exploring what actually happened in the processes could not be opened, and thus the potential of current algebra as a powerful tool for exploring the constitution and structure of hadrons could not be fully realized.

4.2 Saturation: Bjorken's constructive approach

Bjorken's approach was different. In his explorations (1967a [2003], 1967b, c) on how to address the saturation problem, Bjorken frequently appealed to history, heavily relied on the analogy with nuclear physics, and, guided by the insights gained from historical analogy, constructed several models about mechanisms or what might actually happen in various kinematic regions of the deep inelastic electron–proton scattering that would saturate the sum rules and inequalities. Experiments were still the ultimate arbitrator for the validity of hypothetical ideas of sum rules and local current algebra. How could anything else be in physics? But now the attention was guided to specific kinematic regions, for checking the specific predictions made by specific models (mechanisms), rather than looking at the data to see if the sum rules were satisfied or not without insight gained as to what actually happened in the process.

Bjorken's approach was significantly different to Adler's. Satisfied or not, the answer to Adler's questioning would only tell him the validity or invalidity of sum rules, or perhaps also of local current algebra (though not necessarily). But for Bjorken, the enquiry into mechanisms which would saturate sum rules opened the door for the exploration of deeper reality (what actually happened in deep inelastic electron–proton scattering). The significance of such an exploration went far beyond the exploration of local broken symmetry initiated by Gell-Mann, although it was triggered by the latter.

When Bjorken raised the saturation issue to Adler at the Varenna summer school of July 1967, he had already explored deeply into the question. From the very beginning, in his incomplete and unpublished manuscript, written in March 1967 with a title "Inelastic lepton scattering and nucleon structure" (1967a [2003]), he approached the inelastic lepton scattering from the perspective of probing nucleon structure, although he clearly understood that the theoretical foundations of the probing, in terms of conceptual apparatus (local currents and their commutators), predictions and constraints, were laid down by local current algebra and Adler's sum rule, both of which were formal and model-independent, and thus enjoyed more general validity than any specific models he entertained would have. Although it was not published until 2003, nevertheless, it can help us to better understand his approach to the saturation problem in his influential Varenna summer school lecture (1967b) and SLAC talk (1967c) because both of them clearly drew on it, although went much less far – a fact that showed his hesitation and lack of confidence in the tentative ideas at the time he was exploring.

What attracted Bjorken to the issue of saturation mechanisms was his realization that various problems related to it were quite fundamental, dealing with the question of whether there are any "elementary constituents" within the nucleon in what seems to be a direct way, using leptons as a probe. For him, the sum rules[1] – in particular the left-hand side of (3.8) [or the right-hand side of (3.11) and (3.15)] was unexpectedly large, of order of Fermi or Rutherford scattering from a point particle, which was in direct contradiction to the established knowledge about form factors, resonances, etc., as Adler and Gilman had already realized – were so perspicuous that by appealing to history, an interpretation in terms of "elementary constituents" of the nucleon could be suggested; and with the aid of this interpretation more predictions for inelastic lepton scattering could be made, and then checked with existing data.

Using his interpretation or constituent model as a guide to find interesting kinematic regions in which testable predictions could be made, Bjorken clearly realized that all the theoretical ideas he deployed in his explorations were closely related to local current algebra and Adler's sum rules, and thus were very sensitive to the built-in locality assumption therein. Without the locality assumption, one of Bjorken's major claims, that one can learn the instantaneous (electric or weak) charge distribution of the nucleon from the deep (large energy, large momentum transfer) inelastic scattering, would make no sense at all (see below).

Saturation became a problem because Adler's sum rule, derived in a general, formal and model-independent way, when inelasticity ν (or the energy loss from the incipient lepton to nucleon) was large, gave a constant result, independent of momentum transfer q^2; and there was no acceptable answer to the question of where the contributions would be to the sum rule. What was assumed was that the dominant part in the sum over states came from resonances which were produced in the region of low inelasticity. But contributions from resonances could not saturate the sum rule because, as noticed by Adler and Gilman, they were form factor dominated and would fall rapidly with momentum transfer. When the inelasticity was large, but momentum transfer was small, there would be diffractive electroproduction or photoproduction, i.e., coherent production of vector mesons. But Bjorken argued, if the Regge ideas were correct, it could be reasonably assumed that by taking the difference between the absorption of a positive charge on the proton, due to the process of a neutrino going into a lepton, and the corresponding process where a negative charge went over to the proton from antineutrino, that the diffractive contributions would cancel out.[2] In the region where both inelasticity and momentum transfer were large, Bjorken assumed that in the same way that the diffractive processes became unimportant all other processes would also be damped out in that limit. Then, how can the sum rule be saturated and where did the constant come from?

"General truths"

The major inspiration to Bjorken's exploration of saturation mechanisms came from his appeal to history, his taking the historical analogy seriously of deep inelastic lepton–nucleon scattering with elastic and inelastic electron–nucleus scatterings, elastic electron–nucleon scattering, and photoabsorption and photoproduction processes. Bjorken himself had been deeply involved in the theoretical studies of these experiments since 1958 when he was still a graduate student at Stanford working with Sidney Drell, a leading theorist and a major figure in the study of lepton (photon)–hadron processes, and thus knew all the subtleties and implications of the theory–experiment feedback loops. The same experimental and the associated theoretical studies in this area had also provided experimenters at SLAC with a working framework for planning the forthcoming inelastic scattering at SLAC (see the next chapter). Unsurprisingly, therefore, there was a parallel or even great portion of overlapping between Bjorken's theoretical explorations and experimenters' working out the results of their framework. This parallel, overlapping, and, later, active interplay between the two was the major driving force in the late 1960s in advancing current algebra forward and unfolding its potentials for revealing deeper reality at the sub-nucleon level.

Bjorken's analysis started from the "general truths" about a new kind of kinematic formalism – the general form of the cross section, how many form factors and their behavior in elastic, inelastic, quasi-elastic and diffractive production processes – which would enable him to bring to bear the constraints likely to exist by looking at limiting regions of the phase space. As we will see, by studying the kinematics of the diffractive processes and comparing it with the quasi-elastic process, Bjorken found reasons to expect it to extrapolate into the deep inelastic regime, and thus was able to suggest a model for saturation.

The "general truths" were important to Bjorken because, as he pointed out (2003, p. 2), at Stanford "the problems were about to be data-driven: not only were the sum rules, etc. provided by the extant theory needed, but more importantly details of what the actual size and shape of the structure functions" were likely to be; and this was "the most important distinction between what we at Stanford were trying to do, relative to the work done by Adler, which emphasized the opportunities to be found" in the realm of neutrino physics that had no immediate prospect of obtaining experimental data for the creation of the deep inelastic theory as the pursuit at Stanford had.

The process Bjorken examined was an incident electron scattering from a nucleon, through the one-photon exchange, into some final state. It would be experimentally the simplest and theoretically the easiest to get to the "general truths" if one looks only at the electron and ignores what the final state is, the so-called inclusive process. Now let us summarize what "general truths" about the inclusive electron–nucleon scattering process were available to Bjorken in the recent past of nuclear physics in the 1950s and early 1960s.

Schiff pointed out (1949) that high energy electrons available from the new electron linear accelerator at Stanford could be used to probe the structure of the proton itself through the known electromagnetic interaction. Soon afterwards, Marshall Rosenbluth (1950) calculated the elastic differential cross section for an electron of energy E_0 scattering through an angle θ in an elastic collision with a proton, transferring energy $(E_0 - E')$ to the recoil proton of mass M, here

$$E' = \frac{E_0}{1 + \frac{2E_0}{M}\sin^2 \theta/2}$$

is the energy of the scattered electron, and obtained, when only the scattered electron is detected, a simple expression quite similar to the original Rutherford scattering formula:

$$\frac{d\sigma}{d\Omega} = \frac{\alpha^2}{4E_0^2 \sin^4 \theta/2} \cos^2 \theta/2 \frac{E'}{E_0}\left[\frac{G_E^2 + \tau G_M^2}{1+\tau} + 2\tau G_M^2 \tan^2 \theta/2\right]. \tag{4.1}$$

Here $\tau = q^2/4M^2$, G_E and G_M are form factors describing the distributions of charge and magnetic moment, respectively, and are functions of only the momentum transfer, $q^2 = 4E_0E' \sin^2 \theta/2$, which is a measure of the ability to probe structure in the proton. The uncertainty principle limits the spatial definition of the scattering process to $\sim \hbar/q$, so q^2, and therefore E_0, must be large in order to resolve small structures. If the charge and magnetic moment distributions are small compared with $\hbar/q$, then G_E and G_M will not vary as q^2 changes; but if the size of those distributions is comparable with $\hbar/q$, then the form factors will decrease with increasing q^2.

In 1955, Robert Hofstadter and Robert W. McAllister (1955) presented data showing that the form factors in the Rosenbluth cross section were less than unity and were decreasing with increasing q^2, and gave an estimate of $(0.7 \pm 0.2) \times 10^{-13}$ cm for the size of the proton. Thus they showed persuasively that the proton was not a point, but an extended structure. Although the connection between spatial extent and structure was intuitively appealing, nobody was seriously questioning the "elementary" character of the proton at the time, when available electron energies were not yet high enough for the exploration of inelastic scattering from the proton, and only elastic experiments provided clues about the proton's spatial extent. Nevertheless, this achievement marked the beginning of the search for substructure in the proton.

The electron–deuterium scattering was also measured for extracting information about the neutron (Hofstadter, 1956). The form factor for elastic scattering from the loosely bound deuterium nucleus was observed to fall off extremely rapidly with increasing q^2, so the neutron was studied through quasi-elastic scattering, namely the scattering from either the proton or the neutron, which together form the deuterium nucleus. The quasi-elastic scattering reached a maximum near the location of the peak for electron–proton scattering, since the scattering took place off a single nucleon, and was spread out over a wider range of energies than the scattering from the free proton because of the momentum spread of the nucleons in the deuterium nucleus.

Inelastic scattering from nuclei was a common feature of the early scattering data at Stanford. The excitation of nuclear levels and the quasi-elastic scattering from the constituent protons and neutrons of a nucleus were observed in the earliest experiments, but the latter became more evident as q^2 was increased. In 1962, Hand gave an inelastic equivalent of the Rosenbluth formula containing two form factors which are functions of q^2 and $\nu = (E_0 - E')$, the energy loss suffered by the scattered electron; here

$$E' = \frac{E_0 - \dfrac{(W^2 - M^2)}{2M}}{1 + \dfrac{2E_0}{M} \sin^2 \theta/2},$$

W is the mass of the final state of the struck hadron (when $W^2 = M^2$, the elastic kinematics are recovered), W^2, q^2, and ν are relativistically invariant quantities in the scattering process (Hand, 1963).

But the most general form of the cross section in the parity conserving one photon approximation was given by Drell and Walecka (1964) in a form very similar to the Rosenbluth expression,

$$\frac{d\sigma}{d\Omega dE'} = \frac{\alpha^2}{4E_0^2 \sin^4 \theta/2} \cos^2 \theta/2 \left[W_2 + 2W_1 \tan^2 \theta/2\right], \tag{4.2}$$

here the structure function W_1 is the magnetic term which would be, for elastic scattering, $(q^2/4M^2)/G_M^2$, and W_2 corresponds to $G_E^2 + (q^2/4M^2)/G_M^2$, both are functions of both q^2 and v, $W_{1,2}(q^2, \text{v})$. If q^2 and v are fixed, then by varying the scattering angle through varying the beam energy at a given q^2, one can do straight-line plots to isolate W_1 and W_2 in the same manner as for elastic scattering once the radiative corrections have been handled properly. In the limit as $q^2 \to 0$, W_1 and W_2 can be expressed in terms of the total photoabsorption cross section $\sigma_\gamma(\nu)$: $W_1 \to \nu\sigma_\gamma/4\pi^2\alpha$, $W_2 \to q^2\sigma_\gamma/4\pi^2\alpha\nu$.

Historical analogy

At the SLAC symposium, Bjorken (1967c) suggested that the kinematical space of q^2 vs. v to be explored in the lepton–nucleon inelastic scattering, where v is the lepton energy loss or the energy of the virtual photon in the laboratory, had its analogy in the familiar photo-production processes as a function of q^2 and v. For the nicely explored elastic scattering, one could follow a line which is the energy loss $v = q^2/2M$. The line $q^2 = 0$ corresponds to photoabsorption. But the most interesting exploration Bjorken did was with the region to the right of $v = q^2/2M$, starting at small q^2 and v, then working up but keeping q^2 small.

With the increase of photon energy, Bjorken noticed, what first came would be the S-wave threshold pion production. Beyond it, the next interesting object would be the N*(1238), the electroproduction of which was fairly well explored. For any inelastic scattering, when the lepton momentum transfers to the nucleon, to a given state $|N\rangle$, it lies along lines parallel to the elastic line: $q^2 = 2M\nu + M^2 - M_N^2$. The general formula for the cross section for the electroproduction of a given discrete resonant state is expressible in terms of three form factors, which according to Walecka's model (Bjorken and Walecka, 1966), have an interesting qualitative feature that also appears in nuclear physics, namely the threshold, meaning that at low q^2 the form factors might rise like q^2 to some power, whereas at large q^2 their behavior is controlled by the elastic form factor (falling off rapidly). The implication is that there may be a peaking which may or may not occur in the physical region where one observes the scattering.

"Upon continuing on up in energy ν beyond a Bev, the photoabsorption becomes a diffractive process" in which the photon goes into a state and then the state goes coherently into a vector meson. An important feature for the coherent production of vector mesons is that nothing violent happens to the nucleon: the momentum transfer to the nucleon should be less than a few hundred Mev, which corresponds to the nominal size of the nucleon. Bjorken expected that diffractive electroproduction similar to diffractive photoproduction would also happen although the kinematics is different because the "photon" would have a nonzero mass. But by analogy, what actually happens in the electroproduction process, Bjorken argued, should also be approximately a virtual photon going to a vector meson, which then scatters.

Bjorken was also guided by an analogy to nonrelativistic nuclear physics despite the fact that the case of lepton–nucleon inelastic scattering relevant to the saturation of the current algebra sum rule was in an extreme-relativistic situation.

To exploit the analogy, he chose the incident energy sufficiently high at a given scattering angle of the lepton so that in the center-of-mass frame the scattering is still small-angle. "In this limit, the kinematics is such that the energy transfer and longitudinal momentum transfer from the lepton to the hadron tends to zero" and "the scattering proceeds via the charge density and is 'Coulombic' in all respects: only transverse momentum is exchanged between probe and target" (1967a).

Having visualized "the scattering as Coulomb scattering from the Lorentz-contracted charge distribution of the hadron target particle," Bjorken appealed to the historical analogy with nonrelativistic quantum mechanics, in which the elastic scattering (e.g., from atoms or nuclei) measures the time-averaged charge distribution, while inelastic scattering with coarse energy resolution measures the instantaneous distribution of charge. At momentum transfer q^2, the inelastic scattering cross section, integrated over the energy of the final electron and summed over all excited states of the target particle is given by

$$\frac{d\sigma}{dq^2} = \frac{4\pi\alpha^2}{q^4}\sum_i Q_i^2 \tag{4.3}$$

where Q_i are the charges of the constituents which can be seen with the spatial resolution $\Delta x \sim 1/q$. For example, for electron scattering from an atom or nucleus there was the sum rule (Forest and Walecka, 1966)

$$\int_0^\infty dE' \frac{d\sigma}{dq^2 dE'} = \frac{4\pi\alpha^2}{q^4}\left[Z + Z(Z-1)\, f_c(q^2)\right] \equiv \frac{4\pi\alpha^2}{q^4}\left[\sum_i Q_i^2\right] \tag{4.4}$$

where $f_c(q^2)$ is a correlation function which vanishes at large q^2 so that the scattering is incoherent from the individual protons, assuming that they are point-like objects. For $q^2 \to 0$, the scattering is coherent from a single charge Z of the nucleus. As q^2 increases further, the elastic scattering decreases because of the finite proton size, but this decrease is compensated by the onset of meson production.

As q^2 continues to increase, Bjorken conjectured, the total yield will be controlled by scattering from elementary constituents of the nucleon itself. His main argument for the conjecture is the assumed locality of hadron currents in current algebra tradition, which leads to "a picture in which instantaneously in time the charge is localized in space within the nucleon." In principle the magnitude of these charges might be arbitrarily small. However, the existence of a local isospin current density "is an argument that the charge of the constituents is of order 1 because two such constituents differ by a unit of charge." Thus $\sum_i Q_i^2$ in Equations (4.3) and (4.4) should be bigger than or equal to 1.

Bjorken was tempted "to take over the same physical picture for the relativistic scattering at infinite momentum," and use the insights gained from the picture to construct models for saturation of Adler's sum rules or his own inequalities. It was possible "because 1) the scattering goes via the charge density; 2) there is only spatial-momentum transfer, essentially as in the nonrelativistic case; 3) because of Lorentz contraction, the scattering occurs instantaneously in time provided we respect the uncertainty principle and do not look closely at the energy of the outgoing lepton" (1967b).

Two models for saturation

With such a physical picture suggested by historical analogy in mind, Bjorken proceeded to highlight the essential feature of current algebra sum rules and examine how far these rules "go and where they stop, where detailed models are required" before constructing models for saturation.

He started from the "general truths": Equation (4.2) for the general form of the inelastic scattering cross section, the behavior of form factors in elastic, inelastic, quasi-elastic and diffractive production processes, and the respective contributions of these processes in various regions of the q^2–v phase space, aiming to find out where the contributions are to the sum rule.

In analogy with the photoproduction processes, Bjorken noticed that elastic scattering lies along the straight line $q^2 = 2Mv$; inelastic scattering to a given state lies along lines parallel to the elastic line, $q^2 = 2M\nu + M^2 - M_N^2$. Traditionally, it is assumed that the dominant part in the sum over states comes from this region of resonances having relatively low ν and low masses. But, as Adler and Gilman had already realized, their contributions to the sum rules decrease rapidly with increasing q^2. The diffractive production of vector mesons in the region of large ν is separated by the quasi-elastic region, which has a smaller slope and cuts across all resonant states, from the region of asymptotic limits of form factors where little scattering is expected (1967a).

The most interesting feature of Adler's sum rule that Bjorken tried to understand is that as the incident energy goes to infinity, the difference of neutrino and antineutrino cross sections is a constant independent of q^2 (see Equation 3.11). Bjorken noticed that the result would also be true were the nucleon a point-like object because Adler's derivation was general and model-independent. This is compatible with Adler's result that the difference of these two cross sections is a point-like cross section and is big.

How to interpret this sum rule? By analogy with nuclear physics, Bjorken suggested that the nucleon is built out of point-like constituents which could be seen only if you could look at it instantaneously in time, rather than in processes where there is a time averaging and in which the charge distribution of these constituents is smeared out. If we go to the region where q^2 is large and the inelasticity or the energy transfer to the nucleon from the electron is very large so that the assumed point-like constituents of the nucleon could be seen, which is the interesting region because where the right-hand side of Adler's sum rule is assumed to be a constant, then, Bjorken further reasoned, we can expect the neutrino to scatter incoherently from the point-like constituents of the nucleon.

Suppose these point-like constituents had isospin one half. The antineutrino, because it goes to an anti-lepton and must pick up a charge, scatters only from the isospin up constituents. Therefore its cross section is proportional to the expectation value of the number of isospin up objects in the nucleon ($\langle N\uparrow\rangle$) times the point cross section: $\frac{d\sigma^{\bar{\nu}p}}{dq^2} = \frac{G^2}{\pi}\langle N\uparrow\rangle$. Likewise, the same thing applies for the neutrino with isospin up replaced by isospin down. Therefore, what the sum rule says is simply $\langle N\uparrow\rangle - \langle N\downarrow\rangle = 1$ for any configuration of constituents in the proton.

This very clear picture of the deep inelastic scattering process, Bjorken further suggested, would look even better if one looks at it

in the center-of-mass of the lepton and the incoming proton. In this frame the proton is Lorentz contracted into a very thin pancake and the lepton scatters essentially instantaneously in time from it in the high energy limit. Further-more, the proper motion of any of the constituents inside the hadron is slowed down by time dilatation. Provided one doesn't observe too carefully the final energy of the lepton to avoid trouble with the uncertainty principle, this process looks qualitatively like a good measurement of the instantaneous distribution of matter or charge inside the nucleon. (1967c)

Bjorken took this idea over to the electromagnetic processes. Then, by analogy, the limit as the energy goes to infinity of $\frac{d\sigma}{dq^2}$ for lepton–proton scattering should be the Rutherford point cross section times the sums of the squares of the charges of those point-like constituents inside the nucleon:

$$\lim_{E\to\infty} \frac{d\sigma^{ep}}{dq^2} = \underbrace{\left(\frac{4\pi\alpha^2}{q^4}\right)}_{Rutherford} \sum_i Q_i^2. \tag{4.5}$$

It is also big and point-like (1967c).

Assuming the above picture makes sense, then the important question is what inelasticity is needed for a given q^2 in order to see these supposed elementary constituents? The answer to this question would lay down the foundation for his models for saturation.

A. *The harmonic oscillator model*

In order to obtain some idea about the important region of inelasticity for observing the constituents, Bjorken appealed to history and first assumed that at sufficient high momentum transfer q^2, the scattering from constituents is quasi-free and quasi-elastic so that free particle kinematics can be used. Bjorken argued by analogy that a quasi-free or quasi-elastic scattering does not necessarily mean that the constituents be loosely bound or escape the nucleon. For example, for a nonrelativistic charged particle bound in a harmonic oscillator potential, the total scattering at fixed q^2 is always the Rutherford value, although the scattering to any given level is damped exponentially for large q^2. "If you hit it hard the cross section is the point cross section provided you sum over all inelasticities" (1967c).

In the nonrelativistic harmonic oscillator potential model, the important excited states are those which lie near the classical kinematics: $q_0 = \Delta E = q^2/2m$. Using quasi-free kinematics in the relativistic case, within the infinite momentum framework, one finds that the energy loss, just like the elastic scattering, goes like q^2 divided by twice the mass of the constituents, $\nu \simeq q^2/2\overline{m}$, here $\overline{m}$ is the mean mass of the constituent in the nucleon.

Thus, if quasi-elastic scattering is assumed, the essential quantity in determining the important inelasticity ν is the mean mass $\overline{m}$ of the constituents, assuming that ν is high enough at given q^2 to see the constituents. Because elastic scattering occurs for $\nu \simeq q^2/2M_p$, we must have $\overline{m} < M_p$ to make sense of the kinematics. Bjorken felt it attractive to try the constituent quark model in which phenomenological constituent quarks have an effective mass around 300 MeV when they are inside the nucleon and moving nonrelativistically. But he immediately clarified the instrumental nature of the notion of effective mass to avoid any confusion: the effective mass of the constituents as seen in the rest frame "does not necessarily have anything to do with the physical mass of any supposed physical quarks which in turn do not have anything to do with our point-like objects" because the physical quark would be "visualized as a meson cloud with the point like quark inside it" (Bjorken, 1967c).

According to Bjorken's harmonic oscillator model, then, the most reasonable place to look for the saturation of Adler's rule is at inelasticities $\nu \simeq q^2/2\overline{m}$, and when the effective mass $\overline{m}$ is around 300 MeV, the sum rule will be saturated in a region of (q^2, ν), which is roughly along the straight line $\nu = q^2/0.6$.

B. The diffractive production model

In the scattered lepton-energy spectrum, in addition to a quasi-elastic peak, there is a broad continuum arising from diffractive production of ρ^0 and other spin-1 mesons at high inelasticities $\nu = q_0 = \Delta E$ at fixed q^2. In order for this to be a coherent process (nothing violent happens to the nucleon), the minimum momentum transfer to the nucleon should be less than a few hundred MeV which corresponds to the nominal size of the nucleon. The onset of these processes can be estimated on kinematical grounds. When the minimum momentum transfer to the nucleon is small compared to the nucleon size ($\sim$ 350 MeV), we can expect the coherent processes to go efficiently. This occurs when

$$\Delta_{\min} = \frac{q^2 + m_\rho^2}{2q_0} \leq 350 MeV \sim \overline{m} \text{ or } q_0 = \nu \geq \frac{q^2}{2\overline{m}} + m_\rho,$$

which is for large q^2 essentially the same estimate as from the harmonic oscillator model. When ν is moved to the right of this rough line, the diffraction electroproduction will be the important process for saturation. Thus the quasi-elastic peak, if any, may well merge into a continuum coming from the coherent diffractive production (Bjorken, 1967b).

In terms of his picture of constituents, Bjorken found that the coherent processes can be interpreted in a simple way. They are simply the scattering of leptons from the vacuum fluctuations of charge surrounding the nucleon; in the quark model, it is the meson cloud of virtual quark–antiquark pairs surrounding the three constituent quarks. On these grounds, it would be reasonable to expect that the q^2 dependence of the diffractive contribution is again point-like $\sim (q^2)^{-2}$, as in Equation (4.3): $\frac{d\sigma}{dq^2} = \frac{4\pi\alpha^2}{q^4}\sum_i Q_i^2$, and may in all cases obscure any quasi-elastic bump.

Predictions

Bjorken's harmonic oscillator model based on the locality assumption predicts a quasi-elastic scattering result [Equation (4.3)]. If a quasi-elastic peak is observed, the yield integrated over the peak at fixed q^2, in the case of electron–nucleon scattering, measures the sum of squares of the charges of the elementary constituents of the nucleon, which is useful in determining the charges the constituents carry; or if the charges were assumed in some particular model, then the measurement can be used to check the predictions the model makes. For example, in the quark model,

$$\sum_i Q_i^2 = \begin{cases} 4/9 + 4/9 + 1/9 = 1 & \text{proton} \\ 1/9 + 1/9 + 4/9 = 2/3 & \text{neutron} \end{cases} \tag{4.6}$$

But Equation (4.3) itself also has other "strong experimental implications inasmuch as more sensible, conservative estimates of inelastic yields take photoabsorption cross-sections and multiply by the elastic form factors $G^2(q^2)$,

$$G^2(q^2) \leq \frac{const}{(q^2)^4}; \quad \frac{d\sigma}{dq^2} \leq \frac{const}{(q^2)^6}." \tag{4.7}$$

This is much smaller in comparison with (4.3). Of course, the test of this prediction crucially depends on whether the available incident lepton energy is high enough at given q^2 to see the constituents (Bjorken, 1967a).

By using Equations (4.3) and (4.6), Bjorken was able to make another prediction. "Because diffractive processes are expected to be the same strength for proton and neutron, in taking the difference of the yields, they should disappear. We would predict (for large q^2) in the quark model," the difference would be

$$\frac{d\sigma_{ep}}{dq^2} - \frac{d\sigma_{en}}{dq^2} = \frac{1}{3}\left(\frac{4\pi\alpha^2}{q^4}\right). \tag{4.8}$$

What if the diffractive processes do compete with the quasi-elastic processes? Bjorken suggested, "a study of the Z-dependence of the inelastic spectrum would help to disentangle the two contributions" (Bjorken, 1967a).

4.3 Closure: Chew's questioning

Bjorken's rationale was soon to be critically questioned by Geoffrey Chew, the archetypal opponent to atomistic thinking in hadron physics in general, and to local quantum field theory in particular, at the Fourteenth Solvay Conference, held in Brussels, October 1967, although Bjorken himself did not attend the conference. Chew's severer criticisms in his subsequent paper published shortly after the conference, from the perspective of bootstrap philosophy, went even deeper into the conceptual foundations of Adler's sum rule and Gell-Mann's local current algebra (Chew, 1967). Ironically, Chew's questioning and criticism contributed, in a negative way, to clarifying the metaphysical implications of current algebra, which grounded its methodological approach to the constitution and interactions of hadrons.[3]

Exchanges at the Solvay Conference

The confrontation between Chew's bootstrap program and Gell-Mann's local current algebra was clearly displayed at the conference. It began even before current algebra and sum rules were brought into the general discussion session by Adler. Immediately after Chew's report, "S-matric theory with Regge poles," Gell-Mann raised a question to him: "you have implied that someone who wants to formulate theoretical principles governing the electromagnetic and weak interactions of hadrons and relate them to those of leptons (which seem so similar) would be foolish to introduce local operators; can you give a constructive suggestion as to what we should do instead?"

Chew's response was polite but alert, stressing his basic idea of asymptotic behavior: "I have no immediate suggestion for tackling the problem of currents or, equivalently, form factors, except to employ the standard dispersion relations with an open mind about asymptotic behavior, which may be related to the question of locality."

When Heisenberg tried to assert a link between "the bootstrap mechanism and a theory starting from a master-field equation", Chew indicated the incompatibility between the two: in all Lagrangian field theories "the special angular-momentum values selected by the master equation become reflected through a non-democratic particle spectrum." H. P. Durr further elaborated that a master equation destroys the democracy of elementary particles since it has to be written down in terms of fundamental fields and thus "introduces certain elementary particles as distinct from composite particles."

But when Adler brought up the issue of sum rules as tests of local current algebra into the discussion, Chew's questioning of the presumptions of the current algebra project became more and more serious.

At the request of various people at the conference, Adler was scheduled during one of the discussion sessions to give a short talk addressing the issue of the sum rules and their saturation. After having reviewed the basics of local current algebra and the sum rules derived therefrom, Adler argued that the saturation of the Cabibbo–Radicati sum rule by low lying resonances was not relevant for the large q^2 case, and emphasized that for saturation of the neutrino sum rule and "the Bjorken inequality, when q^2 is large ... the inelastic continuum must be of primary importance." Adler then summarized Bjorken's two models for saturation, and pointed out that according to Bjorken, if $q^2 = (1\ \text{BeV}/c)^2$, then to saturate the sum rule, one "must excite states with $q_0 \geq 1.5$ BeV, corresponding to an isobaric mass of $W \geq 1.7$ BeV. So the CEA experiments, which go up to $W = 1.7$ BeV, have not necessarily gone high enough. But we see that according to Bjorken's estimates the SLAC machine will certainly reach the region where the sum rule (or electron scattering inequality) should start to get saturated."

Although Adler's special reading of Bjorken's oscillator model did not pay attention to the quasi-elastic scattering from the elementary point-like ingredients of the nucleon, but focused, as people then usually did, on higher resonances, "the sum rule gets saturated at large q^2 by resonances with spin J_n big enough" to compensate the decrease of form factors, Chew's eyes were much sharper, he immediately put his finger on the point:

> the Borken sum rule is physically understandable only if an elementary substructure for the nucleon really exists. Essentially, the statement is that when you hit a nucleon hard enough (with an electron) you get the sum of the cross sections of certain basic constituents. This kind of relation follows from a completeness relation but without an elementary substructure, particle or field, no meaning is possible for "completeness". In a bootstrap regime I would not expect the Bjorken sum rule to be valid.

Facing Chew's serious questioning, Steven Weinberg tried to save the local current algebra by sacrificing the sum rule: "Local commutation relations are true in a model where the currents are Yang-Mill type fields. The sum rule is not true." But Chew would not accept such a compromise, and carried his questioning to a deeper level:

> Let us take the familiar model of a deuteron as a composite of proton and neutron. If you hit this system with an electron whose momentum transfer is small compared to the reciprocal nuclcon size, then the completeness of the wave functions of the deuteron is all you need in order to get the kind of sum rule we are talking about here. The cross section becomes the sum of the neutron and the proton cross sections. But if you hit the deuteron hard enough to start breaking up the nucleons, then you wonder what constituents determine the cross section. Of course, if there were a sequence of well defined substructures you would at a certain level expect, as a reasonable approximation, the sum of cross sections of whatever made up the nucleon at that level. But if you take the strict bootstrap point of view, and if you say that there is no end to this dividing and no fundamental substructure, then I see no reason for expecting high energy cross sections to approach anything simple.

Chew further pointed out that the roots of his reservation with the sum rules was in the principle of locality. According to him, the bootstrap mechanism "implies that there is no completeness of the Hilbert space in the physical sense." We do "have on mass-shell completeness, but there is no complete set of states which could give a meaning to the sum rules" and that was "the reason why I don't expect those sum rules to hold which come from local currents, requiring you to go off the mass-shell."

When Heisenberg criticized Chew being too radical to abandon one of the pillars of quantum mechanics, the notion of the state vector, Chew replied "I agree that it does." But then he retorted that "there is no state vector that makes any sense in this [bootstrap] picture." It makes sense to work with superposition of asymptotic states, or superposition of amplitudes connecting asymptotic states, "but you have neither completeness nor state vector concepts."

Chew's paper: "Closure, locality and the bootstrap"

Soon after the Solvay conference, Chew published a paper (1967) to elaborate his ideas and arguments raised at the conference and articulate, with an admirable clarity, the incompatibility between the closure property of conventional quantum theory and locality principle underlying local field theory on the one hand, and the true bootstrap idea on the other.

Chew framed the confrontation between the two sides within the general consensus then in the hadron physics community, namely, all hadrons are equivalently composite, or the so-called "nuclear democracy" hypothesis. The urgent issue that bothered Chew the advocator of bootstrap was that it was not clear that "democracy necessarily requires a true bootstrap dynamics." For physicists like Sakata and his followers (cf. Section 2.2), it is conceivable that "beneath nuclear matter lies a basic field, unconnected with special hadrons and obeying a simple 'master equation' motion." Chew admitted, regrettably, "no general demonstration ever has been given that a master equation precludes nuclear democracy." Even worse, "a master field would be inaccessible to direct measurement, so one might despair of ever verifying its actuality."

After having set the stage, Chew quickly moved to his real targets, Gell-Mann and Adler. "Gell-Mann, while eschewing reference to a master equation, has observed that an underlying local structure for strong interactions should manifest itself through corresponding commutation relations for physically measurable currents. Adler, furthermore, has proposed sum rules for observable cross sections to test Gell-Mann's commutation relations."

These sum rules, Chew rightly pointed out, "depend on the 'closure' aspect of conventional quantum theory, i.e., the existence of a 'physical Hilbert space' of states with respect to which the product of two physically meaningful operators may be defined." In pure S-matrix theory, however, "there are no operators besides the scattering matrix itself and correspondingly no general concept of closure." Thus there is no room for sum rules of the Adler type.

Chew's basic argument was that "conventional quantum-mechanical closure implies a distinguishable set of degrees of freedom for dynamical system." Here he used the term "degrees of freedom" to "describe a 'minimum' set of commuting operators through repeated applications of which the entire Hilbert space may be spanned." He stressed that "the degrees of freedom may be infinite or continuous," but closure requires "a sense in which a minimal set can be identified." For example, "a fundamental field (or set of fields) provides such a sense." He further clarified that "the fact that the manifestations of such a field may include an infinite number of composite particles should not be confused with the identifiability of its degrees of freedom." Such identifiability, he claimed, is "physically inseparable from the general concept of closure." Thus, without the notion of degrees of freedom, "the notion of closure (or completeness) of the physical Hilbert space" would be "unnatural."

In contrast, Chew stressed, "the essence of the pure bootstrap idea is to abandon meaning for 'dynamical degrees of freedom.'" Bootstrap allowed the evident freedom "enjoyed by the experimenter in setting up initial configurations of stable particles." But this is quite different from possessing "the capacity to set up a 'quantum state' in the sense of Dirac, where a meaning is attached to subsequent state evolution from one instant of time to another. The capabilities of physical measurement require meaning only for free-particle configurations, and it begs the question to assume that the flexibility available to a nuclear experimenter corresponds to the 'degrees of freedom' of a conventional quantum system."

The bootstrap hypothesis is incompatible with the concept of degrees of freedom because it "postulates a dynamical determination of the experimenter's alternatives." Thus although the experimenter does have a certain choice of initial particles, "we see no conceivable sense in which an initial state could be specified without foreknowledge of the solution to the bootstrap conditions." Chew further related "the conflict between a 'minimal' set of operators and the bootstrap" to locality. He admitted it "possible to define 'off-shell' operators connecting asymptotic states of different energy-momentum." But he stressed that "the locality of any operator has meaning only if a 'minimal' set of operators exists. The bootstrap notion, by foregoing such a minimal set, may preclude locality."

To make his message clearer and stronger, Chew clarified several issues with which many physicists had shown great concern.

The first issue was related with the notion of "on-shell closure." Chew acknowledged that "S-matrix unitarity implies 'on-shell' closure within subsets of states sharing a common value of energy-momentum." Thus "commutation relations for conserved quantities like total charge (or isospin) are correspondingly not being called into question" because they "do not involve the full Hilbert space" and "amount to no more than a statement of the symmetry underlying the conservation law." But "commutation relations for charge or current densities" in local current algebra are different. They "require that 'off-shell' closure have a meaning" and thus "are being questioned here."

The second was related to the notion of "approximate closure" and that of "approximate locality." Chew noticed that "when unitarity sums are truncated in approximate S-matrix calculations, certain particles are in effect given an 'elementary' status and provide a basis for 'counting.'" In such a special situation, Chew admitted that one may speak of approximate closure and correspondingly of approximate locality, "but no basis is provided by S-matrix theory for any general notion of closure."

Chew further explored the consequences of approximate locality and claimed that, although by foregoing a physical Hilbert space certain established principles of quantum mechanics are abandoned, "no experimental successes of quantum theory are undermined." The reason for this is simple. "The S-matrix permits (in fact requires) approximate particle localization, on a scale larger than the particle Compton wavelength," and thus "the description of long range interactions by conventional quantum theory remains unquestioned."

However, Chew stressed that the adoption of approximate particle localization should not be confused with the construction of local fields. "The crucial distinction is that particle localization down to the Compton wavelength forms an adequate basis for nonrelativistic quantum theory but not for a more general quantum theory which conforms to the principles of special relativity" or local quantum field theory.

An illuminating example for the approximate localization, Chew indicated, can be found in the quantum treatment of nuclear physics. "The key here lies in the exceptionally small mass of the pion, whose Compton wavelength provides the scale of spatial dimensions for the nonrelativistic physics of hadrons": all other hadrons have smaller "Compton wave lengths on this scale and thus approximately localized."

The third is related with the notion of a formal complete Hilbert space. Chew admitted that "since S-matrix theory employs superposition of different energy-momentum values, and since unitarity implies a Hilbert space for each individual value thereof," inevitably "there should exist a complete Hilbert space." But, Chew argued, "since the S-matrix lacks elements connecting different values of total energy-momentum, there is no compelling reason to expect physical usefulness of this formal space for defining off-shell operator products."

The fourth one is related with the connection between approximate localization and closure. Chew clarified the issue by further elaborating his discussion on the deuteron at the Solvay conference. If the unitarity condition "is truncated so as to include only the two-nucleon channel, and if nonrelativistic kinematics is employed," the construction of a local two-nucleon wave function satisfying a Schodinger equation would be allowed. "There is then a corresponding closure relation which allows the derivation of sum rules." For example, "the total cross section for the scattering of 'low-energy' electrons by a deuteron, summed over both elastic and inelastic final states, can be shown to approach the cross section of a free point-charge proton." Clearly this is exactly what Bjorken took as a model for the construction of his own model for saturation in the case of electron–nucleon deep inelastic scattering.

But Chew immediately warned that "this evidently is only an approximate sum rule" because "the proton is only approximately a point charge. If the electron energy is increased sufficiently, the sum rule becomes inaccurate for large momentum transfers. The truncation of the unitarity sum is now manifestly invalid (reflecting, e.g., the fact that the proton has structure) and the justification for the two-nucleon wave function disappears." This warning, mutatis mutandis, was clearly applicable to the case of Adler's sum rule and Bjorken's inequalities.

Chew recognized that the "sum rules based on the Gell-Mann commutation relations are supposed to have a more general basis than the approximate sum rule." Still, he believed that they were to "rest on the identifiability of underlying degrees of freedom." Having noticed that the CEA data failed the sum rules,[4] Chew conceded that "it is possible that a careful study of Adler's derivation will reveal loopholes which permit violation of his sum rules without abandonment of locality." However, in Chew's opinion, Adler's "sum rules may allow confrontation between an underlying local spacetime structure for strong interactions and a true bootstrap." For this reason, he concluded that "accurate experimental verification of the sum rules" would "constitute a severe blow to the original bootstrap concept."

4.4 Remarks

Several issues reviewed in this chapter had far-reaching consequences, thus a few remarks are in order for situating them in proper perspective.

Intermediate states

A crucial step in the derivation of the sum rule is to insert a complete set of intermediate states into the matrix element of a commutator of two current operators from one state to another. One of the benefits of having a sum rule is to bypass the difficult question of knowing exactly what kind of intermediate states would contribute to the sum rule. In Bjorken's oscillator model of saturation, one only knows that some hadron states, resulted from the quasi-elastic scattering of the electron from the constituents of the nucleon, the sum of which saturates the sum rule, but not which ones. According to Bjorken (1967b), "the origin and the nature of the [hadron] states which saturate the high momentum transfer sum rule is very obscure." The reason is, as later realized and pursued, these hadron states are produced at the end of scattering through a very complicated process of final state interactions, including the unknown process of hadronization of the struck constituent (parton, quark, or gluon).

Dominant region for saturation

By hindsight, it is detectable that in Bjorken's preliminary models there already were hints of the dominance of a regime where the energy transfer ν grows proportionately to the value of momentum transfer q^2. But the nature of the proportionality here was quite different from the one later in his work on scaling (see Section 5.1). As Bjorken recalled (2003), in order to interpret the quasi-free dynamics, he retreated back from the infinite momentum frame, which was adopted ever since Fubini and Furlan as the simplifying element in understanding dynamics, to the rest frame of the nucleon and used the inferior concept of mass fraction, expressed in terms of the effective mass of the constituents as seen in the rest frame of the nucleon, rather than the momentum fraction in the infinite momentum frame, to interpret what was happening. Two reasons for such a retreat, as Bjorken admitted later. First, he needed Richard Feynman's help to clarify the understanding of inclusive distributions so that he could generalize it from electron to the constituents of nucleon. Second, it required a proper understanding of light-cone quantization of field theories (for more on this see Section 6.3).

Inelasticity and atomicity

If sum rules are not satisfied up to some large inelasticity, then naturally people would assume either states important for saturation have a very large mass, or just the same, there is a subtracted dispersion relation. With Bjorken's oscillator model, however, the focus was shifted from resonance region to quasi-elastic scattering off the constituents. But, as Bjorken (1967c) noticed, the data from CEA had reached into the quasi-elastic region, and penetrated far enough to make him worried because the quasi-elastic model did not look good from the data, although the inelasticity in CEA is too small and not inelastic enough to draw conclusions.

What if no quasi-elastic peak is found at large momentum transfer ? Must one abandon locality? Of course, one can always argue that the inelasticity may not be large enough if the constituents had small effective masses or too small probability of constituents being found inside a nucleon; in those situations, the quasi-elastic peak might not be observable at available energies. But with the diffraction model in mind, Bjorken had another possibility, namely that the diffraction region, which requires larger inelasticity at given momentum transfer and is important for saturation, has not been reached yet. He declared that "it will be of interest to look at very large inelasticity and dispose without ambiguity of the model completely."

The model Bjorken mentioned did not necessarily refer to the quark model which he used frequently in his works. But rather, his use of the quark model was driven by a deep feeling about locality or, rather, atomicity more accurately. At the end of the discussion period of his SLAC talk (1967c), he explicitly wished to "disassociate myself a little bit from this as a test of the quark model. I brought it in mainly as a desperate attempt to interpret the rather striking phenomenon of a point-like behavior. One has this very strong inequality on the integral over the scattering cross section. It's only in trying to interpret how that inequality could be satisfied that the quarks were brought in."

Atomicity and structure

Sum rules had a built-in locality assumption, and Bjorken's models for saturation explicitly assumed the existence of point-like constituents of the nucleon. Thus, although the sum rule neither adopted any master equation for fundamental fields nor formulated in any local field theory language, but utilized, for its derivation, dispersion relations which were in the general framework of S-matrix theory, Chew was justifiably seeing it as having posed a serious challenge to his bootstrap philosophy and exemplified a head-on "confrontation between an underlying local spacetime structure for strong interactions and a true bootstrap."

Chew in his response to the challenge questioned a set of related concepts underlying the sum rule: locality, closure or completeness, degrees of freedom. But its upshot was questioning against the atomistic thinking in hadron physics and defending *a bootstrap version of holism or structuralism: hadrons are placeholders in the network of relations without being independently definable relata.* Indeed, when Chew questioned the conceptual foundations of the sum rule, closure or completeness, and locality, his objection aimed directly at the very notion of atomicity, which, with its connotation of undividability, grounds the notion of point-like locality, and with that of finiteness, grounds the notion of closure, which, according to Chew, "implies a set of degrees of freedom describing a 'minimum' set of commuting operators through repeated applications of which the entire Hilbert space may be spanned."

But atomicity is not a tenable notion, according to Chew, because with impinging energy and momentum transfer sufficiently large, a defining condition for Adler's sum rule, a proton or any other hadron would reveal its internal structure. There is no end to dividing, and thus no fundamental substructure is definable. All are composite, none elementary. That is the true bootstrap philosophy: hadrons are structures without any atomicity involved.

As a physicist, Chew would allow the use of notions of locality and closure and all those notions related with or derived from them, but only in an asymptotic (on-shell) or approximate sense, not in the off-shell and absolute sense. In the true bootstrapping world in which all are dividable, no point-like locality and no closure could be defined.

Chew was right in the sense that no locality and closure could be defined in the absolute sense because, true, no end of dividing could be conceived. But if we take dividing in a broader sense of including divisions into constituents, rather than the sense of bootstrapping in which all those dividable are sitting at the same level, then practically or scientifically, Chew's objection to atomicity missed an important sense. As a result, he also missed an important philosophical sense.

Practically or scientifically, as Chew himself clearly realized, atomicity could be well defined in each specified energy region. For example, the atomicity of hadrons can be defined in terms of the pion's Compton wavelength. Chew would argue that this kind of approximately defined atomicity would collapse once the momentum transfer is large enough to reach a spatial region that is smaller than the pion's Compton wavelength.

That is true. But the philosophically interesting point that Chew missed is that with the collapse of atomicity at the hadron level, another well-defined atomicity emerges from a deeper level, for example, at the level of hadron's constituents.

Chew was right that nothing is undividable, and all is structural. But it is also true that, contrary to the conception of bootstrapping, the true picture of the physical world is that all structures are structured by structureless or atomic constituents at any particular stage of human practice and conception. It is the recognition of this point that has driven physicists to dig into deeper layers of the physical world, fruitfully. In the case of the confrontation Chew addressed, it was the other side of the confrontation that won the battle; the victory soon led to the creation of QCD, and struck "a severe blow to the original bootstrap concept," the result of which was the irreversible decline of bootstrap in general, and Chew's influence in particular, in the hadron physics community, although the sum rule itself remained to be approximately rather than absolutely true: in any case, once quarks and gluons reveal their own internal structure, which is conceivable in the future, it would be unclear how to define the sum rule in that novel situation.

Notes

1 Adler's sum rules for neutrino scattering were special cases of the more general local current algebra sum rule derived by Fubini (1966) and Dashen and Gell-Mann (1966). However, as Bjorken noticed that since the integrand of these general sum rules was not expressible in terms of measurable structure functions, Adler's sum rule in fact was "the only case of which I know where these general sum rules of Fubini *et al.* are at all susceptible to a direct experimental test" (1967c).

2 Bjorken (1967c) noticed that "if the Regge ideas are not correct there might be (in the Regge language) fixed singularities near $J = 1$ for the exchange of isospin one objects. The particle and antiparticle neutrino cross section will then not approach each other at high energies," and there would be no cancellation. But then one of the underlying assumptions in Adler's derivation of his sum rule, the unsubtracted dispersion relation would be violated, and the whole conceptual foundation of Adler's sum rule and its offsprings would collapse (see also Bjorken and Tuan, 1972).

3 Unless otherwise specified, all materials in this section are taken from *Fundamental problems in Elementary Particle Physics* (Proceedings of the Fourteenth Conference on Physics at the University of Brussels, October, 1967), Interscience Publishers, 1968.

4 The confusions caused by the CEA data were reviewed by Adler at the Solvay conference and also by Bjorken at the SLAC conference.

5
Scaling

The notion of scaling conceived and proposed by Bjorken in 1968[1] played a decisive role in the conceptual development of particle physics. It was a bridge leading from the hypothetical scheme of current algebra and its sum rules to predicted and observable structural patterns of behavior in deep inelastic lepton–nucleon scattering. Both its underlying assumptions and subsequent interpretations had directly pointed to the notion of constituents of hadrons, and its experimental verifications had posed strong constraints on the construction of theories about the constituents of hadrons and their interactions.

The practical need for analyzing deep inelastic scatterings planned and performed at SLAC in the mid to late 1960s had provided Bjorken the general context and major motivation to focus on the deep inelastic kinematic region in his constructive approach to the saturation of sum rules derived from local current algebra. The deep inelastic experiments themselves, however, were not designed to test Bjorken's scaling hypothesis. Rather, the experimenters, when designing and performing their experiments, were ignorant, if not completely unaware, of the esoteric current algebra and all those concepts and issues related with it. They had their own agenda. This being said, the fact remains that the important implications of their experiments would not be properly understood and appreciated without being interpreted in terms of scaling and subsequent theoretical developments triggered by it. That is to say, scaling in deep inelastic scattering as an important chapter in the history of particle physics requires for its understanding a proper grasp of the positive interplay between the MIT–SLAC experimenters and Bjorken and other theorists.

5.1 Hypothesis

It was clear to Bjorken that for the analysis of inelastic scattering, the current algebra sum rule machinery was not sufficient. The reason was simple. According to the general "truths" expressed by Drell and Walecka (1964),[2] the major concern for the experimenters was the actual size and shape of the structure functions as a function of their arguments v, which is responsible for the convergence of the sums, and q^2, which is responsible for the asymptotic sum rules to be relevant experimentally. Thus detailed assumptions of the behavior of the structural functions with respect to v and q^2 would be desirable.

In his preliminary models for saturation, the harmonic oscillator model and the diffractive model, the notion of effective mass of the constituents of the nucleon played a crucial role. The underlying picture for the harmonic oscillator model was the quasi-elastic scattering of the electron off the nucleon's constituents with an effective mass $\overline{m}$. The sum rule would be saturated if $v \simeq q^2/2\overline{m}$. In the diffractive model, the underlying picture was the same. The only difference lies in the inelasticity, $v \geq q^2/2\overline{m} + m_\rho$, so that more energy would be available for the production of vector mesons. But for large q^2, there is essentially the same fixed relationship between v and q^2 as from the harmonic oscillator model, that is, $\overline{m} = q^2/2v$.

It is worth noting that in Bjorken's above reasoning for saturation, what was important was not the specific values of v and q^2, but their fixed proportionality when both tend to ∞. Originally, the proportionality was expressed in terms of the effective mass of constituent quarks. But in fact, Bjorken was not particularly fascinated by the constituent quark model for various reasons.[3] This made it easier for him to move away from the constituent quark model as an underpinning picture in his further explorations, while maintaining the desirable proportionality between v and q^2, which is related to the notion of the electron's quasi-elastic scattering off the nucleon's constituents, and in fact characterizes the constituents' behavior in scattering.

One obvious defect in the notion of effective mass, in his exploration of the saturation in terms of the actual size and shape of the structure functions as a function of v and q^2, was that while the structure functions themselves were dimensionless, the effective mass is a fixed dimensional constant. Clearly it was attractive to move from a dimensional constant to a dimensionless variable. Bjorken took this ingenious move, tentatively, first in the spring of 1967 (Bjorken, 1967a), and then, over the summer of 1968, elaborated the move into an transformative notion of scaling (Bjorken, 1968, 1969).

In order to characterize the structure functions as a function of ν and q^2 by a dimensionless variable while maintaining proportionality between the variable ν and q^2, Bjorken introduced a new variable $\omega = -q^2/M\nu$ which is obviously dimensionless. With the help of such a new variable, Bjorken introduced a new notion of limit, a limit later called the "Bjorken limit" in which both ν and q^2 approach infinity, with the ratio $\omega = -q^2/M\nu$ held fixed.[4]

Assuming, as he did before, that the structure functions W_2 and W_1 in inelastic lepton–nucleon scattering could be related to matrix elements of commutators of currents at almost equal times at infinite momentum if the $q_0 \rightarrow i\infty$ method for asymptotic sum rules and the $P \rightarrow \infty$ method of Fubini and Furlan were combined, Bjorken conjectured that if the "Bjorken limit" for these commutators does not diverge or vanish, then the two quantities νW_2 and W_1 become functions of ω only. That is, $\nu W_2(\nu, q^2) \rightarrow f_2(\nu/q^2)$ and $W_1(\nu, q^2) \rightarrow f_1(\nu/q^2)$ as ν and q^2 tend to ∞.

More precisely, Bjorken (1968/69) predicted that

$$\lim_{q^2\rightarrow\infty,\frac{\nu}{q^2}\,fixed} \nu W_2(q^2,\nu) = F_2(-q^2/M\nu), \tag{5.1}$$

and

$$\lim_{q^2\rightarrow\infty,\frac{\nu}{q^2}\,fixed} MW_1(q^2,\nu) = F_1(-q^2/M\nu) \tag{5.2}$$

with

$$F_t(\omega) \equiv F_1(\omega) = \frac{-i}{\pi}\lim_{P_Z\rightarrow\infty}\int_0^\infty d\tau \sin\omega\tau \int d^3x \left\langle P_Z \left| \left[J_x\left(x,\frac{\tau}{P_0}\right), J_x(0)\right] \right| P_Z. \right\rangle \tag{5.3}$$

and

$$\begin{aligned} F_l(\omega) &\equiv \frac{F_2(\omega)}{\omega} - F_1(\omega) \\ &= \frac{i}{\pi}\lim_{P_Z\rightarrow\infty}\int_0^\infty d\tau \sin\omega\tau \int d^3x \left\langle P_Z \left| \left[J_Z\left(x,\frac{\tau}{P_0}\right), J_Z(0)\right] \right| P_Z. \right\rangle. \end{aligned} \tag{5.4}$$

Bjorken's scaling hypothesis, once it was conceived and circulated in the high energy physics community, immediately became a very effective means for the analysis and understanding of experimental results that were quickly accumulated in SLAC and some other laboratories.

5.2 Experiment

By hindsight, the SLAC deep inelastic electron–proton scattering (DIS) experiment was inseparably intertwined with Bjorken's theoretical conjecture of scaling. When it was conceived, planned, designed, and initially performed, however, it was an exploration within its own experimental and theoretical tradition, had its own research agenda, which had nothing to do with the line of thinking in current algebra and its offspring: sum rules, models, or mechanisms for saturating these sum rules, and certain specific behavior patterns derived from one or the other mechanism, such as scaling from the quasi-elastic scattering mechanism. The later history of its being appropriated by the Bjorken scaling and its constraining, not only the research line of current algebra, but almost all theoretical thinking in hadron physics, has convincingly shown that it is wrong to assume that the only or major function of experiment in scientific development is to test the validity of theoretical conjectures, or to assume that the only or major function of theorizing is to understand and conceptualize independently existing and accumulating experimental results. But before drawing the lessons from this piece of history and grasping its implications to the general issue of experiment–theory relationship, let us have a closer look at the experiment first.

Context

At the time the MIT–SLAC group started its deep inelastic electron–proton scattering experiments in the summer of 1967, there was no detailed model of the internal structures of the hadrons known to the experimenters. This, however, does not mean that they started from scratch. On the contrary, the experiments were conceived, planned, and conducted within a very successful tradition of experimentation together with an implied picture, vague though, of what a nucleon looked like and how would it behave in various experimental situations.

There was a long tradition at Stanford of using the electron as a probe to study the nuclear structure and nucleons, a tradition established mainly by Robert Hofstadter and his collaborators. When the updated traveling wave linear accelerator, the MARK III linac, was able to produce a high energy electron beam (up to 225 MeV) in 1953, Hofstadter and his collaborators produced the first evidence of elastic electron scattering from the proton (Hofstadter, Fechter and McIntyre, 1953). Immediately afterwards, the inelastic electron scattering was carried out. Through the inelastic breakup of the deuteron by the scattered electron, the constituent behavior, acting incoherently, of neutron and proton in the deuteron was also observed (Fregeau and Hofstadter, 1955).

Conceptually, as we mentioned before, Rosenbluth realized in (1950) that the momentum transfer q^2 is a measure of the ability to probe the structure of the proton. The uncertainty principle limits the spatial definition of the scattering process to $\sim h/q$, so q^2 and therefore the energy of the electron must be large in order to resolve small structure. His expression of the differential cross section (Equation 4.1), similar to Rutherford's, with two form factors, which are functions of q^2, describing the distributions of charge and magnetic moment, was the starting point for analyzing experimental information about the electromagnetic structure of the proton.

In their pioneering studies of the proton's form factors, Hofstadter and his student found from the elastic electron–proton scattering that the form factors in the Rosenbluth cross section were less than unity, and were decreasing with increasing momentum transfer. On this deviation from the point-charge behavior, they remarked, "if we make the naïve assumption that the proton charge cloud and its magnetic moment are both spread the same proportions, we can calculate simple form factor for various values of the proton 'size.'" From the results of calculations they estimated that the proton had a size of about 10^{-13} cm and a smooth charge distribution (Hofstadter and McAllister, 1955).

These experiments marked the beginning of the search for the substructure in the proton; they also provided a background to the MIT–SLAC group. The discovery that the proton behaves not like a point, but displays a spatial extent, however, did not immediately question the elementary character of the proton. The reason was simple: the only clues about the proton's spatial extent came exclusively from elastic scatterings, rather than inelastic ones. These clues thus suggested only a smooth spatial distribution of charge and magnetic moment within the proton, rather then a discernable structure of the proton's constituents.

In terms of the behavior of form factors, the picture of the proton having a smooth distribution of charge and magnetic moment was first implied and then further supported by Hofstadter's observations, which can be summarized as follows. If the charge and magnetic distributions were small compared with h/q, then the form factors would not vary as the momentum transfer in the scattering changes; but if the size of those distributions was comparable with h/q, then Hofstadter and his collaborators discovered by the mid 1950s that as the momentum transfer in the scattering increased, the contributions of the form factor to the scattering cross section dropped sharply relative to the contributions from a point charge. The rapid decrease of the cross sections as the momentum transfer increased was also the predominant characteristic of the high energy hadron–hadron scattering data of the time, which was compatible with the picture that hadrons were soft inside and would yield distributions of scattered electrons which reflected their diffused charge and magnetic moment distributions with no underlying point-like constituents.

The same behavior pattern was also observed in the scattering from deuterium. The form factor for elastic scattering from the loosely bound deuterium nucleus fell off rapidly with increasing momentum transfer. Thus the neutron could be studied through quasi-elastic scattering from either the proton or the neutron, which together form the deuterium nucleus. The quasi-elastic scattering was spread out over a wider range of energies than the scattering from the free proton because of the momentum spread of the nucleons in the nucleus. It reached a maximum near the location of the peak from electron–proton scattering since the scattering took place off a single nucleon and the recoil energy was largely determined by the mass of the nucleon.

The success in having determined the proton and neutron size, and the hard work on the detailed shape of the form factors, prompted Hofstadter and his colleagues at Stanford to see clearly that the next great advances would require a much larger electron linear accelerator that could provide much higher energies. Such a recognition ultimately led to the construction of SLAC.

Theoretically, the foundation for understanding the nucleon's form factors was first laid down by Yoichiro Nambu (1957). Nambu was the first to suggest that the form factor of the proton and neutron could be explained by the existence of a new heavy neutral isoscalar meson, which he called ρ and is now called ω. That particle was discovered in 1961. The isovector analog of his discussion is the particle now called ρ. Both play a role in the electromagnetic form factors: the ρ dominates the isovector form factors, and the ω dominates the isoscalar form factors. But more important was Nambu's mass-spectral representation of the nucleon's form factor, according to which the nucleon form factors could be written simply as a sum in terms of the masses of the intermediate states, particularly those states with the lightest masses. Nambu's approach was justified, on the basis of the dispersion theory, by Chew and his collaborators (Chew, Karplus, Gasiorowicz, and Zachariasen, 1958), by P. Federbush and his collaborators (Federbush, Goldberger and Treiman, 1958), and by Frazer and Fulco who gave a treatment of the isovector nucleon form factor in agreement with experiment (1959, 1960), and soon became the standard approach to the electromagnetic structure of the nucleon.

It was within this experimental tradition and theoretical context, including the background picture of the nucleon, the behavior patterns of the form factors, and the theoretical understanding of the form factors in terms of the sum of intermediate states (resonances), that the MIT–SLAC group conceived, designed, and conducted its experiments, which, unexpectedly to the practitioners, turned out to be path-breaking events, leading away from the old disciplinary matrix to a new paradigm, with a new picture of the nucleon and other hadrons, searching for new behavior patterns of the form factors, and for a new understanding of the form factors, not in terms of the sum of resonances, but in terms of contributions from the scaling limit of the kinematic region.

Plan

A major objective of the two-mile long electron linear accelerator was the investigation of the structure of the nucleon. The 20 GeV energy of the new machine made both elastic and inelastic scattering experiments possible in a new range of values of momentum transfer q^2. In comparison, the two existing large electron synchrotons, the Cambridge Electron Accelerator (CEA) and the Deutsches Electronen Synchrotron (DESY) had peak energies only of 5 GeV and 6 GeV respectively.

The focus of thinking during the construction period centered on the excitation of resonances and the q^2 dependences of the transition form factors when the nucleon makes a transition from the ground state to the resonant state.[5]

Originally a group of experimenters from Caltech was also part of the collaboration. The Caltech–MIT–SLAC collaboration prepared a proposal that consisted of three parts, elastic scattering measurements, inelastic scattering measurements, comparison of positron and electron scattering cross sections. It is clear that the elastic experiment was the focus of interest at that juncture. But for completeness, the proposal also included an examination of the inelastic continuum region (the deep inelastic region) since this was a new energy region which had not been previously explored.

The inelastic scattering measurements began in August 1967, using a 20 GeV spectrometer, when the elastic part of the project was completed and the Caltech group had withdrawn from the collaboration. The method adopted in the experiments was traditional, consisting of measuring the spectra of inelastically scattered electrons as a function of the momentum and energy transfer to the target system.

What was expected from the inelastic scattering experiments was, understandably, mainly shaped by the research tradition at Stanford summarized above, within which the whole project was conceived.

More precisely, the inelastic studies, initiated by Wolfgang Panofsky and E. A. Alton in 1958, concentrated mainly on measurements of electroproduction of nucleon resonances. For example, in the studies previously carried out at the mark III accelerator of Stanford the (3,3) resonance was excited and studied as a function of four-momentum transfer up to about 0.79 GeV. The CEA experiments observed evidence for the excitation of 1.512- and 1.688-GeV resonances at four-momentum transfer up to about 1.97 GeV. Thus what was expected and what was observed in the initial stage of the experiments, was mainly the appearance of resonances, which now included the resonances at 1.236 Gev, 1.510 Gev, 1.688 GeV, and 1.920 Gev, and the rapid fall off of the resonances with q^2 for $q^2 \sim 3$–4 Gev^2 and above. Since the inelastic cross section contains q^2 dependences which are similar to that of the form factors describing elastic scattering, the counting rates drop rapidly as a function of q^2. Thus the real expectation was simply that the higher energies and intensities available at SLAC would (only) enable both the scope and precision of measurements in the electroproduction of resonances to be greatly enlarged. Clearly this was a very modest and conservative expectation.

Since the main purpose of the inelastic scattering experiments was to study the electroproduction of resonances as a function of q^2, and it was thought that higher mass resonances might become more prominent when excited with virtual photons, it was the intent of the MIT–SLAC group to search for these resonances at the very highest masses that could be reached. The hope was to learn more about each of the observable resonances, and to find new resonances. Initially, however, nothing unexpected was found: the transition form factors fell about as rapidly as the elastic proton form factor with increasing q^2.

There was a change of plan in the early summer of 1968. In a letter sent to the Program Coordinator by Elliott Bloom and Richard Taylor dated May 10, 1968, what was proposed was "a measurement of the total photo absorption cross section $\sigma_{\gamma P}$ from threshold to $W_{\gamma p} \sim 5$ GeV. The measurement would be accomplished by placing the 20 GeV spectrometer at $\theta = 1.5^0$, and hence, observing low q^2 electroproduction from proton" (Bloom and Taylor, 1968). In the letter sent to the program coordinator by Taylor alone dated 26 July 1968, however, the direction of the research began to change, and a new research agenda began to emerge.

Two factors contributed to the change. First an unexpected feature emerged from the data collected. Second, Bjorken's direct intervention.

Taylor noticed at the beginning of his letter an unexpected feature that "the dependence of the deep inelastic scattering (missing mass W > 2 GeV) on q^2 is much weaker than the elastic dependence. This scattering, therefore, becomes the dominant feature of the data even for moderate q^2 (~ 4 GeV^2)." The weak momentum transfer dependence of the inelastic cross sections for excitations well beyond the resonance region was clear cut to the experimenters because the scattering yields at the larger values of q^2 were found to be between one and two orders of magnitude greater than expected.

Within the research tradition in which the MIT–SLAC group conducted its experiments, the weak q^2 dependence was unexpected and demanded an explanation. The experimenters did not realize that the feature was quite understandable in the current algebra sum rule frame. In fact, one of the most distinguished features of Adler's sum rule was that asymptotically, the sum rule is independent of q^2 (see Section 3.2). Still, the appearance of the feature paved the way for the experimenters to assimilate some of the ideas and concepts from the current algebra frame for the understanding of their results, and for a more active interplay between the current algebra approach and the DIS experiments.

Although the main objective of the project was still "a separation of the transverse and longitudinal (or W_1 and W_2) contributions to the cross section for the deep inelastic region $W > 2$ GeV" and "a large amount of resonance information will be collected," Taylor in the letter did report, for the first time, that "we plan to submit a proposal on deuterium inelastic scattering to check Bjorken's sum rules:

$$1)\quad \int_0^\infty d\nu\left[W_{2p}(q^2,\nu)+W_{2n}(q^2,\nu)\right] > \frac{1}{2}$$

$$2)\quad \lim_{q^2\to\infty}\int_0^\infty d\nu\left[W_{2p}(q^2,\nu)-W_{2n}(q^2,\nu)\right] > \frac{1}{3}."$$

Taylor even suggested that "from our present data, we can say that the contribution of W_{2p} to the first integral appears to give a reasonable expectation that the sum rule will be satisfied." The interest in the sum rule now began to be translated into real experimental research interest: "we would like to get a rough idea of the behaviour of the cross sections for Deuterium to help us construct a proposal" (Taylor, 1968).

Bjorken's intervention (for example, through his exchanges with Henry Kendall) involved more than the sum rule. In fact, he suggested a study to determine if νW_2 was a function of ω alone, that is, if his scaling hypothesis was valid or not (Kendall, 1991). Clearly then and even more prominently now by hindsight, Bjorken's intervention had a profound impact on the DIS experiments.

Results

The scattered electron spectra observed in the DIS experiment conducted by the MIT–SLAC group had a number of features whose prominence depended on the initial and final energies and the scattering angle.[6] At low q^2, both the elastic peak and the resonance excitation were large. The elastic cross sections measured at SLAC behaved in much the same way as those measured at lower energies, falling on the same simple extrapolation of the earlier fits as the CEA and DESY data. As q^2 increased, the elastic and resonance cross sections decreased rapidly as expected, with the continuum scattering becoming more and more dominant.

Two features of the non-resonant inelastic scattering that appeared in the first continuum measurements were unexpected. The first was a quite weak q^2 dependence of the scattering at constant W as we mentioned above. The second feature was the phenomenon of scaling.

Kendall (1991) reported that during the analysis of the inelastic data, Bjorken suggested a study to determine if νW_2 was a function of ω alone. To test for scaling it is useful to plot νW_2 for fixed ω as a function of q^2. Constant scaling behavior is exhibited in such a plot if νW_2 is independent of q^2. Note that scaling behavior was not expected where there were observable resonances because resonances occur at fixed W not at fixed ω. Nor was it expected for small q^2 because νW_2 cannot depend solely on ω in this limit. But the earliest data so studied by Kendall and his colleagues in the group for 10 values of q^2, plotted against ω, showed that the Bjorken scaling, $\nu W_2(\nu, q^2) = F_2(\omega)$, to a good approximation, was correct.

Kendall, "who was carrying out this part of the analysis at the time, recalls wondering how Balmer may have felt when he saw, for the first time, the striking agreement for the formula that bears his name with the measured wavelengths of the atomic spectra of hydrogen" (Kendall, 1991).

Bjorken's original inequality for electroproduction (Equation (3.15)), expressed in terms of structure functions, $\int_0^\infty d\nu\left[W_{2p}(q^2, \nu) + W_{2n}(q^2, \nu)\right] \geq \frac{1}{2}$, was also evaluated as planned, and found to be satisfied at $\omega \simeq 5$. This certainly was consistent with the scaling behavior displayed in the data.

More data showed that scaling holds over a substantial portion of the ranges of ν and q^2 that have been studied. That is, νW_2 was shown to be approximately independent of q^2, or to scale in the region $1\ \text{GeV}^2 < q^2 < 20\ \text{GeV}^2$, $4 < \omega < 12$, and $W > 2.6$ GeV. In addition, the deuteron and neutron structure functions showed the same approximate scaling behavior as the proton.

But one of the most remarkable aspects of the earliest results was that the approximate scaling behavior, which was presumed to be a property asymptotically reached in the Bjorken limit, set in at surprisingly nonasymptotic values of $q^2 > 1.0\ \text{GeV}^2$ and $W > 2$ GeV. The theoretical significance of such a precocious scaling will be discussed in Section 5.4.

5.3 Interpretation

For physicists in the category of Bjorken and Chew, the physical significance of scaling was clear cut: the scaling hypothesis presumed and derived from the existence of point-like constituents of the nucleon, and thus its experimental confirmation was none less than giving a direct evidential support to the constituent structure of the nucleon. For others, in particular for experimenters, the experimenters in the MIT–SLAC group included, it took time to digest what the physical significance of scaling was. Thus, when Jerome Friedman, one of the leaders of the MIT–SLAC collaboration, gave the first public presentation of their preliminary results, namely the weak dependence of cross section on the momentum transfer during the scattering, and the scaling feature of structure functions, in a parallel session at the 14th International Conference on High Energy Physics in Vienna, August–September, 1968, no interpretation of their data was given (Friedman, 1968). The reason was simple: the group had reached no interpretation of their data.[7]

By hindsight, a suggestion presented by the director of SLAC, Wolfgang Panofsky in his Rapporteur's plenary talk at the same conference might be viewed as the first public interpretation of the MIT–SLAC results and the first public announcement of a constituent picture of the nucleon when he said, now "theoretical speculations are focused on the possibility that these data might give evidence on the behavior of point like, charged structures within the nucleon" (Panofsky, 1968).

This, however, is too Whiggish a reading of the history. Initially, Panofsky was a member of the collaboration. But when the project moved to the inelastic part, he had already withdrawn from the group. Not even being a member of the group, he could not be authorized by the group to give a public interpretation of the group's data, let alone the fact that the group itself had no interpretation of data at all.

Of course, Panofsky's suggestion was an interpretation of the data. But it was his personal interpretation, not the official interpretation made by the experimenters who produced the data. It was reported at the conference solely because of his status in the particle physics community and his being a Rapporteur for a plenary session of the conference.

As an interpretation, however, Panofsky's suggestion made little impact on the community in terms of understanding the SLAC experiment data. The reason for this is simple. The dominant view about hadrons at the time was the nuclear democracy and bootstrapping, not the constituent model. Of course, the constituent quark model had been around for a few years already, and it had enjoyed some successes in understanding the hadron spectrum. But the failure in the search for the existence of free quark in numerous accelerator and cosmic-ray investigations and in the terrestrial environment required strong interactions among quarks or assigning quarks very large masses, or both, to prevent them to be observed. Nobody had any idea about how quarks interact with each other, and to many physicists assigning large masses was too *ad hoc* to be a satisfactory solution. Thus many physicists, Gell-Mann the initiator of the quark model included, at the time when the SLAC results were first made public still took quarks to be merely a mathematical device of heuristic value, not the real constituents of the nucleon.

But the situation changed drastically soon afterwards, mainly because of the intervention of Richard Feynman.

Feynman paid a short visit to SLAC in the late August of 1968, and had seen the early results that were about to be presented at the Vienna conference. He immediately saw in his own idea of partons an explanation of both the weak momentum dependence and the scaling phenomenon in the SLAC deep inelastic scattering data.

Over years, Feynman had developed a model for hadron–hadron interactions, in which the point-like constituents of one hadron interact with those of another in the infinite momentum frame. Feynman called these point-like constituents of hadrons partons, which were identified with the fundamental bare particles of an unspecified underlying field theory of the strong interactions (Feynman, 1969a, b). It was ideal for Feynman to apply his model to deep inelastic electron–nucleon scattering, because, in this case, both the electron's interaction and structure were known, whereas in his original case of hadron–hadron scattering, neither the structures nor the interactions were understood at the time. Thus in his second visit to SLAC in October 1968, Feynman gave the first public talk on his parton model, which stimulated much of the interpretative attempts to conceptualize the SLAC data.

According to the idea of partons, in the case of electron–proton scattering, the proton in the infinite momentum frame looked like a beam of constituent partons, whose transverse momentum could be neglected and whose internal motion would be slowed down by the relativistic time dilation, so that during the scattering the internal motion and interactions could be ignored. As a result, the incoming electron sees and incoherently scatters from partons that do not interact with each other during the time in which the virtual photon is exchanged, and thus the impulse approximation of quantum mechanics is valid; that is, electrons scattered from partons that were free, and thus the scattering reflected the properties and motion of the partons.

Viewing each parton as the carrier of a fraction of the total momentum of the proton, i.e., $P_{parton} = xP_{proton}$, here x is related to the Bjorken scaling variable ω, $x = 1/\omega$, the phenomenon of scaling could be seen as a natural consequence of considering protons as composed of point-like structures and carriers of the total momentum of the proton, because then it was easy to show νW_2 to be the fractional momentum distribution of the partons within a proton, depending only on $\omega = 1/x$, weighted by the squares of their charges. That is, the scaling function $F_2^e(x) = \nu W_2(x)$ can be expressed as $F_2^e(x) = \sum_N P(N)\left(\sum_i Q_i^2\right)_N xf_N(x)$, where all the possibilities of having N partons is denoted by $P(N)$ and the momentum distribution function of the partons is denoted by $f_N(x)$, the charge of the i-th parton by Q_i; here the functions $f_N(x)$ and $P(N)$ were empirically assumed, and thus were model dependent rather than properties abstracted from current algebra.

In the case of Adler's sum rule for neutrino scattering, which in terms of the scaling variable $x = q^2/(2M\nu)$ reads $2 = \int_0^1 \frac{dx}{x}\left[F_2^{(-)}(x) - F_2^{(+)}(x)\right]$ and in which the q^2 independence is evident,[8] the integral of the structure function over x would be interpreted as the total fraction of the nucleon's momentum carried by the constituents of the nucleon that interact with the neutrino. It is consequential (cf. Section 6.1) because this directly measures the fractional momentum carried by the quarks and antiquarks, and thus the remaining part of the nucleon's momentum could be interpreted as to be carried by the strong gluons that were assumed to have no weak interaction couplings.

As to the weak q^2 dependence, if electron–parton scattering is similar to electron–electron scattering, as two point-like objects collide, rather than one point-like object colliding with an extended object, then clearly whether the collision would occur or not is independent of the momentum transfer.

All these ideas were in perfect harmony with what Bjorken elaborated in his Varenna summer school lecture (1967b), in which deep inelastic scattering was treated as incoherent scattering of the lepton from the beam of point-like constituents, and the sum rule was derived as the incoherent superposition of point cross sections.

With the publications of Bjorkan and Paschos (1969) and Feynman (1969a, b), and mainly because of Feynman's influence and authority in the physics community, the parton interpretation of scaling soon prevailed. By the time of the 15th International Conference on High Energy Physics held in Kiev in 1970, even the very cautious SLAC experimenters felt reasonably convinced that they were seeing constituent structure in their results, and made comparisons with the remaining competing models, the most important of which related to the idea of vector meson dominance (VDM) (see Friedman, 1997).

Indeed when the SLAC data was made public, the existing theoretical climate, as felt by Bjorken (1997), was strongly conditioned by VMD. The general idea of VMD was that the photon converted in the vacuum to a fluctuation of hadronic matter, and that then the fluctuation did the interacting with the nucleon. For a real γ, the VMD hypothesis that the fluctuation was a ρ, ω, or Φ worked reasonably well. For virtual photons, as in the case of deep inelastic electron scattering in the one photon approximation, high mass states must be added. Basically VMD anticipated a short-distance behavior that was softer by a power of q^2 and could make a reasonable guess for the probability of finding a state of mass m in the virtual photon. The idea had a respectable theoretical underpinning called field algebra in which the electromagnetic current is proportional to a spin one field (Lee, Weinberg and Zumino, 1967), and later on it was carried forward vigorously by Jun Sakurai.

Sakurai (1969) in fact made a calculation of the SLAC deep electron–proton inelastic scattering based on VMD, and obtained a result in which the scaling of νW_2 was implied in the large ν and fixed ω limit, and the characteristic diffractive limit that νW_2 equals a constant as $\omega \to \infty$ was given. It also gave a prediction that the ratio between the photoabsorption cross sections of longitudinal and transverse virtual photon, $R = \sigma_s(q^2, W)/\sigma_t(q^2, W)$, had a strong dependence on q^2. The prediction, however, was in gross disagreement with experiment in the kinematic region in which the two cross sections were separately determined.

A related interpretation based on the diffractive model was ruled out by the measurement of the ratio σ_n/σ_p, which also made other non-constituent interpretations, such as those based on the Regge model, the duality model, and resonance models untenable.[9] As to the field algebra itself, it was also ruled out soon through the analysis given by Callan and Gross and the measurement of the ratio R (for more on this see Section 6.1).

Thus, within a year or so, the constituent interpretation, in terms of partons, of scaling prevailed, although what these partons were was clear to nobody.

5.4 Remarks

Bjorken's conjecture of scaling and its experimental confirmation by the MIT–SLAC collaboration was a turning point in the history of particle physics. Before that, the investigations were either phenomenological or speculative. After that, the investigations were reoriented toward the clarifications of the hadron's fundamental constituents and their dynamics that underlie the observed hadronic phenomena. Now, the expectations of experimenters were no longer restricted to the improvement of the scope and precision of their measurements of the phenomena that was shaped by the structureless picture of the nucleon and hadron in general; and the concerns of theoreticians were no longer restricted to the verification of their predictions derived from their model-independent esoteric schemes of current algebra or analytic S-matrix theory. Rather, a vast range of processes and phenomena that were out of reach, meaningless or even invisible, such as the Drell–Yan dilepton production or jet production in hadron–hadron collisions and the hadron production in electron–positron collisions, now through the lens of scaling and its parton model interpretation became meaningful to explore. For a proper appreciation of such a paradigm shift, a few remarks may be of some help.

Historical origin

The whole idea of scaling originated from a historical analogy to nuclear physics, or as Bjorken himself said, "was basically a lot of guesswork" (1997). What was crucial to Bjorken was the realization that the Adler sum rule would be saturated if there were quasi-elastic scattering between lepton with large q^2 and point-like constituents. But this meant that in the infinite energy and momentum transfer limit, the proportion between the two would be fixed. Originally, the proportionality was expressed as the effective mass of the constituents. Later, it was denoted as the scaling variable ω or x, the portion of the proton's momentum carried by the point-like constituent in the parton interpretation of scaling. The connection between the old and new notion is clear: both pointed to the existence of point-like constituents.

With hindsight, it can be seen clearly that scaling was suggested by the q^2 independent character of Adler's sum rules (3.11), as it was clear that the contributions from resonances decrease rapidly as $q^2 \to \infty$. Thus it was hard to see how discrete resonance contributions can give the behavior Adler's rule required for large q^2, and the simplest way to ensure the q^2 independence of the left-hand side of (3.11) as both energy and q^2 approached infinity would be to assume that the scaling limit (5.1) exists and the Adler sum rule can be expressed in terms of the scaling variable.

It is worth noting that the saturation comes from a broad distribution described by scaling and was not modeled as a sum of discrete resonance contributions. The inelastic continuum has a very different structure from resonance contributions, and this structure is simply summarized by scaling. That is to say, scaling can be seen as a simple and natural way to see how sum rule gets saturated.

Dynamic origin

Assuming Adler's sum rule was valid, scaling would be kinematically understandable. But when scaling was conjectured and accepted in the late 1960s, its dynamic origin remained to be a mystery.

In the derivation of the current algebra sum rule, the equal-time commutator was related to asymptotic behavior of the forward Compton amplitude, and quasi-free field behavior was assumed for the commutator. Closely related to this, for obtaining the scaling behavior, quasi-free field behavior was also assumed. But there were no typical quantum field theory calculations that would support such quasi-free field behavior.

In fact, the experimentally confirmed scaling as an inviolable constraint for theory construction was taken by many theoreticians to be a strong signal that no quantum field theory would be suitable to account for the observed strong interactions, which were asymptotically free as they were characterized by the scaling. Even worse, if the scaling and its underlying free field behavior assumption were true, then two colliding protons should break into their constituents and, if the constituents or partons were taken to be fractionally charged quarks, then free quarks would be observable. Thus scaling had posed a serious challenge to experimenters to find free partons, and an equally serious challenge to theoreticians to explain their evasiveness because experimenters failed to find them.

Precocious scaling

One of the most remarkable aspects of the earliest MIT–SLAC results was that the approximate scaling behavior, which was presumed to be a property asymptotically reached in the Bjorken limit, set in already at surprisingly nonasymptotic values of $q^2 > 1.0\ \text{GeV}^2$ and $W > 2.6$ GeV. (Bloom and Gilman, 1970). That is, the region of moderate x was the region in which the scaling set in and the sum rules saturated.

Conceptually, the precocious scaling would be understandable if we took what the scaling measured not as the asymptotic behavior of the genuinely point-like partons, but as that of the dressed partons, or the so-called valence partons. These dressed partons appeared, in the experimentally accessible region of energy and momentum transfer in the SLAC deep inelastic scattering, as if they were structureless and point-like, but in fact they were not really point-like. They were much bigger, and thus could be approached or seen at much lower values of energy and momentum transfer.

In a deep sense, therefore, the precocious scaling has vindicated Chew's notion of approximate or effective closure, and correspondingly the notion of approximate or effective locality.

Scale invariance

Bjorken's scaling brought the notion of scale invariance into the forefront of theoretical thinking in particle physics because the scaling would be trivially satisfied if no scale is defined so that the dimensionless structure functions could only depend on the one dimensionless variable available, the scaling variable ω. More discussion on this will be given in Section 6.2.

Achievements

The most remarkable achievement of scaling was that it revealed the potential of current algebra in exploring the structure of hadrons in terms of their elementary constituents, which paved the way for further identifying what these elementary constituents actually are and what would be the dynamic theory to describe their behavior. Without the achievement of scaling, no further developments in this direction would be conceivable in particle physics.

Secondly, the Bjorken scaling hypothesis and its interpretation in terms of parton model ideas inspired powerful theoretical tools for analyzing deep inelastic scattering experiments. For instance, by assuming the Bjorken scaling, Callan and Gross (1969) derived a proportionality relation between two of the deep inelastic structure functions, which was testable in electroproduction experiments.

In fact, once the scaling interpretation was clear, a lot of other parton model predictions were quickly derived using Bjorken's method, which often went beyond what one could do with just the local current algebra.

Third, scaling marked a shift of physicists' concerns from current algebra to deep inelastic scattering itself. For example, when the scaling function was expressed in terms of partons, $F_2^e(x) = \sum_N P(N) \left(\sum_i Q_i^2 \right)_N x f_N(x)$, the functions $f_N(x)$, which describe the momentum distribution function of the partons, and $P(N)$, which describe all the possibilities of having N partons, therein were empirically assumed. They had nothing to do with current algebra, but were crucial for any parton model treatment of deep inelastic scattering.

Notes

1 It was published first in the form of preprint in September 1968, SLAC-PUB-510 (TH), which was widely read and thus was influential in the particle physics community. But more important for its impact on the evolution of particle physics was Bjorken's direct and effective communications with the experimenters of the MIT–SLAC group who were actually planning and performing the deep inelastic scattering experiments and trying to understand and interpret their results. When it finally was published in *Physical Review* in March 1969, Bjorken scaling had already been regarded as a confirmed notion.
2 See also Equation (4.2) and the explanation that follows.
3 See the discussions after his talk at the SLAC symposium (Bjorken, 1967c).
4 Note that there was a change of notation. The notation adopted in this section was taken from Bjorken's original publication. Later, however, the scaling variable was redefined as $\omega = 2M\nu/q^2$, and the original one played the role of a new variable $x = 1/\omega$.
5 I am grateful to Professor Jerome Friedman for several extensive conversations on this subject. I have benefited hugely from his suggestion that I should examine the original proposal carefully. With the help of Professor Bjorken, the SLAC Archive sent me the proposal and all relevant documents, which provided me with a foundation on which this and the next sections are based. For the help I received in this regard, I would like to express my gratitude to Professor Friedman, Professor Bjorken and Abrahan Wheeler, a Research Librarian at SLAC, and Jean Dekan, an archivist at SLAC.
6 All the results mentioned in this subsection can be found in the Nobel lectures by Taylor (1991), Kendall (1991), and Friedman (1991), in which the detailed references to original publications of the results can also be found.
7 I am grateful to Professor Friedman for sharing with me this information and the information about Panofsky's involvement in the collaboration and his Rapporteur's plenary talk mentioned below.
8 Here the scaling variable x is different from the older Bjorken notation $\omega = -q^2/(M\nu)$, which has an integration range 0 to 2, but is used in all parton model literature. Note that here the convention is spacelike $q^2 > 0$.
9 For more details, see Friedman and Kendall, 1972.

6

Theorizations of scaling

The experimental confirmation of the approximate scaling from SLAC stimulated intensive theoretical activities for conceptualizing the observed short-distance behavior of hadron currents and for developing a self-consistent theory of strong interactions, starting from and constrained by the observed scaling. At first, the most prominent among these efforts was the parton model that was originated from Bjorken's thinking on deep inelastic scattering and Feynman's speculations on hadron–hadron collision, and made popular by Feynman's influential advocacy.

The assumption adopted by the parton model that the short-distance behavior of hadron currents should be described by free field theory, however, was immediately challenged once the scaling results were published. The theoretical framework the challengers used was the renormalized perturbation theory. Detailed studies of renormalization effects on the behavior of currents, by Adler and many others, reinforced the conviction about the limitations of formal manipulations in the PCAC and current algebra reasoning, which made these physical effects theoretically invisible, and discovered, first, the chiral anomaly and, then, the logarithmic violation of scaling. The theoretically rigorous argument for scaling violation was soon to be incorporated into the notion of broken scale invariance by Kenneth G. Wilson, Curtis Callan, and others, which was taken to be the foundation of such approaches as Wilson's operator product expansion and Callan's scaling law version of the renormalization group equation for conceptualizing the short-distance behavior of hadron currents, for giving a more detailed picture of these behaviors than current algebra could offer, and even for a general theory of strong interactions.

Also motivated by scaling, attempts were made to extend current algebra of the equal-time commutators to that of commutators defined on the light cone, in which the infinite momentum frame method adopted by the parton model was canonized and much of the original territory of the parton model, deep inelastic scattering in particular, co-opted. The ambition of the major advocates of light-cone current algebra, Gell-Mann and Harold Fritzsch, was to synthesize the newly emerged ideas of scaling, partons, broken scale invariance, and operator product expansions into a coherent picture of strong interactions.

As we will see in Section 6.3, there was an implicit but deep tension between the basic assumption of light-cone current algebra, "Nature reads books on free field theory" (Fritzsch and Gell-Mann,1971a), and the idea of broken scale invariance. Although Gell-Mann and Fritzsch enthusiastically embraced the idea of broken scale invariance, they had a hard time accepting renormalized perturbation theories as relevant to the general theory of strong interaction. They noticed the fact that scale invariance would be broken by certain kinds of interactions, but did not fully digest the fact that the underlying mechanism for breaking scale invariance was provided by the renormalization effects on the behavior of currents in strong interactions. The tension would not be released until the appearance of further major conceptual developments in the next few years that will be the subject of Chapters 7 and 8.

Now let us have a closer look at the theoretical efforts directly stimulated by scaling.

6.1 The parton model

The initial popularity of the parton model mainly came from the fact that its explanation of scaling was intuitively appealing. The mathematical definition of structure functions was such that each received a contribution at a given value of the scaling variable ω only when the struck parton carried a fraction $x = 1/\omega$ of the total momentum of the proton. Effectively, the structure functions measured the momentum distribution of partons within the proton weighted by the squares of their charges, and their measurement was dependent only upon ω, the ratio of v and q^2, and not upon their individual values. Thus, all of the effects of the strong interactions were contained in the parton momentum distributions, which could be empirically measured, although not theoretically derived.

But even at the intuitive level, a naïve idea of partons as non-interacting constituents of a physical proton had a problem: the problem of binding. The trouble, however, could be bypassed, to some extent, by adopting the infinite momentum frame, which was designed to treat extremely relativistic particles in the same way nonrelativistic quantum theory handled nonrelativistic particles.

That is, in this frame or at very high energies, the rate of internal fluctuations in a proton slowed down while its time of passage through a target did not. Thus the scattering process the parton model intended to explain could be thought as consisting of three stages. First, the decomposition of a proton to parton configuration, which was taken care of by the infinite momentum method; second, interacting with the virtual photon or with a parton in another proton or colliding partner; third, after the interaction, the excited state of emerging partons had to be put back together into the final physical particles, the so-called hadronization of partons.

In terms of treating strong interactions, the parton model was very successful for the second stage of hadron interactions, but said almost nothing, nothing beyond phenomenological descriptions, about the third stage. Thus, as a theory for strong interactions in general, the parton model was far from satisfactory, its intuitive attraction in the interpretation of scaling notwithstanding.

But the real strengths and the real achievements of the parton model, and its real contributions to particle physics, resided elsewhere. That is, its major contributions resided in having provided a framework, with the help of which the constituent picture of hadrons was consolidated by having structurally identified essential properties the constituents must have and, even more importantly, by having identified two kinds of constituents, one kind which was involved in the electromagnetic and weak interactions and the other kind which had nothing to do with both interactions. The identifications just mentioned would not be possible simply by observations and experiments without the parton model analyses and calculations.

When the idea of partons first attracted the attention of physicists through its intuitive account of the scaling, it seemed that a natural candidate for partons should be quarks. In fact, when Bjorken talked about the constituents, he frequently had quarks in mind. But there were very strong reservations in taking quarks seriously among physicists when partons were first introduced to the stage. The reason was simple. Quarks required strong final state interactions to account for the fact that free quarks had not been observed in the laboratories and in cosmic rays. In fact, there had been a serious problem in making the free behavior of the constituents during deep inelastic scattering compatible with this required strong final state interaction before the advent of QCD. If the evasiveness of quarks was explained away by letting them have very large masses, then the difficulty of constructing hadron structure from quarks would appear unsurmountable.

Out of these considerations, some physicists took partons to be bare nucleons and pions and developed a canonical field theory of pions and nucleons with the insertion of a cutoff in transverse momentum, which was the essential dynamical ingredient that guaranteed the scaling (Drell, Levy and Yan, 1969). Later, a fully relativistic generalization of this model was formulated in which the restriction to an infinite momentum frame was removed (Drell and Lee, 1972).

When the idea of partons being bare nucleons and pions was suggested, the first parton model analysis of the constituents by Curtis Callan and David Gross also appeared. Based on Bjorken's formulation of scaling, Callan and Gross showed that depending on the constitution of the current, either longitudinal or transverse virtual photons would dominate the electroproduction cross sections for large momentum transfer, and "the connection between the asymptotic behaviour of photoabsoption cross sections and the constitutions of the current is surprising and clean" (Callan and Gross, 1969).

More specifically, they showed that the ratio $R = \sigma_L/\sigma_T$, where σ_L and σ_T represent respectively the cross section for longitudinal and transverse polarized virtual photons, depended heavily on the spin of the constituents in the parton model. According to their calculations, assuming Bjorken scaling, spin zero or spin-1 constituents led to the prediction of $R \neq 0$ in the Bjorken limit, which would indicate that the proton cloud contains elementary bosons; for spin 1/2 constituents, R was expected to be small.

The experimental verdict arrived quickly. At the 4th International Symposium on Electron and Photon Interactions at High Energy held in Liverpool in 1969, the MIT–SLAC results were presented. These results showed that R was small for the proton and neutron at large values of q^2 and ν (Bloom *et al.*, 1970). This small ratio required that the constituents responsible for the scattering should have spin 1/2 as was pointed out by Callan and Gross, and thus was totally incompatible with the predictions of VMD, and also ruled out pions as constituents, but was compatible with the constituents being quarks or bare protons.

In addition, the asymptotic vanishing of σ_L indicated that the electromagnetic current in the electroproduction was made only out of spin 1/2 fields. Thus, even if the assumed gluons existed, as scalar or vector bosons, they must be electromagnetically neutral.

Another quantity, the ratio of the neutron and proton inelastic cross section σ_n/σ_p, also played a decisive role in establishing the nature of the constituents. The quark model imposed a lower bound of 0.25. In contrast to this prediction, Regge and resonance models predicted that when the scaling variable x was near 1, the ratio would be 0.6; diffractive models predicted the ratio being 1 when x was near 1; while the relativistic parton model with bare nucleons and mesons predicted that the ratio would fall to zero at $x = 1$, and 0.1 at $x = 0.85$. The MIT–SLAC results showed that the ratio fell continuously as the scaling variable x approaches unity: the ratio was 1 when $x = 0$; 0.3 when $x = 0.85$ (Bodek, 1973). The verdict was that except for the quark model, all other models, including the relativistic parton model of bare nucleons and mesons, were ruled out by the results.

Although quarks in the parton model had the same quantum numbers, such as spin, charges and baryon number, as those of quarks in the constituent quark model, they should not be confused with the latter. As point-like mass-less constituents of hadrons, they could not be identified with the massive constituent quarks with an effective mass roughly of one third of the nucleon mass, but only with current quarks.

The whole idea of the quark model became much more sophisticated when Bjorken and Emmanuel Paschos studied the parton model for three valence quarks that were embedded in a background of quark–antiquark pairs, the sea quarks, and when Julius Kuti and Victor Weisskopf included neutral gluons as another kind of parton, which were supposed to be the quanta of the field responsible for the binding of quarks. But for this inclusion to be possible and acceptable, several complicated parton model analyses had to be completed and confirmed or at least supported by the corresponding experimental results. The analyses and results can be briefly summarized as follows.

In the quark–parton model calculation of the electroproduction structure function $F_2^p(x)$, $F_2^p(x)$ was defined by

$$F_2^p(x) = \nu Wp_2^p(x) = x\left[Q_u^2\left(u_p(x) + \bar{u}_p(x)\right) + Q_d^2\left(d_p(x) + \bar{d}_p(x)\right)\right] \qquad (6.1)$$

where $u_p(x)$ and $d_p(x)$ were defined as the momentum distributions of up and down quarks in the proton, $\bar{u}_p(x)$ and $\bar{d}_p(x)$ were the distributions for anti-up and anti-down quarks, and Q_u^2 and Q_d^2 were the squares of the charges of the up and down quarks, respectively, while the strange quark sea has been neglected. Using charge symmetry it can be shown that

$$\begin{aligned} \frac{1}{2}\int_0^1 \left[F_2^p(x) + F_2^n(x)\right]dx &= \frac{Q_u^2 + Q_d^2}{2} \\ \int_0^1 x\left[u_p(x) + \bar{u}_p(x) + d_p(x) + \bar{d}_p(x)\right]dx. \end{aligned} \qquad (6.2)$$

The integral on the right-hand side of the equation, according to the parton model, should be the total fractional momentum carried by the quarks and antiquarks, which would equal to 1.0 if they carried the nucleon's total momentum. Based on this assumption, the expected sum should equal to

$$\frac{Q_u^2 + Q_d^2}{2} = \frac{1}{2}\left[\frac{4}{9} + \frac{1}{9}\right] = \frac{5}{18} = 0.28. \tag{6.3}$$

The evaluations of the experimental sum from proton and neutron results over the entire kinematic range studied yielded

$$\frac{1}{2}\int_0^1 \left[F_2^p(x) + F_2^n(x)\right]dx = 0.14 \pm 0.005. \tag{6.4}$$

This suggested that half of the nucleon's momentum was carried by electrically neutral constituents, which did not interact with the electron.

Then the parton model analysis of the neutrino DIS offered complementary information that, when checked with experiments, provided further support to the existence of non-quark partons.

Since neutrino interactions with quarks through charged current were expected to be independent of quark charges but were hypothesized to depend on the quark momentum distributions in a manner similar to electrons, the ratio of the electron and neutrino deep inelastic scattering was predicted to depend on the quark charges, with the momentum distributions cancelling out

$$\frac{\frac{1}{2}\int_0^1 \left[F_2^{ep}(x) + F_2^{en}(x)\right]dx}{\frac{1}{2}\int_0^1 \left[F_2^{\nu p}(x) + F_2^{\nu n}(x)\right]dx} = \frac{Q_u^2 + Q_d^2}{2}, \tag{6.5}$$

where $\frac{1}{2}\int_0^1 \left[F_2^{\nu p}(x) + F_2^{\nu n}(x)\right]dx$ was the F_2 structure function obtained from neutrino–nucleon scattering from a target having an equal number of neutrons and protons.

The integral of this neutrino structure function over x, according to the parton model, should be equal to the total fraction of the nucleon's momentum carried by the constituents of the nucleon that interact with the neutrino. This directly measures the fractional momentum carried by the quarks and antiquarks because the assumed gluons, which were supposed to be the carrier of inter-quark force only, were not expected to interact with neutrinos.

The first experimental results of neutrino and antineutrino total cross sections produced by the Gargamelle group at the CERN were presented in 1972 at the 16th International Conference on High Energy Physics held at Fermilab (Perkins, 1972). By combining the neutrino and antineutrino cross sections the Gargamelle group was able to show that

$$\begin{aligned}\frac{1}{2}\left\{\int \left[F_2^{\nu p}(x) + F_2^{\nu n}(x)\right]dx\right\} &= \int x\left[u_p(x) + \bar{u}_p(x) + d_p(x) + \bar{d}_p(x)\right]dx \\ &= 0.49 \pm 0.07\end{aligned} \tag{6.6}$$

which was compatible with the electron scattering results that suggested that the quarks and antiquarks carried half of the nucleon's momentum.

When the result was compared with $\frac{1}{2}\int_0^1 \left[F_2^{ep}(x) + F_2^{en}(x)\right]dx$ [see (6.4) and (6.3)], the ratio of neutrino and electron integrals was found to be 3.4 ± 0.7, as compared to the value predicted by the quark model, $18/5 = 3.6$. This was a striking success for the quark–parton model in terms of the charges partons carried.

Thus one of the greatest achievements of the parton model was to bring the notion of gluon back to the picture of hadrons, which notion, after its initial introduction, quite quietly, by Gell-Mann (1962), had been invisible in hadron physics because nobody was able to give it any role to play until the advent of the parton model.

Finally, noticing that partons in the parton model were independent of each other without interactions, and any masses in the parton model were negligible in its infinite momentum frame, meaning that no scale existed in the parton model, the parton model, as a free field theory without scale, gave hadron physicists the first glimpse at a physical theory, not toy models, with scale invariance. In fact, the crucial theoretical developments stimulated by and accompanying the parton model that finally led to the genesis of QCD were strongly constrained by the desire to find a proper way to incorporate the idea of scale invariance as we will see in later sections.

6.2 Broken scale invariance

Both the free field theory of short-distance behavior of hadron currents assumed by the parton model and the scale invariance implied by the parton model were seriously questioned and challenged, even before the parton model was first announced, within the framework of renormalized perturbation theory. The challenge took the forms of anomalies and scaling violations; and then the whole situation was reconceptualized in terms of operator product expansions and the scaling-law version of renormalization group equations on the basis of a new idea of broken scale invariance. The impact of broken scale invariance on the conception of light-cone current algebra, within which QCD was first conceived, and on the conception and justification of QCD itself was immediate and profound, as we will see in later sections.

Anomalies

Intensive investigations into the anomalous behavior of local field theories were carried out along three closely related lines: as a general response to the formal manipulations of equal-time commutators that led to the discovery of anomalous commutators (Johnson and Low, 1966; Bell, 1967); the necessity of modifying PCAC (Veltman, 1967; Sutherland, 1967; Bell and Jackiw, 1969; Adler, 1969); and as part of a study of the renormalization of axial-vector current and axial-vector vertex (Adler, 1969; Adler and Bardeen, 1969).

In terms of anomalous commutators, that is, those that deviated from the canonical one, Julian Schwinger was a recognized pioneer. He showed in (1959) clearly that there must be an extra term in the vacuum expectation of equal-time commutators of space components with time components of currents, which involves space derivatives of delta function instead of a canonical delta function.

The next step was taken by Kenneth Johnson. In an analysis of the Thirring model (1961), he argued more generally that the product of operators defined at the same spacetime point must be singular and in need of regularization. As a result, if renormalization effects were taken seriously, the canonical equal-time commutators would be destroyed if the interaction was not invariant to the symmetry of the theory.

It is worth noting that the argument for the anomalous commutator is much stronger than the argument for the violation of free field behavior. One of the basic assumptions of current algebra was that even if the interaction was not invariant to the symmetry of the theory, if it involved no derivative couplings, then independent of dynamical details, the currents would retain the original structure, and their equal-time commutators, which formally depend only on the structure of currents, would remain unchanged (cf. Section 2.2).

Along the same lines, Johnson and Low (1966), in a study of a simple perturbation theory model, in which currents coupled through a fermion triangle loop to a meson, found that in most cases the result obtained by explicit evaluation differed from those calculated from naïve commutators by well-defined extra terms, and thus argued that free field behavior of currents at short distances would be violated and commutators would have finite extra terms in perturbation theory. The root cause for the anomalies was the same as Johnson discovered, the product of operators in a local relativistic theory must be singular enough in need of regularization.

All these had shown the limitations of formal manipulations, which was uncritically adopted by current algebra and PCAC reasoning. "The unreliability of the formal manipulations common to current algebra calculations" was demonstrated in a very simple example by John Bell and Roman Jackiw (1969). They found the PCAC anomaly in a sigma model calculation of triangle graphs and related it to the observable $\pi \rightarrow 2\gamma$ decay, which had raised enormous interest in and drawn great attention to the existence and importance of anomalies that were investigated in the renormalized perturbation theory.

In his field-theoretical study of the axial-vector vertex within the framework of perturbation theory, Adler showed that the axial-vector vertex in spinor electrodynamics had anomalous properties and the divergence of axial-vector current was not the canonical expression calculated from the field equations. These anomalies, Adler argued, were caused by the radiative corrections to, or the renormalization effects on, the vertex, when the presence of closed-loop triangle diagrams was taken into account. The corrections are purely quantum theoretical in nature and are absent in classical field theory. Technically, Adler pointed out that when a calculation involved a worse than logarithmically divergent integral with two or more limits to be taken, these two limiting processes were not legally interchangeable; in the language of momentum space, an equivalent statement would be that the integral variable cannot be legally shifted. Thus there is a limiting ambiguity, and the extra terms result from a careful treatment of this ambiguity. In an appendix Adler carried the Bell–Jackiw work to its logical conclusion by modifying PCAC by adding the anomaly contribution. This gives a nonzero prediction for the $\pi \rightarrow 2\gamma$ decay rate in terms of the constituent charge structure.

It should be stressed that anomalies are intrinsic effects of quantum processes on the structure of a field theory. Thus different field theories with different structures would have their corresponding anomalies of different characters. For example, the axial anomaly articulated by Adler and Bell–Jackiw is topological in nature, while the scale anomaly we will discuss shortly has nothing to do with topology, although both are caused by renormalization effects rather than by non-symmetrical terms in the Lagrangian or by non-invariant vacuum.

Scaling violation

After Bjorken scaling and the Callan–Gross asymptotic cross section relations for high energy inelastic electron and neutrino scattering derived from it became known to the community, studies of scaling in perturbation theory were conducted, and an anomalous commutator argument of Johnson–Low style was raised to challenge the very idea of scaling, which was based on the assumption that the equal-time commutators were the same as the naïve commutators obtained by straightforward use of canonical commutation relations and equation of motion (Adler and Tung, 1969, 1970; Jackiw and Preparata, 1969).

For example, Adler and Wu-ki Tung did a perturbation theory calculation in a renormalizable model of strong interactions consisting of an SU(3) triplet of spin 1/2 particles bound by the exchange of an SU(3) singlet massive vector gluon, and showed that for commutators of space components with space components, the explicit perturbation calculation differed from the free field theory or from canonical commutators by computable terms. That is, there were logarithmic corrections to scaling and to the Callan–Gross relation. The implication was clear: only free field theory would give exact scaling and actual field theories would have logarithmic corrections to scaling.

Broken scale invariance: early history

The discoveries of approximate scaling and logarithmic scaling violation made it desirable to reconceptualize the whole situation about the idea of scale invariance in the construction of theories of strong interactions. The reconceptualization was achieved mainly by Kenneth Wilson in terms of operator product expansions, and by Callan and Kurt Symanzik in terms of the scaling law version of renormalization group equations. But what they obtained was the result of synthesizing several lines of previous developments.

Historically, the idea of scale dependence of physical parameters appeared earlier than that of a scale invariant theory. Freeman Dyson, in his work on the smoothed interaction representation (1951), tried to separate the low frequency part of the interaction from the high frequency part, which was thought to be ineffective except in producing renormalization effects. To achieve this objective, Dyson adopted the guidelines of the adiabatic hypothesis and defined a smoothly varying charge for the electron and a smoothly varying interaction with the help of a smoothly varying parameter g. Along the same line, Lev Landau and his collaborators (Landau, Abrikosov, and Khalatnikov, 1954a, b, c, d) later developed an idea of smeared out interaction, according to which, the magnitude of the interaction should be regarded not as a constant, but as a function of the radius of interaction. Correspondingly, the charge of the electron must be regarded as an as yet unknown function of the radius of interaction. Both Dyson and Landau had the idea that the parameter corresponding to the charge of the electron was scale dependent, but the physics of QED should be scale independent.

Independently, Urnest Stueckelberg and Andre Petermann (1951 and 1953) developed parallel and even more sophisticated ideas. They noticed that while the infinite part of the counter-terms introduced in the renormalization procedure was determined by the requirement of cancelling out the divergences, the finite part was changeable, depending on the arbitrary choice of subtraction point. This arbitrariness, however, was physically irrelevant because a different choice only led to a different parameterization of the theory. They observed that a transformation group could be defined which related different parameterizations of the theory. They called it the "renormalization group," which was the first appearance of the term in the history. They also pointed out the possibility of introducing an infinitesimal operator and of constructing a differential equation.

In their study of the short-distance behavior of QED, Gell-Mann and Low (1954) exploited the renormalization invariance fruitfully. First, they observed that the measured charge e was a property of the very low momentum behavior of QED, and that e could be replaced by any one of a family of parameters e_λ, which was related to the behavior of QED at an arbitrary momentum scale λ. When $\lambda \to 0$, e_λ became the measured charge e, and when $\lambda \to \infty$, e_λ became the bare charge e_0. Second, they found that by virtue of renormalization, e_λ^2 obeyed an equation: $\lambda^2 \frac{de_\lambda^2}{d\lambda^2} = \psi\left(\frac{e_\lambda^2, m^2}{\lambda^2}\right)$. When $\lambda \to \infty$, the renormalization group function ψ became a function of e_λ^2 alone, thus establishing a scaling law for e_λ^2. Third, they argued that as a result of the equation, the bare charge e_0 must have a fixed value independent of the value of the measured charge e; this is the so-called Gell-Mann–Low eigenvalue condition for the bare charge. The eigenvalue condition in fact was a very strong assumption, equivalent to the assumption that the renormalization group equation would have a fixed point solution; that is, the assumption that a theory is asymptotically scale invariant.

The scale invariance of a theory is different from the scale independence of the physics of a theory or the independence of the physics of a theory with respect to the renormalization scale, as expressed by the renormalization group equations. The scale invariance of a theory refers to the invariance of a theory under the group of scale transformations, which are only defined for dynamical variables, the fields, but not for the dimensional parameters, such as masses. While the physics of a theory should be independent of the choice of the renormalization scale, a theory may not be scale invariant if there is any dimensional parameter.

In Gell-Mann and Low's treatment of the short-distance behavior of QED, the theory is not scale invariant when the electric charge is renormalized in terms of its value at very large distances. The scale invariance of QED would be expected since the electron mass can be neglected in the regime of very high energy, and there seems to be no other dimensional parameter appearing in the theory. The reason for the unexpected failure of scale invariance is entirely due to the necessity for charge renormalization: there is a singularity when the electron mass goes zero. However, when the electric charge is renormalized at a higher energy scale by introducing a sliding renormalization scale to suppress effectively irrelevant low-energy degrees of freedom, there occurs an asymptotic scale invariance. This asymptotic scale invariance was expressed by Gell-Mann and Low, with the help of the scaling law for the effective charge, in terms of the eigenvalue condition for the bare charge, meaning that there is a fixed value for the bare charge, which is independent of the value of the measured charge.

Operator product expansions

Although there was a suggestion by Johnson (1961) that the Thirring model might be scale invariant, and another by Gerhard Mack (1968) concerning the scale invariance of the strong interactions at short distances, the real advance in understanding the nature of scale invariance, stimulated by the discoveries of scaling and anomalies, was first achieved in Wilson's formulation of the short-distance expansion of products of operators (1969).

As a response to the failures of current algebra, formulated in terms of equal-time commutators, in dealing with the short-distance behavior of currents in strong interactions, such as the nature of the Bjorken limit, which were known to many since the work of Johnson and Low, Wilson tried to formulate a new framework for analyzing the short-distance behavior on the basis of two hypotheses. First, as Mack (1968) suggested, the strong interactions become scale invariant at short distances: "This means that scale invariance is a broken symmetry in the same sense as chiral $SU(3) \times SU(3)$." Second, operator product expansions (OPE) for products of two or more local fields or currents near the same point, as an extension of current algebra, was supposed to exist and contain singular functions when the operators were defined at the same spacetime point or one defined on the light cone through the other.

OPE had its origin in detailed studies of renormalization in perturbation theory (Valantin, 1954a, b, c, d; Zimmermann, 1958, 1967; Nishijima, 1958; Hagg, 1958; Brandt, 1967). Yet, Wilson developed it on the new basis of broken scale invariance (Wess, 1960; Johnson, 1961; Mack, 1968), which could be utilized to determine the singularity structure in the expansion.

The point of departure was Johnson's work on the Thirring model, in which the canonical commutators were shown to be destroyed by renormalization effects if the interactions were non-invariant, and the scale dimensions of fields, which are determined by requiring that the canonical commutation rules are invariant, were shown to vary continuously with the coupling constant.[1]

The deviation of the scale dimension of currents from their non-interacting value suggested to Wilson a breakdown of scale invariance caused by the renormalization effects as observed by Johnson, and he made a generalization that Johnson's observation also held in the case of his OPE formulation for the strong interactions.

Wilson argued that his new language of OPE would give a more detailed picture of the short-distance behavior of currents in the strong interactions than one obtained if one only knew equal-time commutators in Gell-Mann's current algebra. According to Wilson, if $A(x)$ and $B(x)$ are local field operators, then

$$A(x)B(x) = \sum_n C_n(x-y)O_n(x), \qquad (6.7)$$

here $O_n(x)$ is also a local field operator, and the coefficients $C_n(x-y)$, which involve powers of $(x-y)$ and logarithms of $(x-y)^2$ and may have singularities on the light cone, contain all the physical information about the short-distance behavior of currents. Note that in comparison to the formulation in current algebra, $[A(x_0,\vec{x}), B(x_0,\vec{y})] = \sum_n D_n(\vec{x}-\vec{y})O_n(x)$, the coefficients functions $C_n(x-y)$ in OPE are not defined at the equal time, so they depend on a four vector, not a three vector; they involve no delta function, but have singularities on the light cone. The expansion is valid for y sufficiently close to x.

The nature of singularities of the functions $C_n(x-y)$ is determined by the exact and broken symmetries, the most crucial of them is broken scale invariance. (Wess, 1960; Mack, 1968). Massless free field theories were supposed to be scale invariant; mass terms and renormalizable interactions would break the symmetry. But, Wilson argued, the ghost of scale invariance would still govern the behavior of singular functions.

In an exactly scale invariant theory, the behavior of the function $C_n(x-y)$ is determined except for a constant by scale invariance. Performing a scale transformation on (6.7), one obtains

$$s^{d_A+d_B}A(sx)B(sy) = \sum_n C_n(x-y)s^{d(n)}O_n(sx). \qquad (6.8)$$

Expanding the left-hand side, one obtains

$$s^{d_A+d_B}\sum_n C_n(sx-sy)O_n(sx) = \sum_n C_n(x-y)s^{d(n)}O_n(sx), \qquad (6.9)$$

which implies

$$C_n(sx-sy) = s^{-d_A-d_B+d(n)}C_n(x-y). \qquad (6.10)$$

Thus $C_n(x-y)$ must be homogeneous of order $-d_A-d_B+d(n)$ in $x-y$. So the strength of the light-cone singularity is determined by the dimension

$-d_A - d_B + d(n)$. C_n can be singular only if $d_A + d_B \geq d(n)$ and becomes more singular the larger $d_A + d_B$ is relative to $d(n)$, and less singular as $d(n)$ increases.

Of particular importance are the fields O_n on the right side of (6.7) of low dimensions, since these fields have the most singular coefficients in OPE. One of them, O_0, namely the operator of smallest dimension, is the dominant operator for short distances, and the short-distance behavior of the corresponding function C_0, the most singular function, can be determined by calculations as if all masses and dimensional coupling constants were zero.

In order to generalize the result to a theory of strong interactions with broken scale invariance, Wilson introduced into the skeleton theory, in which all free parameters were set to zero, a generalized mass vertex which breaks the scale invariance of the theory. According to Wilson, generalized mass terms are the logical choice of interaction when one wants a symmetry of the skeleton to be a broken symmetry of the theory with interaction. Once the generalized mass terms as scale non-invariant interactions are introduced in to the theory, they would destroy the equal-time commutators associated with the symmetry, as Johnson and Low observed earlier, and would produce corrections to expansion functions which are logarithmically more singular than the skeleton terms.

One implication of Wilson's analysis is this. The scale invariance of the strong interactions is broken, not by symmetry-breaking terms in the Lagrangian, nor by a non-invariant vacuum, but, like the anomalous breakdown of γ_5 invariance, only by some non-invariant interactions introduced in the renormalization procedure. This implication was soon intensely explored by Wilson himself (1970a, b, c; 1971a, b, c; 1972), and also by others (Callan, 1970; Symanzik, 1970; Callan, Coleman and Jackiw, 1970; Coleman and Jackiw, 1971). The exploration directly led to the revival of the idea of the renormalization group as we will see shortly.

Wilson's attitude towards the anomalous breakdown of scale invariance, however, was quite different from others'. While acknowledging the existence of a scale anomaly, which destroyed the canonical commutators and was reflected in the change of the scale dimension of the currents, Wilson insisted that all the anomalies could be absorbed into the anomalous dimensions of the currents, so that the scale invariance would persist in the asymptotic sense that the scaling law still held, although only for currents with changed scale dimensions.

It seems that this attitude is attributable in part to the influence on his work by Gell-Mann and Low (1954) on the scaling law of bare charge in QED, and in part to that of the scaling hypothesis of Fisher (1964), Widom (1965a, b) and Kadanoff (1966) in critical phenomena, which gave him faith in the existence of scaling laws, or, to use later terminology, in the existence of fixed points of the renormalization group transformations. According to Wilson, the fixed point in quantum field theory is just a generalization of Gell-Mann and Low's eigenvalue condition for the bare charge in QED. At the fixed point, a scaling law holds, either in the Gell-Mann–Low sense or in Bjorken's sense, and the theory is asymptotically scale invariant.

It is worth noting that there is an important difference between Wilson's conception of the asymptotic scale invariance of OPE at short distances and that of Bjorken. While Bjorken's scaling hypothesis about the form factors in deep inelastic lepton–hadron scattering suggests that the strong interactions seem to turn off at very short distances, Wilson's formulation of OPE reestablishes scale invariance only after absorbing the effects of interactions and renormalization into the anomalous dimensions of fields and currents. But this is just another way of expressing the logarithmic corrections to the scale invariance of the theory that were found in perturbation theory studies of Bjorken scaling.

As a powerful conceptual device, Wilson's OPE has many applications. One example was to determine Bjorken's structure functions in deep inelastic scattering in terms of the behavior of coefficient functions $C_n(x-y)$ in OPE for small distances together with the hadronic matrix elements of the operator $O_n(x)$. The underlying idea behind this application was soon to be absorbed into the framework of light-cone current algebra as we will see shortly.

Scaling law and renormalization group equation

Wilson's insights on broken scale invariance were further fruitfully explored by Callan (1970) and Kurt Symanzik (1970). The explorations resulted in a new version of renormalization group equations, the Callan–Symanzik equation, which, as a general framework for the further studies of broken scale invariance in various theoretical contexts, became central in the ensuing theoretical developments.

At the formal level, Callan, Sidney Coleman and Jackiw (1970) pointed out that scale invariance can be defined in terms of the conservation of the scale current $S_\mu = \theta_{\mu\nu}x^\nu$. Here, the symmetrical stress-energy-momentum tensor $\theta_{\mu\nu}$, as a source of linearized Einsteinian gravity coupled to the gravitational field, was introduced by Gell-Mann (1969), and $\theta = \theta^\mu_\mu$ is proportional to those terms in the Lagrangian having dimensional coupling constants, such as mass terms. The differential dilation operator or charge $D = \int d^3xS_0$ formed from the current S_μ acts as the generator of scale transformations,

$$x_\mu \to x'_\mu = \rho x_\mu, \ \phi(x) \to \rho^l \phi(\rho x)$$

$$[D(x_0), \phi(x)] = -i(d + x \cdot \partial)\phi(x). \tag{6.11}$$

The violation of scale invariance is connected with the non-vanishing of $\theta = \theta^\mu_\mu$, since $dD/dt = -\int \theta d^3x$, $dD/dt = 0$ implies $\theta = 0$.

With the help of the current S_μ and its equal-time commutation relations with fields, $[D(x_0), \phi(x)] = -i(d + x \cdot \partial)\phi(x)$, a standard Ward identity for scale current can be derived:

$$\left[n(d-4) + 4 - \sum p_i \cdot \frac{\partial}{\partial p_i}\right] G(p_1 \cdots p_{n-1}) = -iF(0, p_1 \cdots p_{n-1}) \tag{6.12}$$

where d is the dimension of the field, G and F are defined as follows:

$$(2\pi)^4 \delta\left(\sum_{i=1}^n p_i\right) G(p_1 \cdots p_{n-1}) = \int dx_1 \cdots dx_n e^{i \sum p_i x_i} \langle 0|T(\phi(x_1) \cdots \phi(x_n))|0\rangle$$

and

$$(2\pi)^4 \delta\left(q + \sum_{i=1}^n p_i\right) F(q, p_1 \cdots p_{n-1}) = \int dy dx_1 \cdots dx_n e^{iq \cdot y + i \sum p_i x_i} \langle 0|T(\theta(y)\phi(x_1) \cdots \phi(x_n))|0\rangle.$$

Callan (1970) further elaborated that if $\theta = 0$, so $F = 0$, the Green's functions G satisfy $SG = 0$, where

$$S = \left[n(d-4) + 4 - \sum p_i \cdot \frac{\partial}{\partial p_i}\right], \tag{6.13}$$

and depend only on dimensionless ratio of momentum variables. This is precisely what one expects from naïve dimensional reasoning in the event that no dimensional coupling constants are present in the theory. Thus the scaling law (6.12) says that the matrix elements F of θ act as the source of violations of simple dimensional scaling in the matrix elements G. The reasoning was formal and thus was generally valid, not depending on the details of the theory.

If the full Green's functions are replaced by one particle irreducible Green's functions, denoted by $\overline{G}$ and $\overline{F}$, then in a simple theory in which the only dimensional parameter is the particle mass μ, the scaling law for one particle irreducible Green's functions takes the form

$$\left[\mu \cdot \frac{\partial}{\partial \mu} + n\delta\right] \overline{G}(p_1 \cdots p_{n-1}) = -i\overline{F}(0, p_1 \cdots p_{n-1}). \tag{6.14}$$

Here $\delta = 1 - d$ does not equal zero even for scalar field, whose naïve dimension is 1, because, as Wilson pointed out, when there are interactions it is not guaranteed that the naïve dimension and the dimension defined by the commutator of the generator of scale transformations with the field are the same.

Callan successfully demonstrated that the scaling operator $S = [\mu \cdot \partial/\partial\mu + n\delta]$ suggested by formal arguments on broken scale invariance is distributive and thus is guaranteed to satisfy the scaling law, and it remains so if differentiation with respect to coupling constant λ is added to it. That is, a more general form of S:

$$S = \left[\mu \cdot \frac{\partial}{\partial\mu} + n\delta(\lambda) + f(\lambda)\frac{\partial}{\partial\lambda}\right] \tag{6.15}$$

would also render the particle amplitudes "satisfy a scaling law, albeit one which differs in a profound way from the one suggested by naïve broken-scale-invariance requirements."

In order to explicate the profound difference, Callan turned off the only explicit scale-invariance-breaking term in the Lagrangian he was discussing, the mass term. In this case, the amplitudes satisfy $S\overline{G}^{(n)} = 0$. If S were simply $[\mu \cdot \partial/\partial\mu + n\delta]$, this would imply that the functions $\overline{G}^{(n)}$ are homogeneous functions of their momentum arguments of degree $4 - nd$, with $d = 1 - \delta$. This is what one might call naïve scaling appropriately modified for the anomalous dimensions of the fields. Turning off the mass terms can actually be achieved by taking appropriate asymptotic limits of momenta, and one would expect the Green's functions to satisfy naïve scaling in such limits. But in fact $S = [\mu \cdot \partial/\partial\mu + n\delta(\lambda) + f(\lambda)\partial/\partial\lambda]$, which entails that even though $S\overline{G}^{(n)} = 0$ can be achieved in appropriate asymptotic regions, this does not mean that the $\overline{G}^{(n)}$ satisfy naïve scaling in the same limit. In place of naïve scaling, what one gets is some restriction on the joint dependence of $\overline{G}^{(n)}$ on momenta and coupling constant. The fact that S contains the terms $f(\lambda)\partial/\partial\lambda$, Callan argued, is equivalent to saying that even in the absence of explicit symmetry-breaking terms, scale invariance is still broken by some mechanism.

In his explanation of the nature of this mechanism, Callan heavily relied on Wilson's idea about the variability of dimension. The source for violation of naïve dimensional scaling, according to Callan, was not terms in the Lagrangian having dimensional coupling constants. A term will not break scale invariance only if its dimension is exactly four. But the terms with dimensionless coupling constants are guaranteed to have dimension four only to lowest order in the perturbation expansion; when the effects of interactions are considered, their dimensions will change, as Wilson convincingly argued, and they will contribute to scale-invariance breaking. Of course, these implicit breaking terms could be incorporated in the scaling law by a rather simple change in its form. The resulting scaling law, it turns out, has provided a simple, direct and model independent analytic expression of the effect of this implicit kind of symmetry breaking on scattering amplitudes.

By studying a special asymptotic limit of this generalized scaling law, Callan further argued, the results of the renormalization group can be recovered. As was just mentioned above, whenever the right-hand side of the generalized scaling law $SG = -iF$ could be neglected, one obtained a constraint on the joint dependence of G on momenta and coupling constant rather than naïve scaling. The joint dependence implies a kind of correlation between the asymptotic dependence on momentum and the dependence on coupling constant, and the correlation is typical of renormalization group arguments. That is why Callan's scaling law has also been regarded as a new version of renormalization group equation, which is a powerful approach to computing renormalized Green's functions as we will see in Section 8.1.

The conceptual developments outlined in this section can be summarized as follows. In systems with many scales that are coupled to each other and without a characteristic scale, such as those described by QFT, the scale invariance is always anomalously broken owing to the necessity of renormalization. This breakdown manifests itself in the anomalous scale dimensions of fields in the framework of OPE, or in the variation of parameters at different renormalization scales that is charted by the renormalization group equations. If these equations possess a fixed point solution, then a scaling law holds, either in the Gell-Mann–Low sense or in Bjorken's sense, and the theory is asymptotically scale invariant. The scale invariance is broken at non-fixed points, and the breakdown can be traced by the renormalization group equations. Thus with the more sophisticated scale argument, the implication of Gell-Mann and Low's original idea becomes clearer. That is, the renormalization group equations can be used to study properties of a field theory at various energy (or momentum or spacetime) scales, especially at very high energy scales, by following the variation of the effective parameters of the theory with changes in energy scale, arising from the anomalous breakdown of scale invariance, in a quantitative way, rather than a qualitative way as suggested by Dyson and Landau.

6.3 Light-cone current algebra

Motivated by the experimental confirmation of Bjorken's scaling hypothesis derived from Adler's current algebra sum rule, Fritzsch and Gell-Mann (1971a, b) extended the algebraic system of equal-time commutators to the system of commutators defined on the light cone, or light-cone current algebra. Their aim was to develop a coherent picture of strong interactions based on a consistent view of scale invariance, broken by certain terms in a non-vanishing $\theta = \theta^{\mu}_{\mu}$, but restored in the most singular terms of current commutators on the light cone. The view itself was achieved, as we will see shortly, by synthesizing the ideas of scaling and partons with the newly emerged ideas of broken scale invariance and operator product expansions. They extended this view to all the local operators occurring in the light-cone expansions of commutators of all local operators with one another.

The new algebraic system was formally obtained by abstracting the leading singularity in the commutator of two currents on the light cone from a field-theoretical quark–gluon model, with the resulting algebraic properties similar to those in the equal-time commutator algebra abstracted from the free quark model. Parallel to Wilson's OPE [see Equation (6.7)], the leading singularity, which turned out to be given in terms of bilocal current operators that reduce to familiar local currents when two spacetime points coincide, was multiplied by a corresponding singular coefficient function of the spacetime interval. In dealing with scaling predictions, which were taken to be the most basic feature of light-cone current algebra, the scaling limit in momentum space was translated into the singularity on the light cone of the commutator in coordinate space, and the scaling functions were shown to be just the Fourier transforms of the expected value of the leading singularity on the light cone.

Clearly the notion of light cone was fundamental to the whole project. But why light cone?

From infinite momentum to light cone

As we have noticed in Section 3.1, current algebra found its first applications only when the infinite momentum frame was introduced by Fubini and Furlan. But the infinite momentum frame method was soon canonized in light-cone field theories (Susskind, 1968; Leutwyler, 1968), which made it possible for most of the original applications of the method, those in the deep inelastic scattering in particular, to be co-opted by light-cone current algebra.

Leonard Susskind and Heinrich Leutwyler noticed that the infinite momentum limit had a very simple intuitive meaning: the limit essentially amounted to replacing the equal-time charges $I_i(t) = \int d^3x j_i^o(x)$ by the corresponding charges contained in a light-like surface Σ, characterized by $n_\mu x^\mu = \tau$, $n^2 = 0$, $n_\mu = (1, 0, 0, 1)$,

$$I_i(\tau) = \int_\Sigma d\sigma_\mu j_i^\mu(x) = \int_\Sigma d\sigma \{ j_i^0(x) + j_i^3(x) \}. \tag{6.16}$$

That is to say, in dealing with deep inelastic scattering, for example, the hadron momenta would be left finite instead of being boosted by a limit of Lorentz transformations, and the equal-time surface would be transformed by a corresponding limit of Lorentz transformations into a null plane, with $x_3 + x_0 =$ constant, say zero; in addition, the hypothesis of saturation by finite mass intermediate states should be correspondingly replaced by one in which the commutation rules of currents can be abstracted from the model not only on an equal-time plane, but also on a null plane as well (Leutwyler and Stern, 1970; Jackiw, Van Royen and West, 1970). This way, as we will see, scaling results can be easily recovered.

Since the equal-time commutation rules for the charges in the infinite momentum limit are equivalent to the same form for the light-like generators, all the results obtained in the infinite momentum frame can be expressed in terms of the behavior of the commutators of light-like charges in coordinate space which are integrated over a light-like plane instead of an equal-time plane.

Leutwyler (1968) pointed out that an important gain by using the algebra of the light-like generators was that it left the vacuum invariant, while the equal-time commutator algebra would not. This is important because, in his view, "many of the practical difficulties encountered in current algebras at finite momentum are an immediate consequence."

Assumptions and formula

First, it was assumed that the commutators of current densities at light-like separations and the leading singularity on the light cone of the connected part of the commutator could be abstracted from free field theory or from interacting field theory with the interactions treated by naïve manipulations rather than by renormalized perturbation expansion.

Second, it was assumed that the formula for the leading light-cone singularity in the commutator contained the physical information that the area near the light cone would be the one with full scale invariance in the sense that the physical scale dimension for operators on the light cone in coordinate space is conserved and the conservation applies to leading terms in the commutators, with each current having a dimension $l = -3$ and the singular function of $(x - y)$ also having dimension $l = -3$.

Finally, it was also assumed that a closed algebraic system could be attained, so that the light-cone commutator of any two of the operators is expressible as a linear combination of operators in the algebra.

The simplest such abstraction was that of the formula giving the leading singularity on the light cone of the connected part of the commutator of the vector or axial vector current. (Fritzsch and Gell-Mann, 1971a, b). For example,

$$\left[F_{i\mu}(x), F_{j\nu}(y)\right] \triangleq \left[F^5_{i\mu}(x), F_{j\nu}(y)^5\right] \triangleq \frac{1}{4\pi}\partial_\rho\left\{\varepsilon(x_0 - y_0)\delta\left[(x-y)^2\right]\right\}$$
$$\times\left\{\left(if_{ijk} - d_{ijk}\right)\left[s_{\mu\nu\rho\sigma}F_{k\sigma}(y,x) + i\varepsilon_{\mu\nu\rho\sigma}F^5_{k\sigma}(y,x)\right]\right.$$
$$\left.+\left(if_{ijk} + d_{ijk}\right)\left[s_{\mu\nu\rho\sigma}F_{k\sigma}(x,y) - i\varepsilon_{\mu\nu\rho\sigma}F^5_{k\sigma}(x,y)\right]\right\}. \quad (6.17)$$

On the right-hand side are the connected parts of bilocal operators[2] $F_{i\mu}(x,y)$ and $F^5_{i\mu}(x,y)$, which reduced to the local currents $F_{i\mu}(x)$ and $F^5_{i\mu}(x)$ as $x \rightarrow y$. The bilocal operators are defined as observable quantities only in the vicinity of the light cone, $(x-y)^2 = 0$.

A simple generalization of the abstraction above turned into a closed system, which was called the basic light-cone algebra, in which the bilocal operators commuted, for example, $F_{i\mu}$ (x, u) with $F_{j\nu}$ (y, v), as all of the six intervals among the four spacetime points approached 0, so that all four points tended to lie on a light-like straight line in Minkowski space. Abstraction from the model gave, on the right-hand side, a singular function of one coordinate difference, say $x - v$, times a bilocal current $F_{i\alpha}$ or $F^5_{i\alpha}$ at the other two points, say y and u, plus an expression with (x, v) and (y, u) interchanged, and the system closed algebraically. The formulae were just like Equation (6.17).

In the free quark model, the bilocal currents were expressed as

$$D(x, y, G) \sim \bar{q}(x)Gq(y). \quad (6.18)$$

For a vector current, $G = \frac{i}{2}\lambda_j\gamma_\alpha$; for an axial vector current, $G = \frac{i}{2}\lambda_j\gamma_\alpha\gamma_5$. In the quark–gluon model, an additional factor gB should be attached to G, where B is the gluon field and g is its coupling with quarks.

Light-cone singularity and scaling

It was easy to derive scaling from the light-cone commutator formula because the deep inelastic structure functions are directly related to the commutators of currents at light-like separation. In the scaling region, for example, the structure function

$$W_1 \approx \frac{\nu^2}{q^2}\left(\frac{\sigma^T}{\sigma^T + \sigma^S}\right)W_2 = \sum_n |\langle p|J_x(0)|n\rangle|^2(2\pi)^3\delta^4(P_n - P - q). \quad (6.19)$$

W_1 may also be written as a Fourier transform of a commutator

$$W_1(Q^2,\nu) = \int \frac{d^4x}{2\pi} e^{iq\cdot x} \langle p|[J_x(x), J_x(0)]|p\rangle. \tag{6.20}$$

The interesting question then is, in the deep inelastic limit, what the important values of x_μ are that contribute to the integral.

In order to see this we should only look at the exponential

$$e^{iq\cdot x} = e^{i\nu t - i\sqrt{\nu^2+q^2}z} \approx \exp\left(i\nu(t-z) - \frac{iQ^2}{2\nu}z\right). \tag{6.21}$$

Here one can expand the square root and keep the first term because $\nu^2 \gg Q^2 = -q^2$ in the deep inelastic region. Assuming that the commutator does not oscillate violently, the important contributions come from

$$t - z \leq \frac{1}{\nu}, \ z \leq \frac{2\nu}{Q^2} = \frac{\omega}{M}, \tag{6.22}$$

so z can be quite large for large scaling variable ω. Causality makes the important transverse distances very small, $t^2 - z^2 - x_\perp^2 \geq 0$, because the commutator vanishes outside the light cone. This implies that

$$x_\perp^2 \leq (t-z)(t+z) \sim \left(\frac{1}{\nu}\right)\left(\frac{4\nu}{Q^2}\right) \sim \frac{4}{Q^2}. \tag{6.23}$$

Thus the important distances are concentrated along the light cone for a distance $z \sim \omega/M$ from the tip.

This clearly shows that the current commutator on the light cone is what determines the nature of the structure function W_1. The commutator can be obtained by Fourier inversion of the data, i.e. W_1:

$$W_1 \cong \iint dz dt \exp\left[i\nu(t-z) - i\lambda z\right] C(z,t) \approx W_1(\lambda) \tag{6.24}$$

with

$$C(z,t) = \frac{1}{2}\int_0^\infty dx_\perp^2 \langle p|[J_x(x), J_x(0)]|p\rangle, \tag{6.25}$$

$\lambda = \omega^{-1}$, $M = 1$. Then

$$C(z,t) \cong \delta(t-z) \int_{-\infty}^{\infty} \frac{d\lambda}{2\pi} W_1(\lambda) e^{i\lambda t}. \tag{6.26}$$

With such a structure for $C(z, t)$, the commutator must have a singularity on the light cone:

$$\langle p|[J_x(x), J_x(0)]|p\rangle = \frac{2}{\pi}|t|\delta\prime(x^2)\int_{-1}^{1} d\lambda W_1(\lambda)e^{i\lambda t}$$
$$= \frac{4i}{\pi}|t|\delta\prime(x^2)\int_{0}^{1} d\lambda W_1(\lambda)\sin\lambda t\ [+\text{less singular terms}]. \qquad (6.27)$$

Thus there is a direct connection between the nature of the commutator on the light cone and the data expressed in terms of structure function, or more explicitly, between the leading singularity in the light-cone commutator and the Bjorken scaling.

Puzzlements over method and underlying model

Light-cone current algebra, just as the current algebra of equal-time commutators, was abstracted from a relativistic Lagrangian model of a free spin 1/2 quark triplet. The essential feature in the abstraction was the constraint that turning on certain strong interactions in such a model would not affect the commutators even when all orders of perturbation theory were included. Similarly, mass differences breaking the symmetry under SU(3) would not disturb the commutators.

Are there nontrivial quark field theory models with interactions such that the basic light-cone current algebra can be abstracted and remains undisturbed to all orders of naïve perturbation theory? Of course, interactions change the operator commutators inside the light cone. But it was confirmed (Llewellyn-Smith, 1971; Cornwall and Jackiw, 1971; Gross and Treiman, 1971) that in a quark field theory model with neutral gluons, using formal manipulation of operators and not renormalized perturbation theory term by term, the basic algebra came out all right in the presence of interactions. If such interactions are mediated by a neutral scalar or pseudoscalar gluon field, the leading singularity on the light cone would be unaffected because the only operator series that satisfies the symmetry-breaking constraint ($l = -J-2$) and contains the gluon would not appear in the light cone expansion.

What if the gluon is a vector field? In this case, formally it is still all right except that the correspondence between the bilocal operators involved in the basic light-cone algebra, D (x, y, G), and their quark expressions must be modified by the presence of the factor $e^{ig\int_y^x B\cdot dl}$ where B is the gluon field, g its coupling constant, and the integral is along a straight line.

In contrast, the renormalized perturbation theory treatment of such a model, taken term by term, had revealed various pathologies in commutators of currents. The light-cone commutators do not have the free field structure, at least they do not order by order in perturbation theory, or in the limit of summing certain infinite classes of diagrams, which was implied by the results showing scaling violations (Adler and Tung, 1969; Jackiw and Preparata, 1969; Vainshtein and Ioffe, 1967). There were in each order logarithmic singularities on the light cone, which destroyed scaling, and violations of the Callan–Gross relation in the Bjorken limit. A careful perturbation theory treatment also shows the existence of higher singularities on the light cone, multiplied by the gluon fields. For example, in vector field theory, there is a term of the form $g(x-y)_\alpha \varepsilon_{\alpha\beta\gamma\delta}\partial_\gamma B_\delta/(x-y)^2$ occurring where one would expect from the basic algebra the finite operator $F_0^5(x,y)$. The gluon field strength, having in lowest order dimension $l=-2$, can appear multiplied by a more singular function than can a finite operator $F_0^5(x,y)$ of dimension -3. Such a term would ruin the basic algebra as a closed system.

Most serious is the fact that a term involving the gluon field strength would elevate that operator to the level of a physical quantity occurring in the light-cone commutator of real local currents. That is, if one takes vector gluons seriously, one must deal with the fact that the vector baryon current $F_{0\mu}$ and the gluon exist in the same channel and are coupled, so that a string of vacuum polarization bubbles contributes to the unrenormalized current. But all the currents have fixed normalization since their charges are well-defined quantum numbers, and it must be the unrenormalized currents that obey the algebra if the algebra is right.

Thus if a vector gluon theory is to be brought into harmony with the basic light-cone algebra, its renormalization constant must be finite and the sum of the renormalization expansion yields the special case of a finite theory if the coupling constant of gluon is not zero. Perhaps this finite theory leads, when summed, to canonical scaling and the disappearance of the anomalous light-cone singularities, so that the basic light-cone algebra is preserved.

However, since certain anomalies, like the anomalous divergence of axial vector current, come out only in the lowest order of renormalization perturbation theory and are difficult to fix by the finite theory approach, methodologically, Fritzsch and Gell-Mann argued, order by order evaluations of scaling based on a renormalizaed quark–gluon theory, such as anomalous singularities discussed by Adler and Tung (1969) had to be rejected so that the abstraction of the basic algebra could be preserved. The renormalized perturbation theory revealed no scale invariance or free field theory behavior at high energy, while "Nature reads books on free field theory, as far as the Bjorken limit is concerned" (Fritzsch and Gell-Mann, 1971a).

Fritzsch and Gell-Mann were cautious enough to realize that if "we blindly accept for hadrons the abstraction of any property of the gluon model that follows from naïve manipulation of operators, we risk making some unwise generalizations of the basic algebra." Methodologically puzzled, they said, "It would be desirable to have some definite point of view about the relation of the abstracted results to the renormalized perturbation theory" (Fritzsch and Gell-Mann, 1971b).

Together with this methodological puzzlement they were also puzzled over the underlying model. Among the four possible attitudes toward the conceptual situation: (a) the whole system, including scaling, is wrong in renormalized perturbation theory term by term; (b) a "finite theory" approach is to be used; (c) a naïve approach applied to a model of super-renormalizable self-interaction of quark currents; and (d) the naïve approach is right, and the basic algebra can be abstracted from a quark–gluon field theory model in which "the gluon field is not necessarily directly observable, but its effect is felt indirectly," they clearly inclined to take the last one. That is, they took the quark–gluon model as the underlying model.

But the nature of this underlying model was not clear because there was a profound question unsolved if the vector gluon model was adopted. Fritzsch and Gell-Mann stressed, "in this case, what happens to the Adler anomaly, which formally resembles the corresponding gluon anomalies, but involves the real electromagnetic field and real electric charge, instead of the presumably fictitious gluon quantities?"

The trouble for Fritzsch and Gell-Mann was that the Adler anomaly was well established and, formally, there was a mathematical relation between the low energy anomalous divergence involved in the Adler anomaly and the high energy anomalous singularity which would spoil their generalized algebra: If an anomalous linear singularity on the light cone was introduced, an anomalous divergence term appeared to be introduced into the divergence of the axial vector current; this anomalous divergence has dimension -4, which was not supposed to be present in the divergence of axial vector current to spoil the situation, in which the divergences of currents are contained in the generalized algebra.

It was unclear to them if the mathematical relation "is real or apparent, when operator products are carefully handled. Like Adler term, the anomalous divergence may be obtainable as a kind of low energy theorem and might survive a treatment in which the anomalous singularity is gotten rid of." That is, they hoped that the effect of the Adler anomaly would not spill over to the high energy regime to spoil their asymptotic scale invariant algebraic system.

In addition, when they talked about "the presumably fictitious gluon quantities," the meaning of "fictitious" was not clear at all. Gell-Mann talked about fictitious quarks for years. Its meaning was clear: because of their fractional charges, they would not be observed in laboratories, cosmic rays or any other situations. But the gluons, if they existed at all, according to Callan and Gross, must be electrically neutral because they played no role in electroproduction processes. Thus the gluon being "fictitious" must have other reasons than having fractional charge. As we will see shortly, this unclear intuition of "fictitious" gluon quantities will get its meaning clarified soon in terms of a new quantum number, color.

In sum, within the framework of light-cone current algebra, ontologically, the underlying model was moving from the quark–parton model to the quark–gluon model or even the quark–vector gluon model. Methodologically, formal manipulations remained to be preferred, but renormalized perturbation theory, because of the discovery of the Adler–Bell–Jackiw anomaly, was taken more and more seriously, and the desire was growing to have a definite point of view about the relation between the two.

6.4 Remarks

The activities in theorizing scaling were complicated and rich in implications, and not lacking in ironies. Perhaps the most interesting ironies can be found in the following three cases.

Parton model calculations

As we have noted, the parton model was very good in explaining scaling, but was useless in understanding the hadron spectroscopy because it had nothing to do with phenomenological symmetry of SU(3), nor with any credible dynamics.

Of course, the parton model calculations played a very important role in clarifying the ontological basis for theories of strong interactions, that is, in identifying essential properties the constituents of hadrons must have and in identifying two kinds of constituents, as we discussed in Section 6.1. But the approach adopted in the calculations was a promiscuous one. Basically free field theory in momentum space was used to calculate cross sections, and used wherever it was wanted. The rules were quite vague, and theorists found themselves doing calculations that they could hardly believe they should be doing: calculating strong interactions with free field theory, and getting the right answer.

Scale anomaly and renormalization group

It is interesting to note that while scale invariance is unavoidably broken by renormalization effects in an anomalous way,[3] the renormalization group equation that is underlain by the scale anomaly in fact is the most powerful means to recover potential asymptotic scale invariance. The equation traces the change of coupling constant with momentum scale caused by renormalization effects, thus the very idea of asymptotic scale invariance posed a strong constraint on finding out what kind of coupling would ensure asymptotic scale invariance. Further elaboration of this remark will be given in Section 8.1.

How formal? Why formal?

Gell-Mann and Fritzsch advocated formal, free field theory manipulations. But in fact their algebraic system was accompanied by the *ad hoc* rule that the matrix elements of the operators that appear on the right-hand side of these commutators should be taken as empirical parameters; certainly these matrix elements could not be formally derived from free field theory.

Gell-Mann and Fritzsch preferred formal manipulations and rejected renormalized perturbation theory treatments as irrelevant to the general theory of strong interaction. They embraced the idea of broken scale invariance enthusiastically, and noticed that scale invariance would be broken by certain kinds of interactions, but did not pay enough attention to the fact that the underlying mechanism for the breaking of scale invariance was exactly provided by the renormalization effects on the behavior of currents in strong interactions. This insensitivity perhaps was caused, in part at least, by the absence of adequate theoretical tools in dealing with the rigorous arguments for anomalies of the Johnson–Adler type and the observed scaling or desirable asymptotic scale invariance in a framework of renormalized perturbation theory. But their desire to have "some definite point of view about the relation of the abstracted results to the renormalized perturbation theory" was a testimony that they had already put the construction of such a framework on the agenda for the theoretical particle physics community.

Notes

1 Wilson further corroborated the dynamical view of the scale dimension of field operators, that the value of the dimension of a field is a consequence of renormalization, with an analysis of the Thirring model (1970a) and the $\lambda\phi^4$ theory (1970b).

2 The notion of bilocal operator, first introduced by Y. Frishman in (1970) was taken to be the basic mathematical entities of the whole scheme.

3 That is, the invariance is not violated by explicit symmetry-breaking terms in the Lagrangian, nor by non-invariant vacuum, but through quantum effects or loop corrections.

7
The advent of QCD

The underlying idea of current algebra, light-cone current algebra included, was to exploit the broken symmetry of strong interactions. The idea was pursued through abstracting physical predictions, as the consequences of the symmetry and in the form of certain algebraic relations obeyed by weak and electromagnetic currents of hadrons to all orders in strong interactions, from some underlying mathematical field theoretical models of hadrons and their interactions, more specifically from models of quark fields universally coupled to a neutral gluon field.

When current algebra was initially proposed, the symmetry was defined with respect to static properties of hadrons, such as mass, spin, charge, isospin, and strangeness. However, over the years, especially after the confirmation of scaling, in the changed theoretical and experimental context, the notion of symmetry underlying the current algebra project had undergone a transmutation. A new component of symmetry, that is, a dynamic symmetry, was brought into the scheme. The result was not only some adaptations in algebraic relations and corresponding physical predictions, but also a radical change in the nature of the underlying field theoretical model. More precisely, the result was the emergence of a gauge invariant quark–gluon model for strong interactions, or the genesis of quantum chromodynamics (QCD).

This chapter is devoted to examining the ways in which the transmutation took place and the reasons for such a radical change. Crucial to the genesis of QCD was of course the formation of the notion of "color" as a new quantum number. For "color" to transmute from a global notion to a local one characterizing a non-abelian gauge symmetry, however, a set of conditions had to be satisfied. Otherwise, no such transmutation would be possible, or acceptable if it was prematurely proposed. Lack of the appreciation of the conditions has been the source of confusion. One of the purposes of this chapter is to dispel the confusion by way of clarifying what these conditions are.

7.1 Color

The notion of color has its origin in the constituent quark model. But before it could be integrated into the underlying field-theoretical model of current algebra for further transmutation, from an *ad hoc* device to a notion for a new physical quantity, and then from a global notion to one for a local gauge symmetry, it had to be brought into the framework of current quarks in the first place.

This was by no means a straightforward extension because the system of constituent quarks and that of current quarks were quite different in nature, and the connection between the two was not clear at all in the early 1970s.

Constituent quarks were conceived as nonrelativistic building blocks, with very low probability of having quark–antiquark pairs appear, for low-lying bound and resonant states of mesons and baryons at rest; they were also helpful for understanding certain crude symmetry properties of these states; but they were not treated as quantum fields. Current quarks, in contrast, were part of the quark–gluon field-theoretical model for high energy relativistic phenomena in which a hadron bristled with numerous quark–antiquark pairs. Thus justifications were required for the notion of color to be carried over from the constituent quark scheme to the current quark scheme. Now let us follow the journey of this notion from its genesis to its being integrated into the current quark picture.

The origin of parastatistics

Within the scheme of constituent quarks, attempts were made to combine particles of different spin into the scheme of SU(6) symmetry (Gürsey and Radicati, 1964; Pais, 1964; Sakita, 1964). The idea of SU(6) symmetry led to several striking consequences which were in agreement with known facts, and thus strongly suggested that the baryon was a three-body system of fundamental triplets.

But the success was immediately followed by an embarrassment. Take the $\frac{3}{2}^{+}$ baryon decuplet as an example. It belongs to the 56 representation of the SU(6) group chosen for baryons. A specific property of this representation is that it is totally symmetric in the spin and unitary spin variables. Yet the static SU(6) symmetry applies only when orbital angular momenta are effectively neglected, thus there should be no spin–orbit couplings, and the quarks making up a baryon must be in S-states. This means that the three-quark wave function is symmetric in the orbital variables as well as the other two variables, and that is in direct conflict with Fermi–Dirac statistics or Pauli's exclusion principle.

The embarrassment was ingeniously removed by introducing an *ad hoc* requirement that quarks do not satisfy Fermi–Dirac statistics but a so-called parastatistics of rank three, in which a state can be occupied by three particles (Greenberg, 1964). Physically, the idea of parastatistics of rank three was equivalent to assuming that quarks carry a new additional three-valued degree of freedom, with respect to which the three-quark wave function in the 56 representation of SU(6) symmetry is totally antisymmetric. As a result, the S-state symmetrical wave function was saved because the total wave function, due to the new antisymmetrical component of the wave function, was antisymmetrical now, and thus had no conflict with the exclusion principle.

The Han–Nambu scheme

The original constituent quark model had an unusual feature that quarks had fractional charges. But according to Moo-Young Han and Nambu (1965), it was not necessary. They argued that if we assumed the existence of three triplets of quarks, an I-triplet, a U-triplet, and a V-triplet, then it would be possible to give an integral charge assignment to quarks as follows: for I-triplet quarks, $(1, 0, 0)$; for U-triplet, $(0, -1, -1)$; and for V-triplet, $(1, 0, 0)$.

Nambu pointed out that the reason he preferred this scheme "to the quark model is that fractional charges are unlikely to exist. At least one of the quarks must be stable and could have been observed rather easily among cosmic rays, for example." He dismissed Gell-Mann's signature assumption about quarks and straightforwardly declared that "we reject the idea that quarks are mere mathematical fiction. If they exist, they ought to have a finite mass of observable magnitude." In the spirit of the Sataka school, which was explicitly evoked in his talk at the Coral Gable conference and its published version (1965a), Nambu also stressed: "Moreover, we would have to say that leptons are also composite."

Han and Nambu noticed another advantage of the three triplet model that "the SU(6) symmetry can be easily realized with S-state triplets. The extended symmetry group becomes now $SU(6)' \times SU(3)''$. Since an $SU(3)''$ singlet is antisymmetric, the overall Pauli principle requires the baryon states to be the symmetric SU(6) 56-plet."

They also noticed that just as an electrically neutral system tended to be most stable, the $SU(3)''$ singlet states would be the most stable states. But if quarks of integral charges as $SU(3)''$ non-singlet states were observable as Nambu believed in (1965a), then other non-singlet states, though relatively unstable, must also be observable in this scheme.

It is usually claimed that the Han–Nambu scheme and parastatistics of rank three are equivalent. But as Oscar Greenberg and Daniel Zwanziger (1966) pointed out that they would be equivalent only if the three triplets carried the same charges; that is, if there were three (u, d, s) sets, each with the same old charges (2/3, −1/3, −1/3). In other words, the equivalence would obtain only when the new quantum number associated with the additional SU(3) symmetry made no contribution to electric charge; or, in later terminology, electromagnetism and color are mutually neutral. Clearly the Han–Nambu scheme did not fulfil the condition for equivalence. The non-equivalence is physically significant because it underlies a fundamental difference in conceiving the new physical property: with or without mixing with electromagnetism.

With the proposal of the Han–Nambu scheme, although the exact term "color" had to wait for some years to come, physicists began to talk about red, white and blue quarks: "if the baryon is made up of a red, a white, and a blue quark, they are all different fermions and you thus get rid of the forced antisymmetry of spatial wave functions" (Pais, 1965).

Color for current quarks

The notion of color as a property of current quarks was first proposed and elaborated by Gell-Mann in his Schladming lectures (1972) delivered February 21–March 4, 1972. The move was directly triggered by Rod Crewther's work (1972) on neutral pion decay treated within the framework of broken scale invariance (OPE and light-cone current algebra), although in fact it was the result of Gell-Mann's deliberations over years about the relation between current quarks and constituent quarks.

The exact $\pi^0 \rightarrow 2\gamma$ decay amplitude was first computed in the PCAC limit by Adler. The basis on which Adler derived it was a relativistic renormalized quark gluon field theory treated in renormalized perturbation theory order by order, and there the lowest order triangle diagram gave the only surviving result in the PCAC limit. Adler's result was that in a quark gluon model theory, the decay amplitude is a known constant times $\left(\sum Q_{1/2}^2 - \sum Q_{-1/2}^2\right)$, where the sum is over the types of quarks, the charges $Q_{1/2}^2$ and $Q_{-1/2}^2$ are those of $I_z = 1/2$ quarks and $I_z = -1/2$ quarks respectively. The amplitude agrees with experiment, within the errors, in both sign and magnitude, if $\left(\sum Q_{1/2}^2 - \sum Q_{-1/2}^2\right) = 1$ (Adler 1970). Three Fermi–Dirac quarks (u, d, s) would give 1/3, and the decay rate would be wrong by a factor of 1/9.

This strongly suggested that current quarks used by Adler in his calculations should obey parastatistics of rank three, because then everything would be all right.

However, this derivation, according to Gell-Mann, had to be rejected because it did not lead to scaling in the deep inelastic limit (Fritzsch and Gell-Mann, 1971b). Thus its relevance to current quarks obeying parastatistics was not even an issue for Gell-Mann.

In late February of 1972, unexpectedly, Crewther, who had just finished his PhD with Gell-Mann a few months earlier, completed a work in which Adler's result was derived from short-distance behavior of operator product expansions without assuming perturbation theory. Once this was clear to Gell-Mann, he decided that "we need color!"[1] In the written version of his Schladming lectures, he explained that the reason why he wanted color was that he would like to carry over the parastatistics for the current quarks from constituent quarks, and eventually to suggest a "transformation which took one into the other, conserving the statistics but changing a lot of other things." Before moving to the transformation, however, let us have a closer look at Crewther's derivation of Adler's result.

The derivation was based on algebraic constraints which relate short distance OPE to those of less complicated products, or high energy cross sections to anomalies in low energy theorems.

Crewther started with the consistency condition:

$$f_n(x,y) \underset{x \ll y}{\longrightarrow} \sum\nolimits_m C_m(x) C_{mn}(y) \tag{7.1}$$

where $f_n(x,y)$, $C_m(x)$ and $C_{mn}(y)$ are c-number coefficient functions, containing poles and cuts in the variables, in the expansions:

$$T\{A(x)B(0)C(y)\} = \sum\nolimits_n f_n(x,y) O_n(0)$$

$$T\{A(x)B(0)C(y)\} = \sum\nolimits_m C_m(x) T\{O'_m(0)C(y)\}$$

$$T\{O'_m(0)C(y)\} = \sum\nolimits_n C_{mn}(y) O_n(0).$$

He then applied the consistency condition to the anomalous low energy theorem for $\pi^0 \to 2\gamma$ decay. In terms of the electromagnetic current for hadrons and the third isospin component of the axial vector current, the definition of Adler's anomalous constant S was

$$S = -\frac{1}{12}\pi^2 \varepsilon^{\mu\nu\alpha\beta} \int\int d^4x d^4y x_\mu y_\nu T\left\langle 0 | J_\alpha(x) J_\beta(0) \partial^\gamma J^5_\gamma(y) | 0 \right\rangle. \tag{7.2}$$

PCAC and the experimental value for the decay amplitude, according to Adler (1970), implied $S \simeq +0.5$.

But Wilson had shown (1969) that S could be completely determined by the leading scale invariant singularity $G_{\alpha\beta\gamma}$ of the short-distance expansion

$$T_{\alpha\beta\gamma}(x,y) = T\left\{J_\alpha(x)J_\beta(0)J_\gamma^5(y)\right\} = G_{\alpha\beta\gamma}(x,y)I + \cdots \tag{7.3}$$

The notion of broken scale invariance requires $G_{\alpha\beta\gamma}$ to be a homogeneous function of degree -9 in x and y. In the limit $x<<y$ of Equation (7.3), a consistency condition for $G_{\alpha\beta\gamma}$ could be deduced from expansions

$$T\{J_\alpha(x)J_\beta(0)\} = \frac{R\left(g_{\alpha\beta}x^2 - 2x_\alpha x_\beta\right)I}{(\pi u)^4} + \frac{K\varepsilon_{\alpha\beta\lambda\mu}x^\lambda J^{\mu 5}(0)}{3\pi^2 u^2} + \cdots \tag{7.4}$$

$$T\left\{J_\mu^5(0)J_\gamma^5(y)\right\} = \frac{R'\left(g_{\mu\gamma}y^2 - 2y_\mu y_\gamma\right)I}{(\pi v)^4} + \cdots \tag{7.5}$$

where dots stand for derivatives of operators terms, which break scale invariance, and operators, which have different quantum numbers. The constants R, K, and R' are parameters to be measured experimentally.

According to Wilson (1969), Equation (7.2) was a surface integral in (x, y) space, and could be calculated explicitly. Crewther's calculation involved a limiting ambiguity similar to the one Adler encountered in his calculation, and after a careful treatment of this ambiguity, which he explained to Fritzsch and Gell-Mann in his six-page letter to them, the result he obtained was an exact formula, $3S = KR'$ for the anomalous constant, which demonstrated the intimate relation between PCAC anomalies and processes induced by highly virtual photons. The constant K was given by Bjorken's sum rule.

When Crewther related his result to the light-cone current algebra proposed by Fritzsch and Gell-Mann, he found that the existence of constraints such as Equation (7.2) meant that neither K nor R$'$ could be altered without changing many other coefficients in the algebra. Since the algebra required $K = 1$, $R' = 1/2$, the result of $S = 1/6$ could not be evaded.

The result for Crewther was "unfortunate" because he knew that the PCAC result was $S \simeq 0.5$. But for Gell-Mann, it was far from "unfortunate." He immediately saw the implication of Crewther's work. If Adler's result, which hinted that current quarks might obey parastatistics but was derived in a way inconsistent with scaling, could be derived in a way consistent with scaling in the current quark picture, then the relevance of Crewther's calculation of $\pi^0 \rightarrow 2\gamma$ amplitude to the robustness of the notion "parastatistics" in the context of current quarks was obvious. In fact, Gell-Mann took this as a convincing piece of evidence that current quarks obey the same parastatistics obeyed by constituent quarks.

Gell-Mann even found one more convincing evidence for current quarks obeying parastatistics. It was related to the newly available experimental results about the electron–positron annihilation to hadrons. Gell-Mann argued that if the disconnected part of the current commutator, namely the vacuum expected value of the commutator of currents, also behaved in the light-cone limit as it did formally in the quark–gluon model, namely the same as for a free quark model, then the high energy limit of the ratio for one photon annihilation to hadrons and muons can be predicted immediately and trivially.

The reason was simple. In the case assumed by Gell-Mann, the form of the leading light-cone singularity in the disconnected part of the current commutator can be easily determined from the formal quark–gluon model, leaving only the constant in front to be determined from the model rather than algebra.

According to the quark–parton model, which was consistent with light-cone current algebra, the production of a quark–antiquark pair, like that of a muon pair, should be exactly described by QED, and the rearrangement of the thus produced quarks into hadrons would not change the QED prediction. This means that the only difference between muon and hadron production in the annihilation lay in the electric charges of muons and quarks, and the ratio R of total cross sections for hadron and muon production was expected to be an energy independent constant. The magnitude of this constant was expected directly to reflect the number of quark species and their electric charges. Thus it was easy to predict that the ratio for single photon annihilation at high energy should approach a constant:

$$R = \frac{\sigma(e^+ + e^- \rightarrow hadrons)}{\sigma(e^+ + e^- \rightarrow \mu^+ + \mu^-)} \rightarrow 3\left[\left(\frac{2}{3}\right)^2 + \left(-\frac{1}{3}\right)^2 + \left(-\frac{1}{3}\right)^2\right] = 2 \qquad (7.6)$$

if current quarks obeyed parastatistics, or 2/3 with Fermi–Dirac quarks, or 4 for the Han–Nambu scheme.

The first data on R at energies above 1 GeV were produced by the Italian electron–positron collider, ADONE, and published in 1970 (Bartoli *et al.*, 1970). By 1972, the ADONE data produced with an energy range between 1 and 3 GeV appeared to be in satisfactory agreement with the parton model prediction that if quarks came in three species obeying parastatistics, R should be a constant at a value of around 2 (Silvestrini, 1972).

These new pieces of evidence had consolidated Gell-Mann's conviction that color must be a property of current quarks rather than an *ad hoc* way of dealing with the statistics problem encountered in the scheme of constituent quarks.

Connection between constituent and current quarks

The carrying over of parastatistics from constituent quarks to current quarks, which resulted in the introduction of a new variable, color, for distinguishing three quark triplets, was only part of Gell-Mann's ambition to clarify the connection between the two pictures: the nonrelativistic picture of constituent quarks and the relativistic picture of current quarks. Two pictures underlie two different algebras. One algebra was based on an approximate symmetry group and led to the idea that hadrons are composed of just a few constituent quarks; and the other was an algebra of physically observable currents, which behave, algebraically, as if they were simply bilinear forms in relativistic quark fields. Mathematically the two algebras are similar, physically, they are totally different.

At Gell-Mann's suggestion, his students Jay Melosh and Ken Young tried to take up what Dashen and Gell-Mann himself had been trying to do in the last seven years and develop a unitary transformation to connect the two algebras and two pictures. Part of their results was reported by Gell-Mann in his Schladming lectures (1972).

Melosh's major result was a unitary transformation connecting the two algebras, which was constructed explicitly in the free quark model, and whose algebraic structure was then found to be roughly the same as the unitary transformation constructed in models with interacting quarks. The unitary transformation constructed by Melosh was able to connect current quarks with constituent quarks in a way that an exactly conserved symmetry of constituent quarks would be produced. Young's major result was the observation that "the transformation must bring in quark pair contributions and the constituent quark looks then like a current quark dressed up with current quark pairs" (Gell-Mann, 1972).

Now with the new pieces of evidence from $\pi^0 \to 2\gamma$ decay and the electron–positron annihilation to hadrons summarized above (see also Bardeen, Fritzsch and Gell-Mann, 1972), Gell-Mann suggested adding one more property to the unitary transformation: it should preserve statistics. If this is generally true, then the indication from the constituent quark picture that quarks obey parastatistics of rank three "should suggest the same behaviour for the current quarks in the underlying relativistic model, from which we abstract the vacuum behaviour of the light cone current commutator" (Gell-Mann, 1972).

This way, the notion of color was integrated into the underlying field-theoretical framework of Gell-Mann's light-cone current algebra. At this stage, however, the notion of color remained a global one, similar to isospin and strangeness, having nothing to do with dynamics. The difference between quantum number color and other quantum numbers such as isospin and strangeness lies only in that while isospin and strangeness differentiated different members within a fundamental triplet of quarks, color differentiated different fundamental triplets.

The color singlet assumption

In his Schladming lectures, immediately after the notion of color was introduced, Gell-Mann required that "*all physical baryon and meson states be singlets under the SU(3) of 'color'*" (1972, my emphasis). This means that all baryon and boson states are totally antisymmetric in color. Two months later, Gell-Mann, together with William Bardeen and Fritzsch further emphasized that physical currents and states "are restricted to be singlets under the color SU(3) . . . Likewise all the higher configurations for baryons and mesons are required to be color singlets," although they also acknowledged at the same time that "we do not know how to incorporate such restrictions on physical states into the formalism of the quark gluon field theory model" (Bardeen, Fritzsch, and Gell-Mann, 1972).

In his Schladming lectures, Gell-Mann further explored the implications of singlets restriction: "Now if this restriction is applied to all real baryons and mesons, then the quarks presumably cannot be real particles," they have to be "fictitious" particles. He clarified that "real" meant that they "are detectable in isolation in the laboratory" and "fictitious" meant "they are permanently bound" or "permanently contained inside hadrons" (see also Fritzsch and Gell-Mann, 1972).

The reason for this implication is that "the singlet restriction is not one that can be easily applied to real underlying objects; it is not one that factors: a singlet can be made up of two octets and these can be removed very far from each other such that the system overall still is a singlet, but then we see the two pieces as octets because of the factoring property of the S matrix."

As a comparison, Gell-Mann also considered an alternative scheme without explicitly referring it to Han and Nambu, in which "there are actually three triplets of real quarks, . . . [and] we would replace the singlet restriction with the assumption that the low lying states are singlets and one has to pay a large price in energy to get the colored SU(3) excited."

Two months later, the comparison was made again with a harsher criticism: "in the Han-Nambu scheme, there are nine quarks, capable of being real . . . not only can the analog of the color variable really be excited, but also it is excited even by the electromagnetic current, which is no longer a 'color' singlet . . . it directly excited the new quantum numbers" (Bardeen, Fritzsch, and Gell-Mann, 1972).

At that stage, the color singlet restirction was only an *ad hoc* assumption rather than a convincing explanation of the absence of all unobserved multi-quark configurations in the known array of hadron systems. But as we will see clearly in Section 7.3, this restriction was not only, or even mainly, one on the spectrum of the microscopic world, but had consequence on the dynamics of the underlying quark–gluon field theoretical model of current algebra.

7.2 The rise of non-abelian gauge theory

The idea of using the non-abelian gauge group to deal with nuclear interactions, as we mentioned in Section 2.1, started from Pauli's work (1953). Pauli's aim in developing the mathematical structure of a non-abelian group was to derive rather than postulate the charge-independent nuclear forces of the meson theory from the SU(2) group of isospin symmetry. In comparison to the similar attempt of Yang and Robert Mills (1954a, b), Ryoyu Utiyama's was more ambitious because he, by postulating the local gauge invariance of the system under a wider symmetry group, proposed a more general rule for introducing, in a definite way, a new field that has a definite type of interaction with the original fields. This general rule embodied the substance of the gauge principle set up by Pauli. While Pauli, Yang, and Mills confined their attention to the internal symmetry of isospin, Utiyama's treatment extended to the external symmetry as well, and thus the gravitational field and interactions could also be incorporated into the general framework of a gauge theory (Utiyama, 1956).

In 1957, Julian Schwinger proposed a more concrete gauge theory, based on an SU(2) group, in which the electromagnetic and weak interactions can be accounted for in a unified way. His idea was picked up by many and significantly improved by his disciple Sheldon Glashow (1961), and later by Steven Weinberg (1967) and Abdus Salam (1968).

In the realm of strong interactions, the first to explore the consequences of the gauge principle in a detailed model was Jun John Sakurai (1960). He proposed that the strong interactions should be mediated by five massive vector bosons, which were associated with three conserved vector currents: the isospin current, the hypercharge currents, and the baryonic current.

The mass problem: evasion and solution

Pauli succeeded in his mathematical derivation but was frustrated by the so-called mass problem. His frustration was shared by all his followers, and by Sakurai in particular, until a decade later when a perfect solution to the problem was found. The problem arises from two conflicting requirements. Gauge invariance and the short range of nuclear forces. Gauge invariance requires that gauge bosons have to be massless, as Pauli pointed out. But then these gauge bosons could not be responsible for the short range of nuclear forces. The short range of the nuclear forces requires, as shown by Yakawa, that it be mediated by quanta of nonzero mass, but then massive quanta ruin the gauge invariance and the renormalizability of the theory as well.

Gell-Mann initially was also attracted by the gauge principle in the late 1950s. He even tried to find a solution to the mass problem by suggesting a soft mass mechanism, which would allow the supposed renormalizability of the massless theory to persist, but failed. With no hope of solving the mass problem, Gell-Mann suggested that instead of starting from massless gauge bosons we might take the equations with massive gauge fields as the starting point of the theory, forgetting about the possible connection with the gauge principle. This attitude was satisfactory for all practical purposes, and thus pushed Sakurai into the pursuit of the vector meson dominance model. It also became the starting point of Gell-Mann's own later theorizing.

However, his early fascination with non-abelian gauge theory was not without consequences. In his original paper on the eightfold way, he remarked that "the most attractive feature of the scheme is that it permits the description of eight vector meson by a unified theory of the Yang-Mills type" (1961). Theoretically, his current algebra could be conceived within the framework of a gauge theory, in which vector bosons were coupled with vector currents of fermions. The commutation rules could also be conceived as the consequence of gauge symmetry. By taking away the bosons because of the failure to solve the mass problem, what was left were currents satisfying current commutation rules.

The possible connection between current algebra and a non-abelian gauge theory was keenly perceived and fruitfully exploited in the mid 1960s. For example, Martinus Veltman (1966) derived the Adler–Weisberger relation and other consequences of current algebra by a set of divergence equations, in which ∂_μ is replaced by $(\partial_\mu - W_\mu \times)$, where W_μ denoted a vector boson. Responding to Veltman's work, John Bell (1967) presented a formal derivation of the divergence equations from the consideration of gauge transformations, trying to argue that the success of current algebra might be a consequence of gauge invariance. Perceiving the logical connection between current algebra and the non-abelian gauge theory this way, Veltman (1968) took the success of the Adler–Weisberger relation as experimental evidence for the renormalizability of non-abelian gauge theories, with a hidden assumption that only renormalizable theories could successfully describe what happened in nature. Veltman's confidence in the renormalizability of non-abelian gauge theory was consequential.

But contrary to Gell-Mann's early pessimism about the mass problem, the solution to it was successfully discovered by Philip Anderson (1963), based on Heisenberg's concept of the degeneracy of the vacuum and his nonlinear field theory (1958, 1960), and on the thereby derived notion of symmetry restoring massless meson, inspired by the notion of symmetry restoring collective excitations in Nikolay Bogoliubov's reformulation of the BCS theory of superconductivity (1958) and explored by Nambu and Giovanni Jona-Lasinio (1961a, b) and by Jeffrey Goldstone (1961). Anderson's breakthrough can be summarized in two statements: (i) in certain cases, when gauge invariance is broken by a degenerate vacuum, the restoration of gauge invariance requires the presence of massless collective modes; and (ii) long-range forces carried by massless gauge bosons recombine these massless modes into massive ones. The Anderson mechanism for solving the mass problem of non-abelian gauge theory was elaborated by Peter Higgs and many others in the context of particle physics (1964a, b).[2]

Han–Nambu's double SU(3) symmetry scheme

Based on their three-triplet model, Han and Nambu (1965) proposed a double SU(3) symmetry scheme, in which in addition to the ordinary "$SU(3)'$ group observed among the known baryons and mesons with their strong interactions," "the superstrong interactions responsible for forming baryons and mesons have the symmetry $SU(3)''$ and cause large mass splittings between different representations." Furthermore, "we introduce eight gauge vector fields which behave as (1,8), namely as an octet in $SU(3)''$, but as singlet in $SU(3)'$. Since their coupling to the individual triplets is proportional to λ_i'' [the generators of $SU(3)''$], ... they will be expected to be massive compared to the ordinary mesons."

Again, in his (1965b), Nambu stressed that "we have the group $SU(3)''$... [and] an octet of gauge fields G_μ, $\mu = 1, \ldots, 8$, coupled to the infinitesimal $SU(3)''$ generators (currents) λ_μ'' of the triplets, with a strength g."

That is, each individual member of the old SU(3) triplet is replaced by an $SU(3)''$ triplet, and associated with this new $SU(3)''$ symmetry, they introduced eight vector fields as the carriers of the interactions between quarks in a standard gauge theoretical way.

What was not so standard was their further assumption that $SU(3)''$ symmetry also generated some additional contributions to the electric charge in such a way that now three integrally charged triplets emerge with respective (u, d, s) charges (0, −1, −1); (1, 0, 0) and (1, 0, 0).

There are other problems for the double SU(3) symmetry scheme to be an acceptable non-abelian gauge theory for the fundamental objects and their interactions at a level deeper than hadrons. The only reason for the new degree of freedom with respect to which the new quantum number and new symmetry group were conceived was the statistics problem in the non-relativistic constituent quark model rather than in a relativistic field theory framework. With no other justification, its introduction was somewhat *ad hoc* in nature. But more importantly, when Han and Nambu proposed their SU(3) gauge theory for inter-quark interactions, there was no quantum theory of classical non-abelian gauge field, let alone the renormalization of such a quantum theory. That explains why their proposal left no trace in the further explorations of quark–gluon field theory although their three triplet scheme was quite well known to the physics community; only as an alternative hadron building scheme, of course.

Quantization and renormalization

The quantization problem was first seriously explored by Feynman in (1963). Formally, his subject was gravitation. But for the sake of comparison and illustration, he accepted Gell-Mann's suggestion and discussed another nonlinear theory in which interactions were also transmitted by massless particles, namely an SU(2) gauge theory.

The central theme of Feynman's discussion was quantization, which in his way was equivalent to deriving a set of rules for diagrams, in particular the rules for loops. These rules should be both unitary and covariant so that further investigations on the renormalization of the rules could proceed systematically. However, what Feynman discovered was that the rules for loops derived from his path integral formalism were not unitary. Lack of unitarity means the conservation of probability will be violated, and this is a very serious problem for physical reasoning.

The roots of the rules for loops failing to be unitary were clearly exposed by Feynman. In a covariant formalism which is necessary for renormalization, a virtual line in a loop carries longitudinal and scalar degrees of freedom, in addition to transverse polarization. When it is broken, it becomes an external line, and represents a free particle. That is, it carries transverse physical degrees of freedom only. In order to maintain Lorentz covariance and match the loop with the trees, some extra non-physical degrees of freedom have to be taken into consideration. Feynman then was forced to face an extra particle, a fake, fictitious particle, that belonged to the propagator in a loop and had to be integrated over but now instead became a free particle. Thus unitarity was destroyed, and Feynman felt that "something [was] fundamentally wrong." What was wrong was a seemingly head-on clash between Lorentz covariance and unitarity.

In order to restore unitarity, Feynman proposed to "subtract something" from the internal boson line in a loop. This insightful suggestion was the first clear recognition that some ghost loop had to be introduced so that a consistent formalism of non-abelian gauge theory could be obtained. The idea was picked up by Bryce DeWitt (1964, 1967a, b, c) and others (Faddeev and Popov, 1967; Mandelstam, 1968a, b), and by the summer of 1968, the Feynman rules for massless non-abelian gauge theories were established and were – for some choices of gauge, the so-called renormalizable gauge – those for renormalizable gauge theories, although the price paid for this achievement was the appearance of the ghost particles discovered by Feynman, or the so-called covariant ghosts.

Two more steps were crucial for the renormalization of non-abelian gauge theories.

The first was the establishment of Ward identity, with which the unitarity of the theory could be established because the covariant ghosts would be cancelled by another kind of ghosts, which were introduced by the gauge fixing term in a scheme in which the Feynman rules for general gauge could be obtained (Faddeev and Popov, 1967). Veltman (1970) derived generalized Ward identities for the massive non-abelian gauge theories, which was followed by John C. Taylor (1971) and Andrei Slavnov (1972) for more general cases with more sophisticated techniques. A formal combinatorial derivation of the Ward identities and, using them, a proof of unitarity and renormalizability for non-abelian gauge theories was achieved by 't Hooft and Veltman (1972b).

The second step was to find a gauge invariant regularization scheme for renormalization. A gauge invariant cutoff scheme was invented by 't Hooft (1971a, b) for one-loop diagrams in which a fifth dimension was introduced. A year later, a gauge invariant regularization scheme procedure, which was valid not only up to the one-loop level or to any finite order, but to all orders in perturbation theory, the dimensional regularization scheme, was proposed by 't Hooft and Veltman (1972a), in which the dimensionality of spacetime was allowed to have non-integer values and this continuation was defined before loop integrations were performed.[3]

The work by Veltman and 't Hooft was enormously profound and influential. It was immediately cheered as being one "which would change our way of thinking on gauge theory in a most profound way" (Lee, 1972). In fact it was universally recognized as a turning point for the fate of non-abelian gauge theories. Before their work, many physicists liked the gauge principle or even played with it, but nobody took it seriously. Without valid quantization and Feynman rules and without the proof of its renormalizability, a non-abelian gauge theory, such as the electroweak theory proposed by Glashow, Weinberg, and Salam was hardly noticed by the physics community, let alone recognized as a consistent, feasible, and workable model, before the work by Veltman and 't Hooft had changed the intellectual landscape (Pais, 1986). Only after their work, theoreticians had the tools in their hands to construct realistic non-abelian gauge theories, and the stage was set for non-abelian gauge theories, such as the one for quark–gluon interactions, or QCD, to be proposed by theoreticians and accepted by the physics community. It became much easier to do so because they had already learnt from Veltman and 't Hooft that a quantized non-abelian gauge theory is renormalizable and thus is feasible, workable, and calculable.

7.3 The first articulation of QCD

At the 16th International Conference on High Energy Physics, held in Chicago, September 1972, Gell-Mann gave two important presentations. One was a plenary talk titled "General status: summary and outlook" (Gell-Mann, 1972b), the other was a presentation of his work with Fritzsch titled "Current algebra: quarks and what else?" (Fritzsch and Gell-Mann, 1972) at a session titled "Currents III: scaling phenomena," organized and chaired by Gross.

Gell-Mann's plenary talk

In his plenary talk, when he moved to the area of "the hadrons, their strong interaction, and their weak and electromagnetic currents," Gell-Mann emphatically set the agenda for the hadron physics community, or perhaps mainly for himself, to construct a consistent conceptual framework, or a unified theory to deal with hadron physics as a whole:

I see, close at hand, another marvelous dream – a unified theory of the hadrons, their strong interaction, and their currents, incorporating all the respectable ideas now being studied: constituent quarks, current quarks, the bootstrap, dual resonance methods, Regge poles and cuts, hadronic scaling, weak and electromagnetic scaling, and current algebra. There are all or nearly all compatible, if we get our **language** straight.

Then he moved to summarize the ideas he developed in his work on current algebra in the previous ten years, and in particular ideas he developed in his work with Fritzsch on light-cone current algebra in the previous two years. He discussed the difference between the current quarks picture and the constituent quarks picture, and the unitary transformation connecting them which would preserve parastatistics of rank three; he explained why the notion of parastatistics could be replaced by "an equivalent but simpler formulation" based on the notion of "color."

Once the notion of color was introduced to the framework, Gell-Mann immediately set the color singlet restriction and explored its implications. "Physical particles are required to be singlets under the SU(3) of color," which requirement "rules out the existence of real quarks, detectable in the laboratory." "The reason being that, if quarks were real, such a restriction would tend to violate the factorization of the S-matrix for two widely separated systems." As a result, "the hadron theory must be such that quarks are somehow permanently contained inside hadrons." In this regard, Gell-Mann severely criticized "the Han-Nambu picture" again: "The 'color' variable that distinguishes the triplets from one another is physical in this scheme, and even coupled to electromagnetism. Baryons and mesons that are not singlet with respect to the SU(3) of 'color' must be found some day if the scheme is correct. Moreover, the basic fermions [quarks] themselves could well be real" (Gell-Mann, 1972b).

When Gell-Mann moved to recent works on deep inelastic scatterings, scaling and partons, he paid particular attention to the convincing arguments for the existence of electromagnetically and weakly neutral gluons: "If the ratio of $\sigma^{\nu P}$ to E_{lab} is indeed the asymptotic one, then we can conclude that only about half the momentum of the nucleon is attributable to current quarks and the rest should be ascribed to neutral glue of some kind." The existence of the gluon certainly gave him more confidence in their light-cone current algebra which was "abstracted from a quark and vector gluon picture."

The power of light-cone current algebra was emphatically stressed. "A further abstraction from quark-gluon field theory models permits us to apply light cone current algebra to processes without hadron targets. The ratio $R = \sigma(e^+ + e^- \rightarrow hadrons)/\sigma(e^+ + e^- \rightarrow \mu^+ + \mu^-)$ to order e^2 would approach a definite number at high energy. This number would be 2 for quarks with quark statistics or else for three triplets of quarks." Light-cone current algebra can also be used to "calculate the amplitude for the decay $\pi^0 \to 2\gamma$ using the same assumption as in the case of e^+e^- annihilation."

But nothing concrete was specified in the plenary talk about the underlying "relativistic field model with quarks and neutral (say vector) gluon" from which light-cone current algebra was abstracted. Most importantly, there was a *huge gap* to be filled before the underlying model can be used as a *proper language* for a unified theory to incorporate "all the respectable ideas now being studied." *The two new pieces of evidence for the notion of color applicable to current quarks exclusively came from electromagnetic and weak interactions, while the evidence for the existence of gluons indicated that gluons must have nothing to do with electromagnetic and weak interactions.* Thus a serious question to be answered before a unified theory could be conceived was: *is it possible, and how, to connect the notion of color with the strong interactions mediated by gluons?* Whatever the answer is, it needs to be justified rather than declared.

The answer and justification, and a concrete model for the underlying field-theoretical framework, were outlined in Gell-Mann's presentation of a joint work by Fritzsch and Gell-Mann (1972) at the session chaired by Gross.

The joint paper by Fritzsch and Gell-Mann

The presentation started with a concise statement about the conceptual foundation of Gell-Mann's major project in hadron physics, current algebra:

> For more than a decade, we have been squeezing predictions out of a mathematical field theory model of hadrons that we don't fully believe – a model containing a triplet of spin ½ fields coupled universally to a neutral spin 1 field, that of the "gluon."

The achievements of "the naïve quark model of baryon and meson spectrum and couplings" should be compatible with the underlying model of current algebra because presumably there would be a transformation effectively converting "the current quarks of the relativistic model into the constituent quarks of the naïve quark model."

In the substantial section II, before addressing the crucial point about the colored quarks that are not real and "related matters concerning the gluons of the field theory model," for setting the stage, Gell-Mann introduced the notion of color, "three triplets of quarks, called red, white, and blue, distinguished only by the parameter referred to as color," and made his fundamental assumption about the color singlet restriction, "real particles are required to be singlets with respect to the SU(3) of color."

The implication of the restriction for quarks was clear: "we assume here that quarks do not have real counterparts that are detectable in isolation in the laboratory – they are supposed to be permanently bound inside the mesons and baryons." The paper went further: "we have the option, no matter how far we go in abstracting results from a field theory model, of treating only color singlet operators."

Their conviction about the power of the singlet restriction was clearly displayed:

> We might eventually abstract from the quark-vector-gluon field theory model enough algebraic information about the color singlet operators in the model to describe all the degrees of freedom that are present. For the real world of baryons and mesons, there must be a similar algebraic system, which may differ in some respects from that of the model, but which is in principle knowable. The operator $\theta_{\mu\nu}$ would be expressed in terms of this system, and the complete Hilbert space of baryons and mesons would be a representation of it. We would have a complete theory of the hadrons and their currents, and we need never mention any operators other than color singlets.

How about gluons or the field operators for vector gluons? This was the crucial question, and Gell-Mann had struggled with it for a long time. The notion of gluon was first introduced by Gell-Mann in the first paper on current algebra (1962). Initially it was assumed to be a scalar particle. Over years, he changed his mind and took it as a vector. In his Schladming lecture, he declared that "we stick to a vector gluon picture so that we can use the scalar and pseudoscalar densities to describe the divergences of the vector and axial vector currents respectively" (1972). But a vector gluon had its own problems (see Section 6.3 and below), and that was the major reason for Gell-Mann being so hesitant to take the next step in constructing a consistent quark–gluon field theory to underlie his current algebra project.

Now, however, the color singlet restriction seemed to give him a new source of inspiration. Remember that one of the major difficulties for having a vector gluon field for Fritzsch and Gell-Mann was that a vector gluon and some vector baryon currents might exist in the same channel and coupled (see Section 6.3). But this would be so only when field operators for gluons were color singlets. If this is so, then a more serious question would be raised. That is an asymmetry between quarks and gluons: While quarks were assumed to be color triplets, not color singlets, and thus not real in the sense that they can not be detected in laboratory in isolation, gluons were color singlets, and thus presumably would be real in the sense that they could be detected in isolation in laboratory. Was it possible to remove this asymmetry? The answer seemed to be a simple and clear yes if gluons were assumed to be non-singlets.

At this juncture, two inputs seemed to have played a crucial role. The first came from Fritzsch, the first author of the paper. Fritzsch studied non-abelian gauge theories carefully when he was a college student; his BA thesis was partially inspired by the work of Utiyama.[4] In his desire to further apply the gauge principle to the underlying conceptual framework of his work with Gell-Mann on the light-cone current algebra, the quark–gluon field-theoretical model, there were two SU(3) symmetries available to them. The original (later being called flavor) SU(3) symmetry and the newly conceived color SU(3) symmetry. The flavor SU(3) symmetry was related to electromagnetic and weak interactions. But as parton model calculations had shown, the evidence for the existence of gluons derived from electroproduction and deep inelastic neutrino scatterings indicated that gluons played no role in electromagnetic and weak interactions (see Section 6.1), and thus could not be related to the flavor SU(3) symmetry. It seemed that the only option left for Fritzsch to pursue was to turn to the global color SU(3) symmetry, and try to turn it into a gauge symmetry.

For this reason, the idea of color non-singlet gluons, inspired by the motivation of removing the asymmetry between gluons and quarks in terms of their reality or observability in isolation in laboratory, had opened the door for Fritzsch to conceive gluons as the gauge potentials of the color SU(3) gauge group, which, logically or mathematically, would entail that gluons be color octet.

The other input was acknowledged in note 5 of the paper after the stage setting sentence: "Now the interesting question has been raised whether we should regard the gluons as well as the quarks as being non-singlets with respect to color." Note 5 reads as follows: "J. Wess (private communication to B. Zumino)." In a telephone conversation (April 20, 2006), Julian Wess informed me that sometime in the summer of 1972 when he, Gell-Mann and Fritzsch, were all at CERN, he had spoken to Gell-Mann. In the conversation, he explained to Gell-Mann what he was doing in supersymmetry and Gell-Mann explained to Wess what he and Fritzsch were doing. Then Wess made a remark, which was conveyed to him by Zumino, that the gauge field with an octet gluon might provide a scheme for confinement.

This remark fell into very receptive ears. On the side of Gell-Mann, this was a story very similar to the story about the origin of quarks: "In March 1963, when I was on leave from Caltech at MIT and playing with various schemes for elementary objects that could underlie the hadron. On a visit to Columbia, I was asked by Bob Serber why I didn't postulate a triplet of what we would now call SU(3) of flavor, making use of my relation $\underset{\sim}{3} \times \underset{\sim}{3} \times \underset{\sim}{3} = \underset{\sim}{1} + \underset{\sim}{8} + \underset{\sim}{8} + \underset{\sim}{10}$ to explain baryon octets, decimets, and singlets"

(Gell-Mann, 1997). On the side of Fritzsch, a color octet gluon perfectly fit into the mathematical structure of the color SU(3) gauge group, which he was himself intensively exploring.

Thus, the situation for Fritzsch and Gell-Mann in the summer of 1972 was this. In order to remove the asymmetry between gluons and quarks, gluons had to be color non-singlets. That is, gluons had to be as unreal as quarks were in the underlying field-theoretical model of current algebra. Once gluons being color singlet was ruled out, in particular when an octet gluon was suggested for its nonobservability, for the alerted minds of Fritzsch and Gell-Mann, who were familiar with non-abelian gauge theory and its mathematical structures, the door was open to take the gluon octet as the gauge potentials of the color SU(3) gauge group.

Gluons, since their first introduction into the picture, by definition, were supposed to be responsible for the inter-quark strong interactions, thus *their status as the gauge bosons of the color SU(3) symmetry made them a bridge connecting the notion of color with strong inter-quark interactions. Once this step was taken, the global notion "color" originating from the parastatistics of the constituent quark model had transmuted into a local notion characterizing the gauge invariant inter-quark interactions.*

Now let us return to Gell-Mann's presentation at Gross's session. Immediately after note 5, Gell-Mann continued: "*For example, they could form a color octet of neutral vector fields obeying the Yang-Mills equations [equations for non-abelian gauge fields].*"

He further justified this move by informing his audience of a significant advantage in terms of conceptual consistency:

> If the gluons of the model are to be turned into color octets, then an annoying asymmetry between quarks and gluons is removed, namely that there is no physical channel with quark quantum numbers, while gluons communicate freely with the channel containing the ω and ϕ mesons. (In fact, this communication of an elementary gluon potential with the real current of baryon number makes it very difficult to believe that all the formal relations of light cone current algebra could be true even in a "finite" version of singlet neutral vector\gluon field theory.)

Another advantage, most importantly, was,

> If the gluons become a color octet, then we do not have to deal with a gluon field strength standing alone, only with its square, summed over the octet, and with quantities like $\overline{q}(\partial_\mu - ig\rho_A B_{A\mu})q$ where $\rho's$ are the eight 3×3 color matrices for the quark and the B's are the eight gluon potentials.

This was enough to specify the QCD Lagrangian. Although this was only a verbal description of the QCD Lagrangian without using the word

Lagrangian and Gell-Mann's nickname QCD wasn't coined till 1973, the implication of this description to a knowledgeable physicist, such as Ken Wilson (2004), Rod Crewther (2005), Stephen Adler (2005), as well as the chair of the session, David Gross, was perfectly clear: a new non-abelian gauge theory of quark–gluon interactions was proposed.

With this first articulation of the color octet vector gluon picture, QCD was born.

7.4 Remarks

In the previous sections, we have examined how the transmutation of the notion of color took place, that is, how the notion of color had transmuted from a quantum number characterizing a global symmetry to one characterizing a gauge symmetry for quark–gluon dynamics, resulting in the genesis of QCD. Now some brief remarks seem to be desirable to clarify the general conditions and historical circumstances under which the transmutation took place and QCD was articulated, so that some prevailing confusions, especially the one concerning the role played by Han–Nambu's gauged version of double SU(3) symmetry, can be dispelled. The major point in the following remarks is that the gauge principle has no magic power by itself; its successful application or realization requires a set of stringent conditions to be satisfied.

Level

The gauge principle has been the subject of worship for many scholars, ever since it was proposed in the early 1950s by Pauli and others, mainly because it has been widely misconceived to be a universal principle for uniquely specifying dynamics. In fact, however, it is far from being universal; the range of its applicability, from a contemporary perspective, is severely restricted to a certain level of interactions.[5] Outside of the domain of its applicability, no matter how deep a physicist is thinking, how sophisticated the theoretical resources he is deploying, and how hard he is working, applying the gauge principle is doomed. The failure of Pauli (1953), Yang and Mills (1954a, b), and Sakurai (1960), trying to construct a non-abelian gauge theory (isospin SU(2) and its extension) at the level of hadron physics, was a case in point.

It is amusing to note that the crucial relevance of level to the applicability of physical theories, gauge theory included, was realized by Gell-Mann only decades later after he first articulated QCD in 1972: "I have always recalled describing QCD quite clearly at the session chaired by David [Gross] and then weakening the presentation when it came to preparing the final version.

The motivation was to allow for string since I *did not understand that QCD was correct at one level and string at a higher level, namely that of unified theory*" (private communication, January 29, 2005; my emphasis).

Entities

As opposed to Pauli, Yang and Mills, and Sakurai who tried to apply the gauge principle at the wrong level, when Nambu tried to implement the gauge principle for quark dynamics with his ideas of the SU(3) gauge group relating three triplets of quarks and of the associated eight gauge bosons conveying super-strong forces, he clearly got the level right. Then why would these sophisticated ideas have played no role in the history of QCD while his closely related three triplet models were widely noticed? The reason for this "unfair" reception of his "profound" ideas by the physics community is manifold. At the risk of over-simplification, the reason could be said to have three dimensions: entities, theoretical context, and intellectual leadership. Let us take the entities first.

Nambu's quarks have integral charges, and thus "colored" quarks and hadrons are observable, not confined. Due to the chronic failure in detecting quarks, this conception of the basic entities found few receptive ears in the physics community, especially among those who took Gell-Mann's fractionally charged and thus permanently confined quarks as a more acceptable conception of the sub-hadronic building blocks. More importantly, this charge allocation has consequences for the nature of other basic entities in the gauge theory of quark dynamics, namely gauge bosons themselves, and thus the nature of gauge symmetry itself.

Since the very notion of three triplets, or in some effective sense the notion of "color," was conceived by Nambu purely on the ground of quarks' para-statistics without any other justifications, it thus appeared to be *ad hoc* in nature, without any physical constraints posed on it for its detailed physical traits. As a result, it may or may not mix with electromagnetism. And in the case of Nambu's scheme of integrally charged quarks, the gauge bosons must carry some electromagnetism to connect differently charged triplets. Or amounting to the same thing, the color force and electromagnetism are not mutually neutral. Thus Nambu's wrong conception of gauge bosons had logically spoiled the very nature of gauge symmetry itself. The resultant symmetry, which is not responsible for the color force, but rather for a mechanism mixing color force and electromagnetism, does not exist in nature at all.[6]

In contrast, the existence of gluons for Gell-Mann was not assumed in an *ad hoc* way, but, rather, was painfully demonstrated on the basis of parton model calculations (see Section 6.1). As a result, these bosons could not have any traits that are in conflict with the physical constraints posed on them by experimental observations and the accepted parton model calculations. In particular there is no way for Gell-Mann's gluons, or color gauge bosons as he later conceived them, to play any role in electromagnetic and/or weak interactions.

Theoretical context

The notion of color gauge and the related notion of gauge bosons could be an important component of QCD. But these notions by themselves were not yet a workable conceptual framework of QCD. In Nambu's case, his three triplet models were variations of the constituent quark model, not even field-theoretical models. Without being incorporated into a field-theoretical context, these notions could not play the role QCD plays.

Even if in the field-theoretical context, these notions would not have any use before the difficulties of the massless and massive non-abelian gauge theories, most importantly the difficulties in the quantization and renormalization of non-abelian gauge theories, were cleared (see Section 7.2), let alone taken as a workable conceptual framework. As we have noted, only after Veltman and 't Hooft's work, certainly not before their work, did theoreticians have the effective tools in their hands to construct non-abelian gauge theories such as QCD. Thus the "unfair reception" of Nambu's contributions by the physics community itself was misconceived: there simply was no way for Nambu, or any other physicists, to develop any tenable non-abelian gauge theory for inter-quark forces before Veltman and 't Hooft's work in 1971.

Gell-Mann's intellectual leadership

The articulation of QCD was not a simple application of gauge principle to sub-hadronic entities, but rather the final result of increasingly successful efforts toward a unified theory of the hadrons, their interactions and their currents, incorporating all the respectable ideas that emerged in the 1960s in the description of elementary particles: current algebra, current quarks, constituent quarks, scaling, bootstrap, and others. Gell-Mann's intellectual leadership in the articulation lies in, first and foremost, setting the agenda for straightening the language or conceptual framework for such a unified theory at each stage of the conceptual development in this regard.

Secondly, since Gell-Mann originated many of these respectable ideas, he enjoyed an excellent reputation and special prestige in the physics community, and thus was able to motivate and mobilize many of the most active and most competent physicists, such as Adler, Bjorken, Crewther, Bardeen, Melosh, and Fritzsch, to pursue the agenda he set for the hadron physics community at each stage of the conceptual development since the early 1960s. These pursuits ultimately resulted in the first articulation of QCD by Gell-Mann himself, also on behalf of Fritzsch, in September 1972.

Again, just because of Gell-Mann's intellectual leadership, his articulation of QCD had motivated and mobilized the mainstream of physics community to further pursue the agenda Gell-Mann set for it, that is, to test, justify and improve the conceptual framework of QCD.

For these reasons, Gell-Mann's role, and thus his place, in the construction of QCD was unique. No other physicist, Nambu included, could have played a comparable role or enjoyed a comparable place in the construction of QCD as Gell-Mann did.

An outstanding problem

"Are the quarks really kept inside or do they escape to infinity?" Gell-Mann himself realized the existence and severity of an outstanding problem that remained to be addressed. "By restricting physical states and interesting operators to color singlets only we have to some extent begged that question" (Fritzsch and Gell-Mann, 1972).

Thus Gell-Mann set the agenda again for the community to pursue: "One fascinating problem, of course, is to understand the conditions under which we can have an algebra resembling that for quarks and gluons and yet escape having real quarks and gluons" (Fritzsch and Gell-Mann, 1972). That is, he wanted to have an exact mechanism for color confinement without changing the basic conceptual and mathematical structure of QCD and the light-cone current algebra abstracted from it. This is indeed the major task on the agenda for theoreticians to pursue in decades to come.

Yet it was dubitable to the sceptics of QCD that the unbroken color SU(3) symmetry for quarks could actually confine quarks. Remember that the study of scaling behavior was a central point for the high energy physics community since 1969 when scaling was established experimentally. It seemed to be a miracle that quarks could move freely at high energy and yet they would stick together inseparably inside the hadrons as Gell-Mann suggested. Scaling appeared to be a signal that no quantum field theory with interactions, the color gauge interactions included, would be fit to account for the observed strong interactions.

Experimentally, if scaling were true, two colliding protons, with high enough energy, should break into their constituents. If among these constituents there were fractionally charged quarks, then fractionally charged particles should be observed. In 1972 when QCD was first articulated, the intersecting storage ring (ISR) at CERN, which allowed the highest collision energy in the world (a maximum center-of-mass energy of 62 GeV) in the interaction between protons to be reachable, was just built and available for experiments. Thus the scepticism against QCD had encouraged a new series of experiments to check if quarks remained confined at these energies. These fractionally charged particles were not observed despite the very high energy of the proton–proton collisions. Thus confinement was experimentally established (Zichichi *et al.*, 1977, 1978).

But conceptually the confinement question raised by Gell-Mann remained to be addressed rigorously by theoreticians (more discussions on this in Section 8.3).[7]

Historical role of current algebra

It is noteworthy that after the proposal of QCD was made, current algebra, light-cone current algebra included, if not dying immediately and completely, gradually faded away or at least receded to the periphery in the explorations of hadron physics. How should we understand this historical process?

Current algebra, as a physical hypothesis about the structure of hadrons' interactions represented by a set of algebraic relations among hadronic currents, had provided an opening for the probing of hadrons' internal constitution and dynamics. The reason why current algebra had such a potential to be explored is that the algebraic structure proposed by current algebra, if confirmed, can only be properly understood, ultimately, in terms of hadrons' constituents and their interactions, assuming the validity of causal explanation and the hypothetico-deductive methodology derived from it, as the majority of the scientific community has assumed. This was exactly the reason why Gell-Mann had grounded his current algebra on an underlying quark model, or quark–gluon field-theoretical model for understanding and "fundamental explanation" (Gell-Mann, 1964a).

The relationship between current algebra and QCD was one between a phenomenology and its underlying theory. Once the underlying theory was successfully set in motion, its implications would be, understandably, much richer, wider, and deeper than the phenomenology, which, although historically was suggestive of the underlying theory, in fact was only part of the latter's consequences. This was exactly the case for current algebra and QCD. The very issue of confinement, and many other deep issues eminent in QCD with profound implications, as we will discuss in the next chapter, was completely absent in the framework of current algebra. Understandably, physicists' attention and focus of explorations were inevitably moving from current algebra as a phenomenology to QCD as the underlying theory, leaving the former in the periphery or to be forgotten completely.

Here it seems pertinent to use Gell-Mann's own culinary metaphor, but, obviously, in the reversed sense. In discussing the abstraction of current algebra from a Lagrangian field theory model, Gell-Mann said: "We may compare this process to a method sometimes employed in French cuisine: a piece of pheasant meat is cooked between two slices of veal, which are then discarded" (Gell-Mann, 1964b). Gell-Mann took current algebra as the real meat, and the Lagrangian field theory model as slices of veal, a kind of scaffolding, to be discarded after its service. In fact, the real meat was the underlying Lagrangian field theory or QCD as it turned out to be, and current algebra was only slices of veal in service of preparing the meat. The reversed use of the metaphor, of course, reveals a profound difference in conceiving physical reality, as we will further discuss in Chapter 9.

As far as Gell-Mann is concerned, his conception of physical reality described by physical theory was not fixed. Rather, it evolved with the evolution of physical theorizing itself, as is usually the case for most scientists. When current algebra was "abstracted from a useful but obviously wrong field theory of quarks with a single neutral gluon" (Gell-Mann, 1997), he took an algebraic structure of the world, assumed by current algebra and confirmed by many observations, as more real than the hypothetic world of quarks and gluons. When QCD was at hand, he "clarified" his view and claimed that "the veal was the incorrect single gluon theory, not field theory in general" (Gell-Mann, 1997) certainly not the field theory of QCD. That is, he took the world of quarks and color octet gluons as more real than the algebraic structure abstracted from it.

As far as the fate of current algebra is concerned, it just fits into the general pattern of the evolving status of a phenomenological hypothesis with the development of its underlying theory.

Notes

1 Gell-Mann's letter to Crewther, March 1972, quoted from Crewther's email (February 22, 2005). I am grateful to Professor Crewther for sending me his six page March 11, 1972 letter to Fritzsch and Gell-Mann, as a response to a please explain cable, explaining how to calculate the neutral pion amplitude from the short-distance behavior of <VVA> without assuming perturbation theory.

2 For more details, see Section 10.1 of Cao, 1997.

3 For more details, see Section 10.3 of Cao, 1997.

4 I am grateful to Professor Fritzsch for sending me a copy of his BA thesis, which was left in Leipzig when he escaped from the former GDR and miraculously recovered in 2006.

5 The contemporary view assumes that the proper level for the gauge principle to be applicable is the level of fundamental interactions. However, whether gravity, which is certainly as fundamental as other fundamental forces are, is subject to the gauge principle is a topic for debate. See Utiyama, 1956; Kibble, 1961; Cho, 1975, 1976; for more discussions about the external gauge symmetry, see also Hehl *et al.*, 1976; Carmeli, 1976, 1977; and Nissani, 1984.

6 Gell-Mann abandoned gauging the flavor SU(3) symmetry for the strong interactions in the early 1960s mainly because of the concern with the "clash" it would bring out between the strong interactions and the electromagnetic and weak interactions realized through hadronic currents (Gell-Mann, 1961, 1962). The same concern with the "clash," or communications or mixing, also made him uncomfortable to advocate his underlying field theory of quarks interacting through color singlet gluons. (See Sections 6.3 and 7.2).

7 As late as 1979, Bjorken was still speculating that if the octet of QCD gluons were given a small mass via the Higgs mechanism, which was not completely unreasonable, then the quark may not be confined (Bjorken, 1979).

8

Early justifications and explorations

In spite of the simple appearance of its Lagrangian, QCD is an extremely complicated and rich theory. As a theory of strong interaction, it is subject to various constraints posed by observations. Most important among them are scaling and confinement, which appear to have conflicting implications for the nature and characteristics of QCD. It also has many observational implications, such as the logarithmic violations of scaling, the spectrum of charmonium, and the two-jet and three-jet structure in the electron–positron annihilations. The successes in satisfying these constraints and in explaining and predicting observational events in the early to mid 1970s immediately after the proposal of QCD had provided QCD with much needed experimental justifications and consolidated its status as an acceptable theory in the particle physics community.

As a non-abelian gauge theory, QCD is also subject to some conceptual constraints posed by its symmetry structure, such as the U(1) anomaly (see Sections 8.2 and 8.5 below). The solution of the U(1) anomaly relied on the further explorations of the theoretical structure of QCD, which revealed its richness and great potential for theorizing the complexity of the sub-hadronic world, such as the existence of instantons and the related theta vacuum state. The fruitful explorations in this direction had shown that QCD, in Lakatos's terminology, was a progressive research program, which keeps opening new territory for explorations, and thus was conceived by the majority of the community as worth pursuing.

This chapter will briefly review some of the most important justifications and explorations of QCD that appeared immediately after its proposal. The underlying conceptual framework for most of the justificatory efforts was provided by, in addition to the color SU(3) gauge theory itself, the renormalization group method of Wilson, Callan, and Symanzik, although the topological approach to the global structure of non-abelian gauge theory also played a crucial role in the explorative efforts.

8.1 Asymptotic freedom

While most physicists thought that the Bjorken scaling demanded a radical departure from any quantum field theory in the treatment of strong interactions, Kurt Symanzik was perhaps the first physicist who realized that this might not be a necessity. Symanzik's hope came from his understanding of the nature of renormalization group analysis, of which he himself was one of the inventors of its new version based on Wilson's idea of anomalous scale dimensions of fields (see Section 6.3). Crucial to his hope was the β-function ($\beta(\lambda) = Md\lambda/dM$, or $f(\lambda)$ in Callan's scaling operator, (6.15)), in the renormalization group equation (in Coleman's notation, 1971):

$$\left(M\frac{\partial}{\partial M} + \beta\frac{\partial}{\partial \lambda} + n\lambda\right)\Gamma^{(n)}(x_1 \cdots x_n) = 0 \tag{8.1}$$

which describes how the effective coupling constant evolves with the change of the scale. If a quantum field theory yields a negative β-function when it is subject to the renormalization group analysis for its short-distance behavior, then the theory's asymptotic behavior will be one of a free field theory, and that would be all that is needed for a proper description of the Bjorken scaling. Is it possible to find a quantum field theory with such a property? That was the task Symanzik set for himself.

Symanzik and 't Hooft

Attracted by the idea that theories with a negative β-function would be asymptotically free and thus reproduce parton-like behavior and the Bjorken scaling, Symanzik examined various quantum field theories and found that the only theory with such a property would be the $\lambda\phi^4$ theory with $\lambda < 0$. He reported his work at the Marseille Conference on Particle Physics in June 1972, when the notion of color as a global symmetry was contemplated by Bardeen, Fritzsch, and Gell-Mann, but it was a few months prior to the notion of color being transmuted into a gauge symmetry and QCD being proposed by Fritzsch and Gell-Mann.

It was interesting. But it had problems. The most severe one, as 't Hooft who met Symanzik at the airport before the conference pointed out to him, was that in such a theory the energy would be unbounded from below.

Symanzik's response was that the problem perhaps could be cured by non-perturbative effects, which nobody really knew how to do at that time.

But 't Hooft informed Symanzik of his own finding obtained in 1971 that the β-function for gauge theories was negative. Symanzik was surprised but skeptical: "If this is true, it will be very important." Certainly he understood what was at stake, and thus at the end of his talk urged further study on the β-function for non-abelian gauge theories.

't Hooft did not understand what was at stake. But he knew the answer to the question, and thus he stood up and wrote down on the blackboard the expression he "had derived:

$$\frac{\mu dg^2}{d\mu} = \frac{g^4}{8\pi^2}\left(\frac{-11}{3}C_1 + \frac{2}{3}C_2 + \frac{1}{6}C_3\right) \tag{8.2}$$

for a theory in which the gauge group has a Casimir operator C_1 and where fermions have a Casimir operator C_2. Scalars contribute with the Casimir operator C_3. In the SU(2) case this would mean that with less than 11 fermions in the elementary representation (and no scalars), the β-function would be negative, and for SU(3) the critical number would be 33/2" ('t Hooft, 1998).

Symanzik published his work (1972, 1973), but 't Hooft did not. In addition, aside from Symanzik, 't Hooft' s calculation of the β-function for non-abelian gauge theories attracted nobody's attention at the conference and left no trace in the history until his result was rediscovered by David Politzer, and also by David Gross and Frank Wilczek almost simultaneously (see below), in the new context in which a conceptual framework of strong interactions or the quark–gluon interactions, namely QCD, was proposed and widely circulated among the alerted particle physicists. Only then, retrospectively, 't Hooft's calculation was recognized by many physicists, or at least, or more precisely, by 't Hooft himself, as potentially important to the justification of QCD.

The reason for the unfortunate fate of this piece of 't Hooft's work is threefold. First, he himself did not understand the implications of his work to a field theory of strong interactions, and did not even bother to publish it. Second, when his work was reported to the community, its attention was on the electroweak theory, to which 't Hooft made crucial contributions, but not on strong interaction, where no field-theoretical model was in good shape. Third, and most importantly, detached from a relevant theoretical context, 't Hooft's calculation was only a mathematical exercise, though quite sophisticated, but not part of an active research program. Similar to the case of Nambu's notion of an SU(3) gauge symmetry for the three-triplet quark model, the significance of 't Hooft's calculation can only be properly recognized retrospectively when its result was seen from the new perspective of an active research program of QCD, but not before.

Dimensional transmutation

Detached from any theoretical context, 't Hooft could freely assume that a non-abelian gauge theory, with or without fermions and scalars, could be subject to a renormalization group analysis. But in the context of QCD, in particular in dealing with its short-distance behavior where there is virtually no mass scale and thus should be scale invariant, it was not obviously clear that the renormalization group analysis, which was based on the notion of scale anomaly of fields, could be applied to it. In this regard, the groundwork for the applicability of renormalization group analysis to massless (and thus scale invariant) theories was laid by Sidney Coleman and Eric Weinberg (1973).

The major topic of their work was an investigation of a new mechanism of spontaneous symmetry breaking (SSB), which was a crucial component of the increasingly popular electroweak theory and perhaps would be relevant to the newly proposed QCD. They explored the possibility that the driving mechanism of SSB was not the negative mass term in the Lagrangian for the electromagnetic interactions of a massless charged scalar field, but rather radiative corrections (certain effects of higher order processes involving virtual photons or any other bosons that mediate interactions). Such a mechanism would have all the benefits of the Higgs mechanism for SSB, giving masses to fermions and bosons and making interactions short range, "even in theories *without* fundamental spinless bosons, such as massless spinor electrodynamics, or the theory of non-Abelian gauge fields coupled to massless fermions" (Coleman and Weinberg, 1973). The last category of theory obviously included QCD. As we will see shortly, it had provided a theoretical context for Gross and Wilczek to make sense of their work on asymptotic freedom.

One of the novel results of their work was the discovery of dimensional transmutation: one dimensionless coupling is traded with an arbitrarily chosen renormalization mass in a massless field theory.

More specifically, in order to explore the high order effects to symmetry breaking, a fixed renormalization mass, for virtual photon or other bosons mediating interactions, has to be introduced in a massless theory as a subtraction point for performing the renormalization of the coupling constant. One cannot take this mass to be zero since then the amplitude would be infrared divergent. Clearly the renormalized coupling constant depends on the mass introduced. That is, a new parameter, the renormalization mass as the hidden scale of the massless theory, has entered the theory through the machinery of renormalization.

With the help of the notion of effective potential, Coleman and Weinberg were able to show that the actual mass and coupling constant were all computable from the effective optional V at a renormalization mass $\langle\varphi\rangle$:

$$\mu^2 = \frac{d^2V}{d\varphi_c^2}|_{\langle\varphi\rangle},\ \lambda = \frac{d^2V}{d\varphi_c^4}|_{\langle\varphi\rangle} \tag{8.3}$$

Since the renormalization mass is arbitrary, it was chosen to be the actual location of the vacuum expectation value $\langle\varphi\rangle$ so that the symmetry would be spontaneously broken. The renormalization mass is indeed an arbitrary parameter, with no effect on the physics of the problem. If one picks a different mass, then one simply defines a new coupling constant. It is a change of definition, not a change of physics.

But, a change in the subtraction point is equivalent to a change in the scale of all momenta since it is the only parameter that fixes the momentum scale. In a deep sense, therefore, the renormalization mass here has vindicated the notion of "arbitrary fundamental parameters" Wilson prophetically introduced in 1969:

The strong interactions contain some arbitrary fundamental parameters, not physical masses and coupling constants, they show up explicitly only in the short distance behavior of strong interactions. Implicitly they determine all of strong interactions, masses and constants. They have particular values, but the theory is self consistent for any values of the parameters. (*Wilson, 1969*)

A surprising consequence of the above reasoning, according to Coleman and Weinberg, is that

we have traded a dimensionless parameter, (the coupling constant) λ, on which physical quantities can depend in a complicated way, for a dimensional one, $\langle\varphi\rangle$, on which physical quantities must depend in a trivial way, governed by dimensional analysis. We call this phenomenon dimensional transmutation, and argue that it is a general feature of spontaneous symmetry breaking in fully massless field theories. (Coleman and Weinberg, 1973)

In fact, dimensional transmutation helps to reveal a simple fact that, for a fixed theory, a change in the arbitrary renormalization mass, whose only function is to define the renormalized coupling constant, leads to a change in the numerical value of the dimensionless coupling constants. Thus a change in mass can always be compensated for by a change in coupling constant and a rescaling of the field. The above rationale can be expressed mathematically as the renormalization group equation:

$$\left(M\frac{\partial}{\partial M}+\beta\frac{\partial}{\partial\lambda}+n\lambda\right)\Gamma^{(n)}(x_1\cdots x_n)=0. \quad (8.4)$$

Or one may say that dimensional transmutation justifies the applicability of the renormalization group equation to apparently scale invariant field theories.

In the case of QCD, dimensional transmutation implies that it has a hidden scale even in its short-distance behavior where it is apparently scale invariant, and thus its scale invariance is dynamically broken. This, in turn, justifies the applicability of renormalization group analysis to QCD. The renormalization group approach is the key to asymptotic behavior of QCD, as was soon to be explored by Coleman's graduate student Politzer and by Gross and his graduate student Wilczek.

Politzer and Gross–Wilczek

Politzer heard Gell-Mann and Fritzsch's proposal of QCD, although only indirectly from Sheldon Glashow, but did not pay serious attention to it. He had no intention of proving that QCD is an asymptotically free theory. What he decided to do was "to look into whether the renormalization group had anything to say about the low energy (or ground state) behavior of Yang-Mills theory." In this pursuit, "a key step was to know the Yang-Mills beta function" (Politzer, 2004). He calculated the β-function for non-abelian gauge theories, and obtained the same result 't Hooft obtained earlier.

A negative β-function was useless for understanding low energy phenomena. In order to utilize his result, therefore, Politzer adopted his advisor's result, namely, "the gauge symmetry breaks down spontaneously as a result of the dynamics. Consequently, the fields obtain (in general massive) particle interpretation." However, he acknowledged, "as yet, nothing is known about the particle spectrum, the low-energy dynamics, or particles describable only by composite fields" (Politzer, 1973). Judging from these concerns of his, it seems that a hidden agenda in his query was about the quark–gluon model of hadron physics or QCD, although nothing of this kind was explicitly mentioned in the paper.

"But it did not take long to go from dismay" over the uselessness of the negative β-function to excitement over the possibility of having reliable perturbative results for strong interactions.[1] The possibility was suggested by the asymptotic freedom, a property of some non-abelian gauge theories which was revealed by Politzer's calculation that their β-function had a minus sign.

Gross also heard of Gell-Mann and Fritzsch's proposal of QCD. In fact, he was the chair of the session at which Gell-Mann first articulated the joint proposal of QCD by Fritzsch and himself (see Section 7.3). As a seasoned theoretical particle physicist, who had been actively engaged in Gell-Mann's research program of current algebra and had made important contributions to it, Gross fully understood what was proposed by Fritzsch and Gell-Mann. However, deeply impressed by the experimentally confirmed Bjorken scaling, Gross was convinced that no quantum field theory, the newly proposed QCD included, would be able to accommodate the Bjorken scaling, because no renormalizable quantum field theory could yield negative β-function and thus be asymptotically free, which was considered to be necessary for accommodating the Bjorken scaling.

Although there existed a general model-independent argument for the positivity of β-function, mainly based on the positivity of the spectral function, Gross was determined to prove, through detailed calculations of β-function of all kinds of renormalizable quantum field-theoretical models, that it must be the case. This way, he could decisively prove that the Bjorken scaling really demanded a radical departure from quantum field theory (Gross, 1997).

For this purpose, Gross mobilized his graduate student Wilczek to do the tedious calculations. Since all renormalizable quantum field theories, with the exception of non-abelian gauge theories, were known to be not asymptotically free (Gell-Mann and Low, 1954; Parisi, 1973; Callan and Gross, 1973; Zee, 1973), Gross and Wilczek's work understandably focused on non-abelian gauge theories. Their calculation resulted in what they expected, a positive β-function, when Politzer obtained a negative one and compared notes with them through Coleman. But they soon found a mistake, corrected it and submitted a paper to *Physical Review Letters* for publication on their great discovery of asymptotical freedom in some non-abelian gauge theories, the crucial ingredient of which was the negative β-function they obtained in some non-abelian gauge theories.

The reason why the same result, that the β-function in some non-abelian gauge theories was negative, which in the case of 't Hooft was only a mathematical exercise, but in the case of Politzer, and particularly in the case of Gross and Wilczek, was a great scientific discovery is roughly threefold.

The first is related with interpretation and understanding. 'T Hooft, while vaguely feeling his result might be relevant to scaling, was far from understanding its significance for the theory of strong interactions. He did not even consider his result "relevant for partons. Since I could not understand their problems, I turned away from the strong interactions that had to be infinitely complicated" ('t Hooft, 2001).

For Politzer, his result seemed to suggest a much more general claim that the perturbation approach for non-abelian gauge theories would be reliable at the high energy domain. For Gross and Wilczek, it almost entailed a justification of a non-abelian guage theoretical framework for strong interactions. For them, the asymptotic freedom they discovered explained "Bjorken scaling and the success of naïve light-cone or parton-model relations" (Gross and Wilczek, 1973a), and allowed one to "derive all the sum rules and relations previously derived in the parton or light-cone models. In addition there will be logarithmic deviations from Bjorken scaling which can be calculated" (Gross and Wilczek, 1973b).

The last claim of Gross and Wilczek was especially important. It indicated that with asymptotic freedom, QCD would no longer be a purely conceptual framework, but would rather became a practical physical theory, which could make calculations and predictions that would be checkable with experiments. Gross and Wilczek substantiated their claim by having actually calculated the logarithmic deviations from Bjorken scaling: they applied renormalization group techniques to Wilson's operator product expansion,

$$J(x)J(-x) \underset{x^2\approx 0}{\sim} \sum_n C^{(n)}(x^2, g)x^{\mu_1}\cdots x^{\mu_n}O^{(n)}_{\mu_1\ldots\mu_n} \tag{8.5}$$

and calculated, to the lowest order of perturbation, the anomalous dimension of $O^{(n)}$, from which the logarithmic deviation from Bjorken scaling was obtained (Gross and Wilczek, 1973b). Furthermore, considering the difficulties in determining the q^2 behavior of the moments of the structure functions, they even made a practical suggestion: "More than likely the most practical place to look for violations of Bjorken scaling is in the vicinity of threshold ($x \approx 1$). There we expect to see the structure functions decreasing like ever-increasing power of $\ln(-q^2)$" (Gross and Wilczek, 1973c). This was perhaps the first checkable prediction made from perturbative QCD, and indicated that QCD was indeed a workable physical theory.

The above discussion already points to the second reason for the difference: the role played in advancing an active research program in particle physics. While 't Hooft's calculation as an isolated event played no role at all and left no trace in history, the work of Politzer, Gross, and Wilczek played a great role in the history of particle physics. It provided QCD with a convincing justification in terms of accommodating a large number of the successes in particle physics (all those related with the current algebra project surveyed in earlier chapters). Furthermore, it also provided QCD with a workable perturbation approach to high energy phenomena. For these reasons, their work greatly advanced QCD as a theory for strong interactions.

Finally, in terms of acceptance. 'T Hooft's result was accepted by nobody aside from Symanzik, who however was not without concern. But the work of Politzer, and of Gross and Wilczek was widely accepted by the community. The acceptance was immediate and even sensational. It was reported that Gell-Mann, "upon hearing of the news of asymptotic freedom, declared that he now considered the color octet vector gluon picture 'established'" (Hill, 2005). Crucially for such a sensational acceptance of their result was that it perfectly fitted into the scheme of QCD and provided it with the most desirable justification. Without such a color octet vector gluon picture, or QCD, as a conceptual framework for strong interactions, the fact that some non-abelian gauge theories might yield negative β-function would mean almost nothing for the particle physics community.

Questions answered and unanswered

Before the discovery of asymptotic freedom, it was not clear if any renormalizable field theory could satisfy one of the major experimental constraints on the theory of strong interactions, namely the quasi-free behavior of sub-hadronic ingredients revealed by Bjorken scaling. The discovery clearly answered the question positively if the theory was a non-abelian gauge theory with certain limited number of flavors, and thus lent strong support to QCD that as a non-abelian gauge theory it is compatible with scaling. The further claim based on asymptotic freedom by Coleman and Gross (1973) that "any acceptable field theory of the strong interactions must involve non-abelian gauge fields" increased the chance of QCD being a theory of strong interactions.

But there were important questions that remained unanswered by the discovery of asymptotic freedom.

First, it was not clear to Gross and Wilczek if the SU(2) $\times$ U(1) electroweak theory, which was also a non-abelian gauge theory, was similarly asymptotically free: "we note that theories of the weak and electromagnetic interactions, built on semi simple Lie groups, will be asymptotically free if we again ignore the complications due to the Higgs particles" (1973a). That is, the resolution of the ambiguity related to the electroweak theory critically relied on the understanding of the contributions of the Higgs particles to β-function.

Closely related to "the complications due to the Higgs particles" was another profound question remaining to be answered. The attitude of Gross and his collaborators toward the nature of color SU(3) symmetry was ambiguous. They gave no clear answer to the important question: "Was the symmetry exact or must it be broken spontaneously?"

The ambivalent attitude to the question was clearly displayed in their publications. On the one hand, Gross and Wilczek claimed that "in order to construct realistic non-abelian gauge models of the strong interactions one must confront the issue of symmetry breaking" (1973b). Coleman and Gross further asserted: "The strongest interactions (like the weak, though for completely different reasons) must be described by a spontaneously broken gauge field theory" (1973).

Their desire for symmetry breaking was based on legitimate worries about the infrared behavior of the asymptotically free gauge theory. If the gauge symmetry is exact, "it would imply the existence of massless strongly coupled vector mesons," which in turn would imply long-range force, which would cause problems with the spectrum of states (the particle content) and spoil S-matrix in perturbation theory. Thus in the construction of realistic physical models of the strong interactions, "the major problem remaining in these gauge theories is how to break the gauge symmetry and provide masses for the vector mesons" (1973b).

Understandably, Gross and Wilczek devoted much of their attention to the issue. At first, they tried to use "the standard means of breaking the gauge symmetry," namely by introducing "scalar mesons explicitly into the Lagrangian." But after detailed analysis of the contributions of scalars to the beta function, they found that "it is very difficult to preserve asymptotic freedom while incorporating scalar mesons, and perhaps impossible to include enough scalars to completely break the gauge symmetry." This led them to the conclusion that "the main problem which requires investigation is whether one can obtain an infrared sensible theory without explicit Higgs mesons" and suggested that "one must pin one's hope on the possibility of dynamical symmetry breaking." The trouble with dynamical symmetry breaking was that even "in that case one would still be faced with the fact that there is no experimental evidence for the existence of such neutral vector mesons, colored hadrons and especially quarks" (1973b).

On the other hand, perhaps because of the troubles they had with acceptable ways of treating symmetry breaking in the case of asymptotically free theories, they also considered another possibility that the gauge symmetry is exact. They cited the work by Bardeen, Fritzsch, and Gell-Mann (1972), and in particular the grounding work in QCD by Fritzsch and Gell-Mann (1972), and commented, "the proponents of 'red, white, and blue' quarks as a mathematical abstraction argue that the color SU(3) group should be exact, and that all noncolor singlets should be suppressed completely. One clearly requires a dynamical explanation of such a miracle." They immediately declared that "It might very well be the violent infrared singularities of an asymptotically free gauge theory provide the requisite dynamical mechanism" (1973b).

The claimed explanation can be seen clearly by looking at the fine-structure constant,

$$\alpha_c(q^2) = \frac{12\pi}{(33 - 2n_f)\ln q^2/\Lambda^2}, \text{ for } \frac{q^2}{\Lambda^2} \gg 1,$$

which can be derived easily in the asymptotically free gauge theory they just obtained. It decreases with q^2 as long as the number of flavors (n_f) is less than seventeen. Clearly, when q^2 decreases, $\alpha_c(q^2)$ increases, and so does $\alpha_c(r)$. Suppose this continues to hold all the way down to low q^2 or, equivalently, large r, then the attractive quark potential would keep growing with r. If this is the case, they reasoned: "One might expect, on physical grounds, that the infrared singularities induced by the gauge charges (color) are so strong that they must be completely shielded, so that only objects neutral under the gauge group could exist. This is an exciting possibility which might provide a mechanism for having a theory of quarks without real quark states," which means "that the colored gauge mesons as well as the quarks could be 'seen' in the large-Euclidean momentum region but never be produced as real asymptotic states" (1973b).

This was a direct response to the question posed by Fritzsch and Gell-Mann in their founding paper of QCD: "One fascinating problem, of course, is to understand the conditions under which we can have an algebra resembling that for quarks and gluons and yet escape having real quarks and gluons" (1972; see also Sections 7.4 and 8.3).

However, Gross and Wilczek were not committed to this exciting possibility. They continued, "whether this can be realized or whether the theory will exhibit dynamic symmetry breaking deserves much attention." The deep reason for their hesitation was that "the infrared instability of the origin indicates the unreliability of the classical (free) approximation to this potential. Thus whether or not the theory exhibits symmetry breaking is a difficult dynamical question, requiring nonperturbative calculations" (1973b).

Thus the nature of the color symmetry, broken or not, did not get clarified by the discovery of asymptotic freedom. It required further arguments to settle the issue.

8.2 "Advantages of the color octet gluon picture"

With the discovery of asymptotic freedom, Gell-Mann, and perhaps also a sizable portion of the particle physics community, felt that the color octet gluon picture could be considered established. But the ambivalent attitude toward the nature of color gauge symmetry displayed in the initial papers of the asymptotic freedom, as we have just discussed, which was perhaps also shared by many other particle physicists, required a clarification. The paper titled "Advantages of the color octet gluon picture" by Fritzsch, Gell-Mann and Leutwyler (1973) served this need perfectly in this context. In fact, it went farther and provided an overall justification of the picture as a conceptual scheme, and even opened a new territory for further theoretical exploration of the picture.

All the advantages that justified the scheme or picture were derived from one claim: "Color is a perfect symmetry." This implied that only color-gauge invariant states, or colorless states or color singlets, are observable. Thus the authors declared, "we do not accept theories in which [colored single] quarks are real, observable particles; nor do we allow any scheme in which the color non-singlet degrees of freedom can be excited" (1973). All color-nonsinglets were taken to be fictitious (not real) because they were not color-gauge invariant, and thus were not observable.

In order to make the advantage of the color octet gluon picture over the color singlet gluon picture clearer, the authors put down two Lagrangian densities explicitly:

$$L = -\bar{q}[\gamma_\alpha(\partial_\alpha - igB_\alpha\lambda_o) + M]q + L_B \tag{8.6}$$

where M is the diagonal mechanical mass matrix of the quarks and L_B is the Lagrangian density of the free neutral color singlet vector field B_α, and

$$L = -\bar{q}[\gamma_\alpha(\partial_\alpha - igB_{A\alpha}\chi_A) + M]q + L_B(Yang{-}Mills), \tag{8.7}$$

where χ_A is the color SU(3) analog of λ_i, and $B_{A\alpha}$ is a color octet ($A = 1\ldots8$).

The most obvious advantage of (8.7) over (8.6), according to the authors, is that "the gluons are now just as fictitious as the quarks," and "does not communicate with any physical channel, since the physical states are all color singlets." To the major worry of Gross and Wilczek, namely the long-range force, produced by massless gluons in the exactly symmetrical gauge theory, and its observational implications, the paper responded that since the color octet "gluon is unphysical, we have no objection to the occurrence of long range forces in its fictitious channel, produced by massless gluons." Furthermore, they claimed that "these fictitious long-range forces and the associated infrared divergencies could provide a mechanism for confining all color nonsinglets permanently. They would not be present in physical hadronic interactions, where long-range forces are known to be absent."

It is worth noting that here the ground for the infrared divergencies, which was utilized as a mechanism for the confinement of color nonsinglets, was different from the renormalization group analysis of an asymptotically free gauge theory, namely a running coupling constant, whose evolution was dictated by the negative beta function, growing indefinitely when the distance between interacting particles was getting larger and larger, as Gross and Wilczek argued in their work. Here it was taken to be simply the result of massless mediating bosons and the related long-range forces, which, however, occur only in a non-physical channel because the color octet gluons themselves are non-physical.

The second advantage is that it gives "a hint as to why Nature selects color singlets"[2] and can avoid the clash between the color gauge and "the electromagnetic gauge or the Yang-Mills gauge of unified weak and electromagnetic interactions." While the non-decisive hint itself did not have much persuasive power, the "no-clash" argument, that color octet gluons would not clash with "the weak and electromagnetic currents [which] form color singlets," was deployed to reject the "Han-Nambu picture in which color nonsinglets can be physically excited by electromagnetism and in which there are three triplets of real quarks with integral charges that average to 2/3, −1/3, and −1/3."

Another important advantage is that it is asymptotically free, and thus is able to explain both the Bjorken scaling and its logarithmic violation.

The last advantage the authors "wish to stress" was a subtle one, the exploration of which in fact touched on a deep layer of the theoretical structure of QCD, as we will see briefly below, and more in Section 8.5.

In the scheme of current algebra, there is an anomalous divergence of the axial vector baryon current $F^5_{i\alpha}$ (cf. Cao, 1997, Section 8.7). While for the other eight axial vector currents $F^5_{i\alpha}(i = 1\ldots8)$ we simply have $\partial_\alpha F^5_{i\alpha}(x) = D\left(x, x, i\gamma_5\{\frac{1}{2}\lambda_i, M\}\right)$, the divergence of $F^5_{0\alpha}(x)$ is (Fritzsch and Gell-Mann, 1971a):

$$\partial_\alpha F^5_{0\alpha}(x) = D\left(x, x, i\sqrt{\frac{2}{3}}M\gamma_5\right) + \sqrt{6}g^2\left(8\pi^2\right)^{-1}G_{\mu\nu}G^*_{\mu\nu} \tag{8.8}$$

where D(x,y,G) is the physical operator that corresponds in a free quark model to $\bar{q}(x)Gq(y)$, $G_{\mu\nu} = \partial_\mu B_\nu - \partial_\nu B_\mu$ for the color singlet case, while $G_{A\mu\nu} = \partial_\mu B_{A\nu} - \partial_\nu B_{A\mu} + gf_{ABC}B_{B\mu}B_{C\nu}$ for the octet case.

According to a previous work of Fritzsch and Gell-Mann (1971a), in the color singlet case, the anomalous divergence term in (8.8) is necessarily associated with an anomalous singularity in the bilocal current $F^5_{0\alpha}(x, y)$ as z^2=(x−y) tends to zero:

$$F^5_{0\alpha}(x, y) \triangleq 3i\left(2\pi^2\right)^{-1}gG^*_{\alpha\beta}z_\beta\left(z^2\right)^{-1}. \tag{8.9}$$

They noticed that such a term "would destroy the light cone algebra as a system since one of the bilocal currents arising from commutation of two physical currents would be infinite on the light cone" (1971a). They could not tolerate such a term because they assumed the correctness of the full light-cone algebra. Thus "how to get rid of the anomalous singularity in $F^5_{0\alpha}(x, y)$, while retaining the anomalous divergence term for $\partial_\alpha F^5_{0\alpha}(x)$ given by triangle diagram" was posed as a puzzle.

The color octet gluon picture solves the puzzle. The anomalous divergence term in $\partial_\alpha F^5_{0\alpha}(x)$ is unchanged, except for replacing $G_{\mu\nu}G^*_{\mu\nu}$ by $G_{A\mu\nu}G^*_{A\mu\nu}$, but it is now associated with a singularity as $z^2 \to 0$ not in $F^5_{0\alpha}(x, y)$, but in a different formal quantity, the corresponding color octet operator:

$$F^5_{0A\alpha}(x, y) \triangleq 3i\left(2\pi^2\right)^{-1} gG^*_{\alpha A\beta} z_\beta \left(z^2\right)^{-1}. \tag{8.10}$$

"Since $F^5_{0A\alpha}(x, y)$ is not a physical operator, being a color octet, we can have no objection to its being singular on the light cone."

So the last advantage the authors advocated was that the color octet gluon picture helped to "get rid of the unacceptable anomalous singularity" (8.9) in $F^5_{0\alpha}(x, y)$. As a result, now one "can believe and make use of the anomalous divergence terms" in (8.8) to resolve "a great mystery" in Gell-Mann's major research project, which included PCAC, current algebra and quark model and now appeared as the color octet gluon picture or QCD (as a synthesis of all these respectable ideas).

The mystery is the following situation related to PCAC. In applying the PCAC reasoning (see Cao, 1997, Section 10.1) to the quark–gluon picture (8.6) or (8.7), "turning on the quark bare masses with $m_u \approx m_d \ll m_s$, we would have four nearly massless pseudoscalar mesons instead of three, in bad disagreement with observation." Here what the authors referred to was the fact that the fourth meson (η-meson) observed (549 MeV) was considerably heavier than three pions (135 MeV).

Similarly, "we do not have in the limit $M \to 0$ the conservation of nine axial vector currents and the existence of nine massless pseudoscalar mesons." Here what they referred to was the observation of the X^0 meson, which was renamed as η' meson (958 MeV), that was much heavier than the other eight mesons (Bollini *et al.*, 1968).

They further reasoned, "to put in another way, as m_u and m_d tend to zero, we would have U(2) × U(2) conservation and four massless pseudoscalar mesons."

Theoretically, then, the mystery can be stated as follows. According to the PCAC reasoning, the existence of three or eight light pseudoscalar mesons was the result of the spontaneous breaking of the chiral SU(2) × SU(2) or SU(3) × SU(3) symmetry followed by the explicit breaking due to the presence of the quark bare masses. When the quark bare masses tend to zero, however, the picture [(8.6) or (8.7)] should have the chiral U(2) × U(2) or chiral U(3) × U(3) symmetry instead of the chiral SU(2) × SU(2) or chiral SU(3) × SU(3) symmetry together with a single U(1) symmetry for baryon number conservation. That is, there should be an additional axial U(1) current that is approximately conserved, with an additional light pseudoscalar meson as a consequence. The heaviness of the observed additional pseudoscalar meson indicated that the additional axial U(1) current is not approximately conserved. Thus the mystery the authors tried to address is also called the U(1) anomaly. What was the cause of the mystery of the U(1) anomaly?

The authors claimed that "the mystery might appear to be resolved" in the new picture: "since the anomalous term in (4) [namely (8.8) above] breaks the conservation of $F^5_{0\alpha}$ even in the limit $M \to 0$ and so in that limit it looks as if there need not be a ninth massless pseudoscalar meson, and in the limit $m_u \to 0$, $m_d \to 0$ it looks as if there need not be a fourth one."

But the trouble was, as the authors immediately pointed out, the anomalous term "is itself a divergence of another (non-gauge invariant) pseudovector, and thus as $M \to 0$ we still have the conservation of a modified axial vector baryon charge; we must still explain why this new ninth charge seems to correspond neither to a parity degeneracy of levels nor to a massless Nambu-Goldstone boson as $M \to 0$."

So what the authors had shown was that "it is important to find the explanation," rather than an achievement of an actual explanation of the mystery on the basis of the color octet gluon picture. In addition, the exact reason why color nonsinglets should be unphysical and thus confined permanently was also unclear. Do such states carry infinite enerey, or is it enough to say that they are not gauge invariant? The answer, and also the explation of the U(1) anomaly, remained to be approached by further explorations of the picture.

Still, by stressing that the color symmetry is an exact symmetry, the paper had presented a consistent conceptual framework, within which various questions, puzzles and mysteries, such as scaling, confinement and the U(1) anomaly, can be successfully analyzed, approached, and explored, as we have seen in the last section, and will see again and again in the next section and Section 8.5. Without such a consistent conceptual framework, for example, if the framework was conceived as based on a spontaneously broken symmetry, as many initially inclined to conceive, then all the researches and explorations would be misguided into a wrong direction, with results that would certainly be radically different from what particle physicists had actually achieved in the last 35 years.

8.3 Early signs of confinement

The fact that quarks had failed to appear as separate particles in a final state raised a serious challenge to the quark model and demanded an explanation. With the proposal of QCD, this outstanding problem of the quark model was generalized to the problem of color confinement: is it possible to give a rigorous theoretical proof that QCD, as a self-consistent model, would not allow any colored degrees of freedom to appear as separate particles in a final state?

Gell-Mann was optimistic in addressing this problem. His solution to the problem was based on the idea that color was an exact symmetry. Thus colored states would not be observable because they are not gauge invariant. Of course, he realized that the singlet restriction for the observable states, which had its origin in the bootstrap philosophy that Gell-Mann respected greatly as far as hadron physics was concerned, amounted to nothing but the requirement of gauge invariance, which in fact had begged the question (Fritzsch and Gell-Mann, 1972). At a deeper level, Schwinger warned a long time before that the general requirement of gauge invariance could not be used to dispose of essentially dynamical questions (Schwinger, 1962a).

With the discovery of asymptotic freedom, Gell-Mann found that QCD now could solve the confinement problem on a dynamical ground rather than on the formal requirement of gauge invariance. A gauge invariant QCD had the welcome "infrared divergencies [which] could provide a mechanism for confining all color nonsinglets permanently. They would not be present in physical hadronic interactions, where long-range forces are known to be absent" (see Section 8.2).

It was not clear, however, that Gross's major worry was properly addressed: even if colored components were bound into a color singlet by long-range forces permanently, why couldn't they be seen separately if they were bound at a long distance? More serious was the possibility that the welcome infrared divergencies might not occur. They might be cancelled by some mechanism similar to that in the case of QED. Turning to the renormalization group argument that led to asymptotic freedom would not be of much help. The renormalization group analysis indicated that for low momenta or large distances, Green's functions reflected a large effective coupling constant. But this only meant that perturbation theory would not be a reliable tool for investigating widely separated quarks, rather than a rigorous proof of the color confinement. Thus, although heuristically Gell-Mann's ideas were interesting and encouraging, they had to be fleshed out in more dynamical terms.

In order to circumvent the unreliable perturbation theory in dealing with strong coupling at large distances, Wilson, who intuited that the color confinement involved in large distance physics must be the consequence of complex dynamics and iteration from small to large scales, proposed a lattice version of QCD to address the confinement problem, with a new expansion scheme. While the idea of iteration derived from his own study of renormalization group flow, the dynamical side of his intuition was inspired by Schwinger's deep insights on gauge invariance, vacuum fluctuations, and the mass of gauge bosons. In fact, Schwinger's insights had provided the foundation for all works on the confinement problem. In order to properly understand conceptual developments in attacking the confinement problem, therefore, it is crucial to have a clear idea about Schwinger's insights.

Schwinger's insights

In his classic papers on "Gauge invariance and mass: I, II" (1962a, b), Schwinger claimed that "the gauge invariance of a vector field does not necessarily imply zero mass for an associated particle if the current vector coupling is sufficiently strong" (1962a). The claim offered a perspective radically different from the one with which the confinement problem was perceived and addressed by Gross and Gell-Mann.

The key notion to Schwinger's claim and argumentation, as it was to all of his works on renormalization in the late 1940s, (see Section 7.6, Cao, 1997) was that of vacuum fluctuations or vacuum polarizations. If the coupling constant is large enough, the vacuum will become polarized in such a way that the vacuum fluctuations are composed of two parts. One is associated with a pure radiation field, which is transverse and has no accompanying current, and thus is related with a massless photon. The other is directly related to corresponding current fluctuations, which will neutralize or totally screen external charges, and as a result "the massless Bose particle ceases to exist" (1962b). If charges are neutralized as the result of complete screening, the long-range effects will be eliminated. This already suggests the disappearance of massless bosons. But Schwinger had a more concrete mechanism for the existence of massive bosons. That is, in the case of electrodynamics, "the creation of a unit angular momentum positronium state." This bound state, held together by having strongly coupled to vector current, shifts the photon pole in its propagator and thus is massive "in the current vector's spectrum of vacuum fluctuations" (1962a).

But how strong a gauge coupling would be sufficiently strong for the phenomenon (gauge bosons have nonzero mass and long-range effects are cancelled out in a gauge invariant theory), the so-called Schwinger phenomenon, to occur? Schwinger suggested that it had "a critical dependence upon the coupling constant." Below "the critical coupling strength," the photon mass would be zero and above the critical coupling strength, the photon mass would be nonzero and vary with the coupling constant (1962b). The very existence of such a critical point suggested the existence of certain types of phase transitions, from the normal phase to the Schwinger phase.

The difference between the Schwinger phenomenon and the spontaneous symmetry breaking (SSB) was duly noticed in the discussion of confinement (Kogut and Susskind, 1974). First, the mechanisms cancelling the long-range effects in the propagator of the gauge boson are different. While for SSB the cancellation is caused by a massless scalar bound state in the proper vacuum polarization tensor, in the Schwinger phenomenon it is the vector positronium-like state that shifts the pole in the propagator. Second, the Schwinger mechanism eliminates the massless gauge boson and the related long-range Coulomb-like force, but leaves the vacuum as a singlet of the gauge symmetry; while in the case of SSB, the vacuum would no longer be a singlet of the symmetry. Third, in Schwinger's case, the screening of external charges is complete, so the total charge of any isolated system must vanish; while this is not the case in the electroweak theory whose symmetry is spontaneously broken, in which flavored states exist (Casher, Kogut and Susskind, 1973; Kogut and Susskind, 1974). The last point was particularly pertinent to the confinement problem: if a Schwinger scheme could be found in QCD, then no colored states would exist.

Thus the investigations on the possibility of a Schwinger mechanism within the framework of QCD became a strategic step in attacking the confinement problem. In this pursuit, Wilson's lattice theory occupied a prominent place; but some works on strings preceded it, and parts of their positive results were integrated into Wilson's work.

String

Heuristically it is widely acknowledged that "the common theme of a string-like structure stretched between quarks has become the standard model of confinement" (Susskind, 1997).

The string picture (Nambu, 1969, 1970; Susskind, 1969, 1970; Nielsen and Susskind, 1970; Goto, 1971) first emerged from a special interpretation of the dual resonance model (Dolen, Horn and Schmidt, 1967; Veneziano, 1968;

Rosner, 1969). In this picture, mesons behave as little stretches of strings with quarks as their end points. As noticed by Tryon (1972), the energy of an elastic string was proportional to its length in string theory, or equivalently, as noticed by Nambu and Goto, action for a world sheet swept out by a moving string was proportional to its area.

The proportionality squared very well with the observed linear relation between the mass-squared and the angular momentum of the heavy mesonic resonances, which was expressed in terms of Regge trajectories (see Section 2.1). But as soon realized by Holger-bech Nielsen (1971), the dual string picture had to be abandoned because it was incompatible with the parton model supported by the results obtained in deep inelastic scattering experiments. Hadrons in the parton model were composed of free partons, which carried their momentum and other quantum numbers in discrete finite bits. This was clearly inconsistent with the continuum string picture.

In the new context of an abelian Higgs model, however, Nielsen and Poul Oleson (1973), and independently Zumino (1973), found that the model allowed the existence of a magnetic vortex line, with monopoles at the end points, that shared several features with the dual string: it was unbreakable and its mass-energy content was entirely due to the tension force. One interesting feature of this model is that it was almost identical to models about superconductor materials in which the Meissner effect was active: the vortices were just the Abrikosov vortices formed by magnetic fields that succeed in penetrating into a superconductor.

The model was generalized by ‘t Hooft to the non-abelian case (1974; see also Polyakov, 1974), and the result was that vortices in this case can break and, as a result, pairs of magnetic monopole and antimonopole can be created. This had opened a new line of reasoning. If the coupling is strong enough in a model considered by ‘t Hooft in which free magnetic monopoles exist, a kind of Schwinger phenomenon would occur and the monopoles occurring in the vacuum fluctuations may condense in the same sense that Cooper pairs condense in the BCS ground state. If this happens, then the vacuum becomes a superconductor with the roles of magnetic and electric fields (or chromomagnetic and chromoelectric fields) interchanged. The electric field was replaced by the condensate and the chromoelectric Meissner effect would occur. That is, the chromoelectric lines of force would be squeezed into a narrow tube and the energy of the tube would be proportional to its length. In this case, the role of monopoles would be played by the quarks, which would be held together by these lines of chromoelectric flux or string-like vortices.

In this model, a line of chromoelectric flux breaks only when a pair of quark and antiquark is created. This heuristically explained why hadrons were always kept color neutral and no colored state could be observed, an explanation in exactly the manner described by the dual string model. The difference from the dual string model, however, is profound: now the strings are not a mathematical device to reproduce what was observed in the hadron's spectrum and amplitudes as in the case of the dual string model, but objects composed of the gauge fields: the chromoelectric flux lines were squeezed into tube-like configurations by dynamical nonlinearities of the non-abelian gauge field, or, more exactly, by the interactions between chromoelectric lines through the exchange of chromomagnetic quanta in the gluon field's vacuum fluctuations when the gauge coupling is strong enough.

The plausibility of this model in explaining the confinement had suggested a general two-step scheme for solving the confinement problem in terms of strings within a gauge invariant theory like QCD. First, in the computation of interactions between quarks by summing up the effects of the non-abelian gauge field, one should be able to find a long-range force between unscreened quarks mediated by flux tubes, which implies that the energy needed to remove a quark from a color neutral system is proportional to the length of the flux tubes or the separation distance between the quarks. Second, one should be able to find the pair production of a quark and an antiquark by the non-abelian gauge field that provides new end points and allows the flux tube to break and hadrons to form as in 't Hooft's model. This process screens the long-range force and, as a result, only a short-range force remains within the gauge invariant theory.

It is worth noting that what eliminates the long-range force and renders the propagator finite can only be the classic Schwinger mechanism: the relevant bound state in 't Hooft's theory was a fermion pair held together by a flux tube, which plays the same role a bound electron pair played in Schwinger's one dimensional QED.

Lattice

One of the major merits of Wilson's lattice gauge theory is that it is able to deal with large distance dynamics (or, equivalently, strong coupling dynamics, according to renormalization group analysis) in a reliable (even in a quantitative) way. Without such an ability it would be impossible to treat the confinement problem effectively.

In Wilson's formulation, the color SU(3) gauge theory is defined on a Euclidean four dimensional hypercubical lattice with equal lattice spacing a in four directions between two neighboring sites of the lattice. This means that the theory has a built-in ultraviolet cutoff. Since it can be shown that classical action is finite and proportional to the area of the loop, and thus is not sensitive to the cutoff, the actual force law is also cutoff insensitive in the regime in which the color confinement is investigated.

On the lattice the fields are $\psi_n, \overline{\psi}_n$, and $A_{n\mu}$, where n is a four-vector with integer components referring to points on a simple hypercubic lattice. A gauge transformation on the lattice can be defined as follows:

$$\psi_n \to e^{iy_n g}\psi_n, \ \overline{\psi}_n \to e^{-iy_n g}\overline{\psi}_n, \ A_{n\mu} \to A_{n\mu} - \left(y_{n+\hat{\mu}} - y_n\right)/a, \tag{8.11}$$

where $\hat{\mu}$ is a unit vector along the axis μ, g is the coupling constant and y_n is arbitrary. The terms $\overline{\psi}_n\psi_{n+\hat{\mu}}$ and $\overline{\psi}_n\psi_{n-\hat{\mu}}$ are not invariant to this transformation; the corresponding invariant expressions are $\overline{\psi}_n\psi_{n+\hat{\mu}}\exp(iagA_{n\mu})$ and $\overline{\psi}_n\psi_{n-\hat{\mu}}\exp(-iagA_{n-\hat{\mu},\mu})$. As to the gauge field A_μ, a gauge invariant lattice approximation for $\nabla_\mu A_\nu - \nabla_\nu A_\mu$ is $F_{n\mu\nu} = \left(A_{n+\hat{\mu},\nu} - A_{n\nu} - A_{n+\hat{\nu},\mu} + A_{n\mu}\right)/a$, which can be rescaled as $f_{n\mu\nu} = a^2 g F_{n\mu\nu} = B_{n\mu} + B_{n+\hat{\mu},\nu} - B_{n+\hat{\nu},\mu} - B_{n\nu}$, where $B_{n\mu} = agA_{n\mu}$. Thus a simple lattice action for the gauge field which preserves periodicity (which is the lattice version of the gauge invariance) is $A_B = \frac{1}{2g^2}\sum_{n\mu\nu} e^{if_{n\mu\nu}} \approx \frac{1}{4}a^4 \sum_{n\mu\nu} F^2_{n\mu\nu}$ (when a is small), and the full gauge invariant action can be written as

$$A = -c\sum_n \overline{\psi}_n\psi_n + K\sum_n\sum_\mu \left(\overline{\psi}_n\gamma_\mu\psi_{n+\hat{\mu}}\exp(iB_{n\mu}) - \overline{\psi}_{n+\hat{\mu}}\gamma_\mu\psi_n\exp(-iB_{n\mu})\right) + \frac{1}{2g^2}\sum_{n\mu\nu} e^{if_{n\mu\nu}} \tag{8.12}$$

where $c = m_0 a^4$, $K = a^3/2$, $a^4\sum_n$ replaces the space time integration of the continuum theory, and $\left(\psi_{n+\hat{\mu}} - \psi_{n-\hat{\mu}}\right)/2a$ replaces ∇_ψ. The action reduces to the usual continuum action for $a \to 0$; for finite a it is gauge invariant and periodic in the gauge field (Wilson, 1974).

A physical picture offered by the above formulation is this. The elementary particles of the theory are quarks that are purely anticommuting objects living at lattice sites. The gauge variables are links (or string bits) located on space-like nearest neighbor sites of the lattice, and must be unitary matrix from the gauge group itself. There is a separate gauge invariance of the theory for each lattice site. There are nearest neighbor couplings of the quark variables with a coefficient denoted K, and plaquette-type couplings of the gauge variables with a coefficient of order $1/g^2$. It is immediately obvious that it would have a reliable strong coupling expansion because the coefficient of the gauge plaquettes term was $1/g^2$ rather than g^2, and this strong coupling expansion becomes accurate at large distant scales.

In this theory, the gauge invariant configuration space consists of (i) a collection of strings with quarks at their ends, which are lines of non-abelian electric flux and can be used to describe the dynamics in the strong coupling limit; (ii) the gauge field plaquettes, in which every index is matched at each site; the match is a non-abelian equivalent of the continuity of electric flux lines in the lattice theory.

In the simple limit in which all purely spacelike terms with small coefficients are dropped altogether from the Lagrangian, only gauge field plaquettes in one space and the time direction and nearest neighbor quark couplings in the time-like direction would remain, and some significant results can be derived. First, all non-gauge invariant states, such as single quark states, have infinite energy. Second, the masses of gauge invariant composite multi-particle states are sums of the masses of each constituent, with the mass of a quark constituent being $-\ln(K)$ and the mass of a constituent string bit being $-2\ln(g)$. For K and g small enough, both these masses are large, indeed much larger than the inverse of the lattice spacing, which is taken to be unity in these formulae. This rule also includes the appearance of string built of string bits whose total mass is proportional to the length of the string (Wilson, 1974, 2004).

Wilson examined the quark confinement by considering the current–current propagator

$$D_{\mu\nu}(x) = \langle\Omega|TJ_\mu(x)J_\nu(0)|\Omega\rangle \tag{8.13}$$

whose Fourier transform determines the e^+e^- annihilation cross section into hadrons, where the currents are built from quark fields. The propagator in the Feynman path-integral picture "is given by a weighted average over all possible classical quark paths and all possible classical values of the gauge field. The currents are thought of as producing a quark pair at the origin which later annihilate at the point x: One has to sum over all paths joining the point 0 and x for each of the pair of quarks." Here is the origin of the now well-known Wilson loop. "All possible loops must be summed over" (Wilson, 1974).

The weight associated with a given quark path or set of paths includes a factor of $\exp\left[ig \oint A_\mu(x) \times ds^\mu\right]$, where $\oint \cdots ds^\mu$ is a line integral or a sum over line integrals for each of the quark–antiquark loops joining the points 0 and x. Independently of the quark paths there is another weight factor, namely the exponential of the free action for the gauge field. The combined weight factor is then averaged over all quark paths and all gauge fields to give the current–current propagator. Thus in the strong coupling expansion for the propagator, there is a double expansion: one in the coefficient K is needed to define quark loops on the lattice; the g^{-2} expansion is needed to define the surfaces filling in the quark loops, the sum of the g^{-2} expansion is a sum over all such surfaces. Such a structure of double summation over all quark paths and over all surfaces joining quark paths is reminiscent of relativistic string models of hadrons.

Then Wilson set the criterion for the confinement: in order for quarks to exist as separate final state particles, it must be possible to have quark–antiquark loops with well-separated quark and antiquark lines. Otherwise no detector will see a quark or antiquark in isolation. This means that any argument for the confinement has to be able to show that large size loops are impossible. And this is exactly the strategy Wilson adopted.

In the strong coupling limit ($g \to \infty, K \to 0$), Wilson was able to show that the gauge field average of $\exp\left[ig \oint A_\mu(x) \times ds^\mu\right]$ behaves as $\exp(ic'A)$, where A is the enclosed area of the loop. "This heavily suppresses large loops." As a result, "vacuum expectation values [the current–current propagator (8.13)] decrease rapidly at separations of only a few lattice sites." This corresponds to the existence of masses much larger than the cutoff, a Schwinger mechanism which was crucial for Wilson's argument for the confinement. The existence of a Schwinger-like mechanism entails that quark paths will not separate macroscopically, and thus there will be no quarks among the final state particles. But this also entails that when the coupling constants varied from small to large, a phase transition, from the free phase to the confinement phase, must occur in the lattice QCD.

By considering perturbations around the strong-coupling limit when the fermion piece of the Hamiltonian is accounted for in the Hamiltonian formulation of Wilson's lattice gauge theory, it can also be shown that "the fermion term considered as a perturbation describes processes in which a $q\bar{q}$ pair and its flux line is created. If that flux line overlaps with a link of the original flux line connecting the initial quarks, it may leave that link in an unexcited state." Thus "these process allow the original string to break, i.e., they screen the long-range interquark forces." However, they do not allow free quarks to escape since each segment in the process must be colorless. This situation gives another example of the Schwinger mechanism for eliminating long-range forces, which in this particular theoretical context also confines quarks (Kogut and Susskind, 1975).

In sum, Wilson's lattice formulation of QCD was able to calculate, in a reliable way, effects that lay beyond the reach of perturbation theory, such as confinement. Although the confinement was not rigorously proved within this framework, now physicists had compelling reasons for believing that quarks and gluons were confined in lattice gauge theory. If a link in the lattice (or a color vortex in a more physical term) ever broke, it could only be caused by the quark–antiquark pair formation, and as a result, hadrons were always kept color neutral. A vortex breaking without pair formation would require large scale structures (namely low energy excitations) of a kind that directly contradict the result obtained in the lattice theory or by the lattice simulations (Creutz, Jacobs and Rebbi, 1979). For this reason, with the progress made along the line of Wilson's formulation of lattice gauge theory, the confinement problem was no longer taken to be a deep puzzle in the quantum field theory of quarks and gluons as it had been ever since the quark model was proposed in 1964.

8.4 Early successes in phenomenology

Within a few years after it was first articulated by Gell-Mann in 1972, QCD enjoyed a series of predictive and explanatory successes. The foundation of these successes was its powerful perturbative analysis of hadronic processes. As an asymptotically free gauge theory, QCD, taken in conjunction with some features from the quark model and the parton model, most important among them the ideas about hadrons' constituents and their behavior in high energy processes, was able to provide a new perturbative analysis of hadronic processes for its prediction and explanation of various hadronic phenomena.

The first prediction the QCD theorists made was the logarithmic corrections to the quark–parton model result of scaling: "the structure functions decreasing like ever-increasing power of $\ln(-q^2)$" (Gross and Wilczek, 1973c). Since a logarithmic function is a slowly varying one of its argument, the checking of the prediction requires experiments over a large range of q^2 and measurements of sufficient precision. For this reason, it was not until 1978 that enough data were accumulated for reasonable claims to be made about the validity of the prediction (Fried, 1979).

Chronologically, however, QCD's first success came from its explanation and prediction in charm phenomenology (Appelquist and Politzer, 1975). It was followed by the confirmation of logarithmic corrections to scaling, and by the prediction and observation of the three-jet events. The three-jet events were unanimously taken to be unambiguous evidence for the existence of gluons in the particle physics community (Ellis, Gaillard and Ross, 1976; Wiik, 1979; Wu and Zobernig, 1979).

Charm phenomenology

The first novel and successful application of QCD was to offer a theoretical model for the "new particles," the hidden- and naked-charm particles. From the achievements of this model sprang a new QCD-inspired tradition of hadron spectroscopy.

The new particle at 3.1 GeV, named J by the MIT group led by Samuel Ting, and ψ by the SLAC–LBL group led by Burton Richter and Gerson Goldhaber, was discovered in November 1974.

Ting, equipped with expertise in the detection of e^+e^- pairs and encouraged by a strong conviction that there should be more massive vector mesons other than the discovered ones, the ρ, ω, and ϕ mesons, assembled a group at MIT for pursuing this idea. Their strategy was to look for peaks in the distribution of total energy of e^+e^- pairs produced in hadronic collisions because the peak is a signature for a resonance decaying into e^+e^- pairs. In the reaction $p + Be \rightarrow e^+ + e^- + anything$ studied at the 30 GeV AGS (alternating gradient synchrotrons) in Brookhaven by Ting's group, the data taken in the late summer of 1974 had revealed a sharp enhancement at 3.1 GeV. The observed large number of e^+e^- events narrowly around 3.1 Gev was naturally ascribed by Ting's group to the production and decay of a new vector meson (Aubert *et al.*, 1974).

Ting's group did not publish the result until the SLAC Mark I experiment found the same particle at the electron–positron storage ring SPEAR in early November (Augustin *et al.*, 1974). The SLAC–LBL group observed a huge but narrow peak at 3.1 GeV in the total cross section for e^+e^- annihilation to hadrons in the colliding beam processes $e^+ + e^- \rightarrow$ hadrons studied at SPEAR. Peaks in cross sections were usually ascribed to particle production, and thus the observed peak was interpreted as evidence for the production of a vector particle with a mass of 3.1 GeV.

The outstanding mystery calling for explanation was the unusual longevity of the J/ψ particle. The conventional wisdom was that a hadronic resonance of 3 GeV should be expected to have a width of several hundred MeV, but the observed width of J/ψ was much narrower, around 0.06 MeV.

An explanation was handily offered by Appelquist and Politzer (1975). They applied the perturbative QCD to the bound system of charm quark and charm antiquark, which they termed charmonium, [suggested by Alvaro de Rujula to them, (Politzer, 2004)] and claimed that the J/ψ particle was only one of its modes. Through the application of perturbative QCD, they successfully explained the longevity of the J/ψ particle and, further, predicted the existence of many new (hidden and naked) charm particles and their properties and behavior.

The term "charmonium" first appeared in the article of Appelquist and Politzer, but the idea of a bound system of charm quark (c) and charm antiquark ($\bar{c}$), or hidden charm states, was well in the air since the summer of

1974, prior to the discovery of the J/ψ particle. In a widely circulated Fermilab preprint "Search for Charm" by Mary Gaillard, Benjamin Lee and Jonathan Rosner (1974),[3] all the physics of hidden-charm states and charmed particles was expounded in detail, and most was correct except for one mis-prediction (about the branching ratio of charm mesons to $K\pi$) and one omission (about the discoverability of the hidden charm states in electron–positron collisions) (Yoh, 1997).

The charmonium model of the J/ψ particle derived much of its explanatory power from its prototype positronium which, as the atomic bound states of an electron–positron pair first observed by Martin Deutsch in (1951), was effectively a hydrogen atom with the proton being replaced by a positron, except for its instability: the electron and its antiparticle positron in positronium can annihilate one another to photons. Similarly, charmed quark (c) and charmed antiquark ($\bar{c}$) in charmonium can also be annihilated to gluons. This annihilation, as we will see shortly, turned out to be crucial for its explanation of the longevity of the J/ψ particle.

The narrow widths of hidden charm states was originally understood qualitatively by Gaillard, Lee and Rosner in terms of the so-called Zweig rule: hadron decays involving quark–antiquark annihilation were suppressed, by a factor of 10^2 in the case of $\phi \rightarrow \rho\pi$, relative to decays in which the constituent quarks maintained their identity. Thus, if a hidden charm state lies below the threshold for the production of a pair of charmed hadrons, then energetically it is impossible for it to decay into a pair of them, and the only decay channels open to it would be those in which the Zweig rule is violated. But in order to have a quantitative agreement with the observed longevity of the J/ψ particle, a "super-Zweig rule" with a suppression factor "20 times stronger than for [the ordinary case of] ϕ" would be needed (Gaillard, Lee and Rosner, 1975).

The required super-Zweig rule naturally arises from the charmonium model proposed by Appelquist and Politzer. According to the model, the mass of charm quark (m_c) is assumed to be large, hence its Compton wavelength is small. Since binding holds particles at a small distance compared to that length, it should be much smaller than that of all other known mesons. This means that one is in the short-distance QCD regime, $\Lambda/m_c \ll 1$, and perturbative QCD should apply because of the asymptotic freedom.[4] In addition, since the heavy bound particles move with low mean velocity, nonrelativistic quantum mechanics should also apply. Such a situation in the analysis of charmonium implies that QCD yields a potential that is Coulomb-like at short distances, $V(r) = -4\alpha_c/3r$, with a small running coupling constant α_c,[5] here the Coulombic interaction is mediated by weakened color gluon exchange (see also Gaillard, Lee and Rosner, 1975).

According to perturbative QCD, the decay of charmonium to conventional states is governed by the annihilation of the $c\bar{c}$ pair into three gluons, which is the minimum number consistent with color, charge, and parity conservation, and then three gluons rematerialize as conventional quarks and antiquarks which would recombine to be ordinary mesons. Thus the first stage of the decay leads to an intermediate state with three gluons, the wave function of which is proportional to the sixth power of the coupling constant, and the decay rate for the process is proportional to the twelfth power of the running coupling constant, which is small because the gluons created by the massive hidden charm states have large energy and momentum and thus small effective coupling with quarks according to the asymptotic freedom. This analysis provided a mechanism for the operation of a "super-Zweig rule," and thus convincingly explained why the decay rate of charmonium was dramatically suppressed. As a result, the longevity of the J/ψ particle was no longer to be a mystery.

The charmonium model successfully passed severe tests. In addition to the explanation of the longevity of the J/ψ particle and the subsequently discovered ψ' (3.7 GeV) particle, with the help of non-relativistic inter-quark potential and the wave function extracted therefrom, the energy spectrum of charmonium was computed as was done for positronium and the hydrogen atom, and many of the details of the other bound states were predicted. For example, in analogy with the established nomenclature of the hydrogen atom, Appelquist and Politzer claimed that "the existence of paracharmonium with this width [$\Gamma_h(para) \approx 6\,\text{MeV}$] and a mass on the order of 3 GeV is one of our most unambiguous predictions and it is important to look for it experimentally" (1975). In standard atomic notation, J/ψ and ψ' are the 1^3S_1 and 2^3S_1 states of $c\bar{c}$; the existence of other hidden-charm particles with different spins and parities, corresponding to other states of the positronium spectrum, was predicted, and then confirmed in 1975 and 1976 at SPEAR (SLAC) and DORIS (DESY) (Wiik, 1976; Appelquist, Barnett and Lane, 1978; Quigg and Rosner, 1979).

Thus it is not without justification to see charmonium as the hydrogen of QCD. By the time of the 18th International Conference on High Energy Physics, which was held in Tbilisi in July 1976, the charmonium model was generally accepted as the explanation of the new particles (Wiik, 1976; De Rujula, 1976),[6] and its success strongly reinforced the belief in QCD.

Scaling violation

Bjorken scaling observed in deep inelastic lepton–hadron scatterings can be understood in terms of, and derived from, the parton or light-cone models.

The latter, as free field theories, however, could offer no explanation as to how quarks were bound together, and thus could not be taken as realistic models for hadron physics. In fact, once interactions were taken into consideration, scaling was violated. As we mentioned in Section 6.2, immediately after the observation of scaling, Adler and Tung (1969) did a perturbative calculation in a model where quarks were bound by SU(3) singlet massive vector gluons, and the result was that there appeared logarithmic deviations to scaling.

As an asymptotically free theory, QCD with color octet gluons was able to offer more reliable perturbative calculations, and thus able to "derive all the sum rules and relations previously derived in the parton or light-cone models. In addition there will be logarithmic deviations from Bjorken scaling which can be calculated" (Gross and Wilczek, 1973b). Gross and Wilczek actually calculated, to the lowest order of perturbation, the logarithmic deviations from Bjorken scaling, but noticed that "it is an open question whether such a picture is consistent with experiment" (Gross and Wilczek, 1973c). They were cautious because they realized that the deviations "from exact scaling are quite small, and one would need rather large variations of q^2 to see any marked change." Thus they declared that verification of logarithmic deviations would be strong evidence for QCD, or even "a crucial test" of the validity of QCD (Gross and Wilczek, 1973c).

The deviations were first observed in 1975 at SLAC (Taylor, 1975) and at Fermilab (Chang *et al.*, 1975). But the data on the deep inelastic electron–proton scattering from SLAC lacked the range of q^2, and the data on the deep inelastic muon–proton scattering from Fermilab lacked the precision required to determine the form of the scaling violations. Thus further improvement in the range of q^2 and precision was needed to clearly establish the validity of QCD in comparison with the parton model.

The situation improved quickly, only three years later, at the 19th International Conference on High Energy Physics held in Tokyo in August 1978, Richard Field was able to show impressive and convincing agreements between the predictions of QCD and a whole range of experimental data. He examined the existing electron–proton and muon–proton scattering data from SLAC and Fermilab, displayed detailed QCD fits to the data, and claimed that the data manifested logarithmic violations of scaling "in precisely the manner expected from QCD" (Field, 1979).

Field went further and examined the latest measurements of scaling violations in neutrino and antineutrino scatterings, which were carried out by the BEBC (Big European Bubble Chamber) collaboration at CERN. The CERN data were precise enough for the calculations of the moments of the structure functions, which were weighted integrals of the structure functions and their ratios were predicted first by Gross and Wilczek (1973c) with no free parameters at all. Field displayed a plot of the CERN moment ratio and then stressed its remarkable agreement with the QCD predictions "at about the 10% level. This is one of the most impressive tests of QCD to date" (Field, 1979).

But to call the compatibility of data with the predictions of QCD a "test" is perhaps too strong a claim. It was immediately realized by Taizo Muta, at the same Tokyo conference and afterwards (Muta, 1979a, b) that the test might be impressive but not clear cut.

It was realized widely among theorists that both scaling and scaling violations were asymptotic claims. In order to see how close the existing energy regimes were to be effectively asymptotic, it was necessary to examine the scaling violations beyond the leading-order approximation which dominated only asymptotically. The calculations of higher order effects showed that they were not negligible. But their calculations were very complicated and sensitively dependent upon the computational schemes (especially the renormalization schemes) employed, as was pointed out by Muta. The implications were clear: the agreements or disagreements between data and the QCD prediction were not without some arbitrariness.

The confused situation was examined carefully by Gaillard (1979) and De Rujula (1979). They pointed out that many terms contributed non-asymptotically to the QCD predictions of scaling violation were not all computed, some of them – those arising from confinement effects – were not even perturbatively computable. Clearly there was a possibility that when all terms relevant at contemporary energy regimes were included, the results might be quite different from the currently adopted approximations. Taking this possibility into consideration, the agreements between data and the current QCD predictions might be accidental.

Possible failure in future, however, could not damp the enthusiasm of QCD theorists, or its reception by the high energy physics community. A remarkable fact was that the QCD prediction continued to fit the data, including the new data from high quality neutrino and muon beams available at CERN. The historical fact was that the successes of QCD in predicting logarithmic deviations to scaling contributed heavily to physicists' early perception of QCD as a research program with progressive problem-shifts, in Imre Lakatos's term (1970).

Observing gluon through three-jet events

The idea of jets, namely a number of hadrons all traveling in roughly the same direction, originated from the parton model.

In the process of electron–positron annihilation into hadrons investigated in the e^+e^- colliders, such as SLAC's SPEAR or DESY's DORIS and PETRA, the resulting virtual photon was conceived by the parton model theorists as first rematerialized into a pair of parton and antiparton, which moved back to back with equal momentum because of the conservation of energy and momentum. When the pair had separated by no more than one Fermi, it was expected that non-perturbative effects would be operative and the process of hadronization or fragmentation would take over through strong interactions, such as the breaking-up of the string, so that the fractional charges of quark-partons obtained from their birth would be shielded, and two bunches of outgoing ordinary hadrons would appear. No matter what the exact mechanism of hadronization would be,[7] the final state hadrons would remember the direction of the parent parton, and would come grouped in jets. Thus, "a double-jet structure for the emerging hadrons is natural in a parton picture" (Bjorken, 1973a; see also Drell, Levy and Yan, 1969; Cabibbo, Parisi, and Testa 1970)

In the process of proton–proton collisions investigated in CERN's ISR, the two protons were scattered off each other at high energy; they would heat up and ultimately produce a variety of hadrons that would then scatter in the same direction as the incoming protons, that is, their transverse momentum should be small. In 1973, however, hadrons with large transverse momentum were frequently observed. The phenomenon was handily explained by the idea of parton: they must emerge from the fragmentation of partons that have collided violently. Since the secondary hadrons always move along the direction of the struck parton, as the parton idea suggested, these hadrons with big transverse momentum could never by produced by themselves, but always as parts of a jet, and thus the final hadron state in such a case should always consist of four jets. Bjorken quickly interpreted the phenomenon this way: "the secondary is the decay product of a parent, both parent and secondary possess approximately the same laboratory angle. A jet is formed. Jet is a necessary consequence of parton model" (1973b).

Since 1969 and 1970, partons in the electron–positron annihilation were expected to dress up as jets of hadrons, and hadrons were expected to show a two-jet structure. The jet structure in the electron–positron annihilation was first observed in 1975 at SPEAR of SLAC, and the spin 1/2 behavior of the parent parton was verified by the observation of an azimuthal asymmetry in inclusive hadron production, which justified identifying parton with quark. Since in the quark–parton model, these jets were interpreted as products of quarks, the observation of these quark-jets was taken to be indirect observation of quarks.

But the distinguishing feature separating QCD from the quark–parton model was the gluons. One possible effect of the gluon would be the broadening of a quark jet in two-jet events as the center-of-mass energy increases. In 1976, John Ellis, Mary K Gaillard and Graham G Ross (1976) went further and made a prediction of three-jet events by invoking a gluon bremsstrahlung process. They conjectured that in the electron–positron annihilation process when one of the primary quarks emitted a hard gluon carrying a large momentum transverse to that of the emitting quark, the gluon would convert into a third jet. The rate of emission could be determined by perturbative QCD, which should be less frequent than two-jet events. They claimed that although "the first observable effect [of gluon] should be a tendency for the two jet cigars to be unexpectedly oblate... Eventually, events with large P_T [transverse momentum] would have a three jet structure" (1976).

The three-jet structure was observed in the electron–positron collision experiments in June 1979 at PETRA of DESY (Wu and Zoberning, 1979), and its interpretation was not difficult. First, two starting fermions could not become three fermions, so these three jets could not all be the products of quarks and antiquarks, and a new particle should be responsible for one of the three jets. Second, since this new particle, similar to quarks, also fragmented into a jet of hadrons, it could not be a color singlet. Color singlets would leave a track if charged, or decay into well-defined final states, but not metamorphose into jets. This reasoning strongly suggested that the gluon or the color non-singlet carrier of strong forces should be responsible for the appearance of the three-jet events through the hard gluon bremsstrahlung process $e^+e^- \to q\bar{q}g$, as predicted by Ellis, Gaillard, and Ross three years before. The interpretation was readily accepted by most high energy physicists and the observation of the events was duly regarded as a "direct" observation of the gluon.[8]

As Björn Wiik announced, immediately after the discovery, in his plenary talk at the Bergen Conference of Neutrino 79:

> If hard gluon bremsstrahlung is causing the large $P_\perp$ values in the plane then a small fraction of the events should display a three jet structure. The events were analyzed for a three jet structure using a method proposed by Wu and Zobernig... A candidate for a three jet event observed by the TASSO group at 27.4 GeV, is showing in Fig 21 viewed along the $\hat{n}_3$ direction. Note that the event has three clear well separated jets and is just not a widening of a jet. (1979)

A few days later, at the 1979 EPS (European Physical Society) Conference on High Energy Physics in Geneva, Bjorken argued that since a gluon with large angle relative to the quark jet was emitted at very short distances, and therefore calculable in perturbation theory, this short-distance process should lead to a distinct three-jet final state, and thus declared that these "gold-plated"[9] three-jet events should be regarded as "an especially clean test of QCD perturbation theory and an independent determination of α_5 [the running coupling constant of QCD]" (Bjorken, 1979).

8.5 Theoretical probing into deep structures

The axial U(1) anomaly, as recognized by Fritzsch, Gell-Mann and Leutwyler (1973), could not be explained away by the Adler–Bell–Jackiw anomaly, $\partial_\mu J^5_\mu = \frac{-ig^2}{16\pi^2} G^a_{\mu\nu} \tilde{G}^a_{\mu\nu}$, where J^5_μ is the axial vector current, $G^a_{\mu\nu}$ the gluon field, and g the strong coupling constant. The reason, as they clearly indicated (see Section 8.2), was that the anomaly "is itself a divergence of another (non-gauge invariant) pseudovector," namely, the divergence of a Chern-Simons current K_μ, $-iG^a_{\mu\nu}\tilde{G}^a_{\mu\nu} = \partial_\mu K_\mu$. Although K_μ is not gauge invariant, it can be used to construct a modified axial U(1) current, $\tilde{J}^5_\mu = J^5_\mu - \frac{g^2}{16\pi^2} K_\mu$, whose charge would be conserved. For this reason, the axial U(1) problem persists even in the presence of anomalies.

't Hooft (1975) once argued that since K_μ was not gauge invariant, it would not obey the boundary conditions required to allow partial integrations in the derivation of the conservation of the modified axial U(1) charge. Thus the axial U(1) problem could be evaded. But the real solution to the axial U(1) problem came from deep explorations of the special topological structure of QCD, centered around the notion of instanton.

Instantons

Motivated by the search for classical solutions with nontrivial topology in analogy with the 't Hooft–Polyakov monopole, which was thought to be relevant in addressing confinement (see Section 8.3), Alexander Belavin, Alexander Polyakov, Albert Schwarz, and Yu Tyupkin (1975) discovered "pseudoparticle" solutions of non-abelian gauge theories. These were classical solutions to the Euclidean equations of motion, characterized by a topological quantum number, and responsible for tunneling events (or gauge rotations) between degenerate classical vacua in Minkowski space associated with the solutions.

If the Lagrangian is $\mathcal{L} = -\frac{1}{4} G^a_{\mu\nu} G^a_{\mu\nu}$ with the field strength $G^a_{\mu\nu} = \partial_\mu A^a_\nu - \partial_\nu A^a_\mu + g\varepsilon_{abc} A^b_\mu A^c_\nu$, then the topological quantum number is

$$n = \frac{g^2}{32\pi^2} \int d^4x G^a_{\mu\nu} \tilde{G}^a_{\mu\nu}, \tag{8.14}$$

with $\tilde{G}^a_{\mu\nu} = \frac{1}{2}\varepsilon_{\mu\nu\alpha\beta}G^a_{\alpha\beta}$. Here n is an integer for all field configurations in Euclidean space that have the vacuum or a gauge transformation thereof at the boundary; it is called the winding (or Pontryagin) number because it classifies all maps from the 3-sphere S_3 (corresponding to $|x| \to \infty$) into the gauge group and indicates how many times the group manifold is covered. The "pseudoparticle" solution with $n = 1$ in Euclidean space is

$$A^a_\mu(x)^{cl} = \frac{2}{g}\frac{\eta_{a\mu\nu}(x - x_0)^\nu}{(x - x_0)^2 + \lambda^2}, \tag{8.15}$$

here the center of the solution x_0 is free because of translation invariance, the scale of the solution λ is a free parameter, and η is a mapping tensor.

Mathematically it is not difficult to see that the solution connects two classical solutions at $x_4 \to \pm\infty$, thus allowing tunneling between topologically different vacua. In the presence of light fermions with N flavors, from (8.14) it is also clear that the Adler–Bell–Jackiw anomaly $\partial_\mu J^5_\mu = \frac{-iNg^2}{16\pi^2} G^a_{\mu\nu}\tilde{G}^a_{\mu\nu}$ is related to the topological quantum number. As 't Hooft noticed, "a configuration in Minkowsky space with $n = 1$ would be associated with a violation of axial charge conservation: $\Delta Q^5 = 2N$" ('t Hooft, 1976a).

These solutions were localized regions of spacetime with very strong gauge field, which entailed a topologically nontrivial localized field configuration in four dimensional Euclidean space with a fixed finite action. The finite action is determined by the corresponding topological charge. In the case of A^{cl} $(n = 1)$, the action is $S = \int \mathcal{L}\{A^{cl}\}d^4x = -\,8\pi^2/g^2$. As 't Hooft (1976a) pointed out in his classic article on the subject, because of "a 4D rotational symmetry, the solution is not only localized in three-space, but also instantaneous in time." For this reason, he coined the term "instantons" to refer to these objects.[10]

In the same article, 't Hooft also suggested "a simple heuristic argument that explains why these solutions are relevant for describing a tunneling mechanism in real Minkowskian spacetime, from one vacuum state to a gauge rotated vacuum (a gauge rotation that cannot be obtained via a series of infinitesimal gauge rotations)." The core of the argument was a careful examination of the tunneling amplitude, through which several important properties of instantons were discovered.

The WKB approach in nonrelativistic quantum mechanics suggested that the best tunneling path is the solution with minimal Euclidean action for connecting different classical vacua. Along the same line, 't Hooft claimed that the leading exponential of the tunneling amplitude in non-abelian gauge theory should be exp(S), here S is the minimal action for a path from one vacuum to a topologically different vacuum, and is determined by the topological charge. Thus for the one instanton sector, he pointed out that "the exponent equals $\exp(-8\pi^2/g^2)$, which explains why we get results that are unobtainable through perturbation expansions with respect to g^2" ('t Hooft, 1976a).

In his calculation of the semiclassical tunneling rate, 't Hooft discovered a crucial property of instanton: in the presence of light fermions, the Dirac operator has zero mode solutions with vanishing action. "There is one such solution for each of the N flavors. They are chiral solutions very much like the fermion bound states" formed in the instanton field that act as a potential well. The contribution of zero modes to the vacuum-to-vacuum tunneling amplitude turns the amplitude to zero. 't Hooft interpreted this result as being a consequence of $\Delta Q^5 = 2N$: "we must not sandwich the functional integral expression between two vacuum states, because the initial and final chiral charges Q^5 should be different" ('t Hooft, 1976a).

When a source term was inserted into the Lagrangian, however, 't Hooft found that what made the zero modes' contributions to the amplitude vanishing was replaced by an expression with a factor which could be "exactly reproduced by the $2N$ propagators that connect the sources with the instanton. Note that the sources have to switch chirality. This explains why the instanton gives $\Delta Q^5 = 2N$" ('t Hooft, 1976a).

Intuitively, this suggests that quarks can hop from one instanton to another, flipping their chirality as they pass through an instanton. Or more precisely, this suggests the existence of physical processes in which every instanton absorbs one left-handed fermion of each species and emits a right-handed one (and vice versa for anti-instantons). The result can also be interpreted in terms of Dirac's hole theory: "during the tunneling process, one of the energy levels produced by the Dirac equation switches the sign of its energy. Thus, chiral fermions can pop up from the Dirac sea, or disappear into it. In a properly renormalized theory, the number of states in the Dirac sea is precisely defined, and adding or subtracting one state could imply the creation or destruction of an antiparticle" ('t Hooft, 1998).

The appearance of this factor in the tunneling amplitudes is crucial for the final result that only those tunneling amplitudes in which the instanton creates or destroys chiral fermions are unequal to zero.

If chirally charged fermions are created and destroyed by instantons, then instantons should be properly understood as sinks and sources for the chiral current, which violated the conservation of chiral current. Thus the discovery of instanton solutions in QCD solved the axial U(1) anomaly.

The structure of the true QCD vacuum

The study of tunneling amplitudes caused by instantons has shown that in the absence of massless fermions, the gauge fields may freely perform transitions from one gauge configuration into a topologically rotated one. Since these gauge rotations are an exact invariance of the system and form the additive group of integers, **Z**, the vacuum state must be a representation of this gauge group, which is characterized by an angle θ. The analysis of the effects of zero modes on the tunneling amplitude has also shown that if there are massless fermions, the effects of this angle disappear ('t Hooft, 1976a, b; Jackiw and Rebbi, 1976; Callan, Dashen and Gross, 1976).

In the absence of fermion fields and Higgs fields, the classic vacuum corresponds to configurations with zero field strength. For non-abelian gauge fields, however, this does not imply that the potential has to be constant. Rather, it only limits the gauge fields to be "pure gauge," $A_i = iU(\vec{x})\partial_i U(\vec{x})^+$. In order to enumerate the classical vacua, one has to classify all possible gauge transformations $U(\vec{x})$, namely to study equivalence classes of maps from the 3-sphere S_3 into the gauge group. In practice, this can be done by looking at the special case of $U(\vec{x}) \to 1$ as $x \to \infty$ (Callan, Dashen, and Gross, 1978). As indicated earlier, such mappings can be classified using an integer called the winding (or Pontryagin) number, which counts how many times the group manifold is covered, $n_w \approx \frac{1}{24\pi^2}\int d^3x\varepsilon^{ijk}Tr\left[\left(U^+\partial_i U\right)\left(U^+\partial_j U\right)\left(U^+\partial_k U\right)\right]$. In terms of the corresponding gauge fields, the number is the related Chern–Simons characteristic $n_{cs} \approx \frac{1}{16\pi^2}\int d^3x\varepsilon^{ijk}\left(A_i^a\partial_j A_k^a + \frac{1}{3}f^{abc}A_i^a A_j^b A_k^c\right)$.

The nontrivial nature of the finite gauge transformations caused by instantons reveals the fact that the naïve vacuum of QCD is unstable, the instanton allows tunneling between all possible vacua labeled by winding number n. Thus the true QCD vacuum must be a superposition of an infinite set of vacua $|n\rangle$, each belonging to a different homotopy class labeled by an integer n. Since they are topologically different, no continuous (infinitesimal) gauge transformations could connect one with another.

The effect of a gauge transformation caused by one instanton, U_1, is to shift the winding number n by one, $U_1 : |n\rangle \to |n+1\rangle$, so the effect of U_1 on the true vacuum can change it only by an overall phase factor. This fact can be used to fix the coefficients of various vacua $|n\rangle$ within the true vacuum as follows: $|vac\rangle_\theta = \sum_{n=-\infty}^{\infty} e^{in\theta}|n\rangle$. Expressed this way, the effect of U_1 on the true vacuum is indeed only to generate a phase shift: $U_1 : |vac\rangle_\theta \to e^{-i\theta}|vac\rangle_\theta$. Thus the presence of instantons causing tunneling between degenerate vacua[11] implies that the true vacuum can be parameterized by an arbitrary number θ.

Strong CP violation

The effects of the θ dependence on, or equivalently instantons' contributions to, the physics of QCD could be expressed by writing down an effective Lagrangian, $\mathcal{L}_{eff} = \mathcal{L} + \frac{\theta}{16\pi^2} TrG_{\mu\nu}\tilde{G}^{\mu\nu}$, $\tilde{G}^a_{\mu\nu} = \frac{1}{2}\varepsilon_{\mu\nu\alpha\beta}G^a_{\alpha\beta}$. The presence of the extra term, $\frac{\theta}{16\pi^2} TrG_{\mu\nu}\tilde{G}^{\mu\nu}$, in the effective Lagrangian does not give any perturbative effects since it is a total derivative, and hence never enters into the perturbation theory. Nonperturbatively, however, it has important effects on the physics of QCD. It violates the conservation of axial charge as indicated above. In fact, it aptly produces the mass terms for the η particle and η' particle, which has de-mystified the observational manifestation of the "mysterious" axial U(1) anomaly (Veneziano, 1979; Witten, 1979a, b). However, the extra term explained one problem away, only to raise another problem, the strong CP problem.

The appearance of $\epsilon_{\mu\nu\alpha\beta}$ in the extra term of the effective Lagrangian, indicates that parity is violated by the strong interaction. T is also violated, so there is a violation of CP. Although CP invariance breaking is predicted by QCD, the amount of breaking being related to the value of θ is theoretically unpredictable because no understanding about the source of CP violation is available. Since explicit CP violation also occurs elsewhere in particle physics, notably in the kaon system, the choice that $\theta = 0$ is not an option: virtual kaons will renormalize θ into a value that is different from zero.

Then the question is how to determine the value of θ from experimental observations. The most severe limit on CP violation in the strong interaction came from the electric dipole moment of the neutron, which is known experimentally to obey $d_n \leq 10^{-24}$ e-cm. This serves as an experimental constraint on the parameters of QCD, which gives us the bound on theta: $\theta < 10^{-9}$. This is the strong CP problem: if instanton effects contribute an extra parameter to QCD, why is it so small?

A popular proposal for solving the strong CP problem is to introduce a new spontaneously broken U(1) symmetry of the quark and the Higgs sector, the Peccei–Quinn symmetry (1977), whose Nambu–Goldstone boson, called axion, a new massless pseudoscalar particle coupled to θ, would absorb all strong CP violating effects by a shift. Steven Weinberg (1978) and Frank Wilczek (1978) had shown that the axion in fact fills the role of θ. Thus the nonzero value of θ would be related to the nonzero mass of axion that is related to the Adler–Bell–Jackiw anomaly of the Peccei–Quinn symmetry.

The axion has been sought for a long time, without any success. Experimentally, the invisible axion, if it exists, should have a mass between 10^{-6} and 10^{-3} eV. In the realm of cosmology, the axion is taken to be a candidate for the missing mass or dark matter, and thus acts as a link between the two domains of fundamental physics: elementary particle physics and cosmology.

8.6 Remarks

Conceptual developments centered around the emergence of QCD, which itself was a process full of ambiguities, were extremely complicated. In order to properly assess the roles played by various lines of development, some remarks seem to be desirable.

Asymptotic freedom and QCD

A widely circulated opinion holds that QCD started with the discovery of asymptotic freedom (Pais, 1986). Pickering (1984) elaborated on the point: "once asymptotic freedom had been discovered, two questions remained to be decided in the construction of a realistic candidate gauge theory of the strong interaction: what were the fundamental fields which appeared in the Lagrangian of the theory, and under what symmetry group was the Lagrangian invariant?" The decision was to pick up quarks and gluons as the fundamental fields and the color SU(3) group as the symmetry group, and the result was QCD. This misleading opinion was based on an improper understanding of the logical sequence of the conceptual developments leading to the genesis of QCD and its early justifications, and thus has to be rectified by looking more carefully at the sequence.

Logically, once the Adler–Bell–Jackiw anomaly was taken seriously because of its relevance to the notion of color (see Section 7.1), the strong interactions had to be taken explicitly in Gell-Mann's current algebra program. Since the underlying framework of current algebra was the quark–gluon field theory, thus historically the primary concern of Gell-Mann and his collaborators for this purpose was understandably to fix the nature of gluon. Was it a scalar particle or vector particle? Was it a color-singlet or color-octet particle? Was it a gauge boson of an exact symmetry or a gauge boson of a spontaneously broken symmetry? Proper answers to these questions resulted in the emergence of QCD, as we recounted in Section 7.3, but the search for the answers was experimentally constrained by scaling and nonobservability of quarks and gluons. This constrained search determined the internal logic of conceptual development, and thus all contributions to the search have to be assessed with respect to this internal logic.

Judging from such a perspective, we find that what the discovery of asymptotic freedom (that certain non-abelian gauge theories might have negative β-functions) contributed to the search was to reveal that QCD as a non-abelian gauge theory in fact satisfied the scaling constraint. This was an important contribution to justifying the answers to the key questions provided earlier by Gell-Mann and his collaborators, and thus its reception by the particle physics community was understandably sensational.

But the discovery in itself was not decisive for the emergence of QCD, nor for its justification. In fact, it was based on the existing framework of QCD,[12] and some heuristic allusions aside, it did not really provide any convincing argument to address the confinement problem, which, of course, was not its intended goal.

But more important were the confusions, concerning the nature of gauge symmetry (exact or spontaneously broken),[13] that clouded the discussion when asymptotic freedom was first discovered. Gross and Wilczek (1973a, b) painstakingly tried to bring the Higgs sector into the discussion, arguing that since the ultraviolet behavior of the Higgs sector was that of the underlying symmetric theory, so its effect on symmetry breaking and related mass generation would have no impact on the asymptotic behavior. The effort was motivated by the desire, which was also shared by Politzer, of having a model "which is both asymptotically free and, via the Higgs mechanism, has mass terms for all, or all but one, of the gauge fields. The search has been strenuous and fairly systematic but by no means exhaustive and, furthermore, has provided no insight as to why the two phenomena are apparently incompatible" (Politzer, 1974).

If one took the discovery of asymptotic freedom as synonymous to the birth of QCD, then one might have to take the electroweak theory – which is also a non-abelian gauge theory having a negative β-function, and its Higgs sector, as argued by Gross and Wilczek, may not affect the ultraviolet behavior of the theory or the asymptotic freedom of the theory – as a candidate theory for the strong interactions. But this would be nonsense.[14]

QCD and the quark model

Another widely circulated conception is to take QCD as a marriage of gauge theory and the quark model (Pickering, 1984). Here the quark model refers to the constituent quark model rather than the notion of current quarks. This is clear in the following problematic assertions: "from 1976 onwards, the properties of the new particles [referring to hidden or naked charmed particles] were held to demonstrate the reality of quarks. In its composite aspect at least, the new physics was established – although its gauge theory component, especially QCD, was still open to dispute" (Pickering, 1984).

The notion that the two components of QCD can have separate validity or reality (the marriage metaphor itself was based on the notion of separation) was a misleading one. No reality of quarks can be demonstrated without the valid argument provided by QCD. This is especially true and clear in the case of the charmonium and other charmed particles, in terms of the super-Zweig rule, as we discussed in Section 8.4. Thus the triumph of the new particles was also a triumph for gauge theory. If the validity of QCD was open to dispute, then the charmonium model would be in trouble, and the reality of the charm quark at least cannot be said to be convincingly demonstrated.

It should be stressed that the reality of quarks and the validity of QCD stand and fall together. Nobody could have one without the other. More generally, this symbiotic relationship also applies to all unobservable entities and the theory in which they were conceptualized, as we will further argue in Chapter 9.

Partons, bootstrap, and QCD

The idea of partons remained alive in QCD, at least in the perturbative regime of QCD (deep inelastic lepton–hadron scatterings, electron–positron annihilation and jets). The QCD theorists tried to accommodate the idea by reasoning that in QCD "we may have to assume that the propagation of gluons is somehow modified at high frequencies to give the transverse momentum cutoff" (Fritzsch, Gell-Mann, and Leutwyler, 1973). It is easy to see that the accommodation was closely related with the desire to satisfy the scaling constraint.

The idea of bootstrap was also alive in QCD, mainly in the form of the permanent confinement of quarks and gluons, or equivalently in the form of the color-singlet restriction. Thus in QCD "a modification at low frequencies may be necessary so as to confine the quarks and antiquarks permanently inside the hadrons" (Fritzsch, Gell-Mann, and Leutwyler, 1973).

The impact of bootstrap on Gell-Mann's conception of hadron physics was deep and permanent. For him, it was a constraint any realistic theory of the strong interactions had to satisfy. Such a conviction may explain why the color-singlet restriction (which was equivalent to the exact color SU(3) symmetry and was crucial for conceiving gluons as the gauge bosons of the color symmetry, which was accordingly conceived as a gauge symmetry) was seriously considered by Gell-Mann, but was not taken seriously by other physicists such as Nambu (see Section 7.1). The relationship between bootstrap and QCD conceived by Gell-Mann and his collaborators was most clearly stated in the same text quoted above: "The resulting picture could be equivalent to that emerging from the bootstrap-duality approach (in which quarks and gluons are not mentioned initially) provided the baryons and mesons then turn out to behave as if they were composed of quarks and gluons" (Fritzsch, Gell-Mann, and Leutwyler, 1973).

The **R**-*crisis and new particles*

It is well-known (Pickering, 1984; Pais, 1986) that the discovery of the ψ and ψ' particles was preceded by months of confusion, the so-called R-crisis. As we mentioned in Section 7.1, the ratio R of total cross sections for hadron and muon production in the electron–positron annihilation was expected to be an energy independent constant, which was confirmed by the ADONE data produced with an energy range between 1 and 3 GeV. However, the situation began to be confusing in 1973, and at the London conference of July 1974, Burton Richter of Spear at SLAC reported that above about 2 GeV total energy, the data showed that R kept rising linearly as a function of s (the square of the center-of-mass energy). This precipitated a crisis of confidence amongst theorists, who talked about "the gravity of this theoretical debacle" (Ellis, 1974).

The intensity of interest in understanding the anomaly was reflected by the number of theoretical efforts: twenty-three models were presented at the conference, each of them made special predictions, ranging from 0.36 to infinity; none of them fitted the data.

The discovery of the ψ and ψ' particles (which was immediately interpreted as the discovery of a new quark flavor which had to be taken into account in the calculation of R) indicated that the apparent linear rise of R was faked by the two resonances, and the R-crisis was over. As a result, theorists' attention quickly switched from R to the new particles themselves, and a series of charmed particles (hidden or naked) were discovered.

The switch of attention from R as an expression of structural relations among various entities (electron, positron, muon, photon, quarks of various flavors) to the underlying entities themselves was very methodical because the former as a necessary window, through which one could have access to deeper layers of physical reality, enjoys much less explanatory power than the underlying entities themselves; or more precisely, structural relationships themselves have to be understood in terms of underlying entities.

Evidence and theory-acceptance

As we have shown in this chapter, a few years after the first articulation of QCD, there was an accumulation of evidence in its favor: the discovery of asymptotic freedom and its immediate application to the calculation of scaling violations; the evidence for confinement provided by Wilson's lattice gauge theory; the prediction and confirmation of charmonium spectrum; the observation of three-jet events as a direct verification of gluon nonlinear couplings; the revelation of the topological structure of QCD (the discovery of instanton solutions and the θ vacuum structure, and 't Hooft's resolution of the axial U(1) anomaly). To be sure, none of them alone was decisive, but taken together they were impressive enough to convince people of the correctness of QCD: they would have falsified QCD if they had not been verified.

As a result, by 1978, most of the experimental and theoretical researches on the strong interactions were structured around QCD, which acted as a backbone of high energy physics.

QCD and cosmology

Instantons as sinks and sources for the chiral current implies that the conservation of the baryon number is violated. The cosmological implication is that the observed baryon excess in the present Universe could be explained by the asymmetry in instanton-related creation and destruction of baryonic matter during the time in the early Universe when bubbles of true vacuum configurations appeared in a supercooled but symmetrical pseudo-vacuum and bubble walls were rapidly expanding.

Another speculation about QCD's cosmological relevance comes from the idea of the axion as a solution to the strong CP problem. The very elusive nature of the axion makes it possible to take it as a candidate for the similarly elusive dark matter, whose existence, however, seems to be necessary for the consistency of cosmology.

Notes

1 The title of his published paper was "Reliable perturbative results for strong interactions?" (Politzer, 1973).

2 In "the crudest nonrelativistic, weak-coupling approximation" to (8.7), they found "a potential $g^2(2\pi)^{-1}\sum_{i\neq j} r_{ij}^{-1}C_{iA}C_{jA}$, where the C_{iA} are the color octet SU(3) charges of the various quarks, antiquarks, and gluons. Then it is easy to envisage a situation in which the only states with deep attraction would be the color singlets."

3 The idea of a fourth quark was first introduced by Yasuo Hara (1964) to achieve a fundamental lepton–baryon symmetry in the weak and electromagnetic interactions, and to avoid the undesirable strangeness-changing neutral currents. It became known as the charm quark owing to the paper by Bjorken and Sheldon Glashow (1964) and was made central in the investigations of the unified theory of electroweak interactions in the early 1970s by the paper of Glashow, John Illiopolos, and Luciano Maiani (1970).

4 Here Λ is the hidden scale of QCD, which is related to a fixed mass of virtual gluon, at which the renormalization is performed. The zero mass of real gluons ($\Lambda^2 = 0$) cannot be taken because there the amplitude is infra-red divergent.
5 Since the potential $V(r) = -4\alpha_c/3r$ does not confine, it was soon to be suggested to use $V(r) = -4\alpha_c/3r + \sigma r$ instead, α_c and σ to be fixed by experiment (Harrington, Park, and Yildiz, 1975; Eichten, Gottfried, Kinoshita, Kogut, Lane, and Yan, 1975).
6 The new particles also include D mesons, which were discovered in 1976 at SPEAR (SLAC), and were immediately interpreted as naked charm particles (Goldhaber *et al.*, 1976; Peruzzi *et al.*, 1976).
7 Feynman suggested an outside–inside cascade mechanism, while Bjorken suggested an inside–outside cascade mechanism, which was further elaborated by Casher, Kogut, and Susskind (1973). See also Susskind, 1997 for these suggestions.
8 A more philosophical discussion on the criteria for a "direct" observation of a seemingly unobservable entity like quark and gluon will be given in Chapter 9.
9 According to Sau Lan Wu, when she went to Bergen to see Wiik and show him the just obtained result, "it turned out that Ellis was there too; after seeing my event, he described this event as 'gold-plated'" (Wu, 1997).
10 Another term "Euclidean-gauge solitons" was also used in the paper to refer to the same objects.
11 Although topologically different vacua are energetically degenerated, in the presence of light fermions, they are distinguishable by physical observable, the axial charge. The change in axial charge between two adjacent vacua connected by one instanton is, as we indicated earlier, $\Delta Q^5 = Q^5(t = +\infty) - Q^5(t = -\infty) = \int d^4x \partial_\mu j^5_\mu = 2N$.
12 This was testified by the original text: "the proponents of 'red, white, and blue' quarks [there were references to Bardeen, Fritzsch and Gell-Mann, 1972, and to Fritzsch and Gell-Mann, 1972] as a mathematical abstraction argue that the color SU(3) group should be exact, and that all noncolor singlets should be suppressed completely. One clearly requires a dynamical explanation of such a miracle. It might very well be the violent infrared singularities of an asymptotically free gauge theory provide the requisite dynamical mechanism" (Gross and Wilczek, 1973b).
13 The Higgs mechanism was attractive because the mass term for the gauge bosons generated thereby would prevent long-range force from occurring, and thus help to address the confinement problem.
14 I am grateful to Heinrich Leutwyler for bringing this observation to my attention in an interview on October 30, 2005.

9

Structural realism and the construction of QCD

The importance of underlying fundamental entities in theoretical sciences, which in most cases are hypothetical, unobservable, or even speculative in nature, such as quarks and gluons in QCD, from the realist perspective, and also from the hypothetic-deductive methodology, is clear and understandable. It has deep roots in human desire for explanation. However, it might be argued that the stress on the importance of underlying entities, which ground a reductive analysis of science, is in direct opposition to the holistic stance of structuralism. According to this stance, the empirical content of a scientific theory lies in the global correspondence between the theory and the phenomena in the domain under investigations at the structural level, which is cashed out with mathematical structures without any reference to the nature of the phenomena, either in terms of their intrinsic properties, or in terms of the underlying unobservable entities. Thus for structuralists no ontological interpretation of structure would be possible or even desirable. A structuralist, such as Bas van Fraassen (1997), would argue that if you insist to interpret the mathematical structure anyway, then different ontological interpretations would make no difference to science. That is, no ontological interpretation should be taken seriously.

Van Fraassen is a structural empiricist. How about structural realists, such as epistemic structural realist John Worrall, or ontic structural realists Steven French and James Ladyman? When these structural realists take issues with Kuhn's challenge to scientific realism, that the ontological discontinuity across theory-change has clearly undermined the realist claim that science gives us true knowledge of what exists and happens in the world, they usually stay at the structural level and take refuge in the continuity of mathematical structures without confronting the Kuhnian argument head-on, at the ontological and explanatory level. For Worrall, the ultimate nature and content of the world is unknowable; for French and Ladyman, structures, but nothing else, are the ultimate nature and content of the world. That is, for both of them, scientific revolutions as radical shifts in ontology don't exist. Thus their confrontation against the Kuhnians degenerates into a fake one by rejecting the very existence of revolutions in the history of science.

In my view, taking such an escapist strategy would not help, and the only way to effectively defend realism is to confront the Kuhnian argument at the ontological level, to give an acceptable account of the apparent ontological discontinuity while arguing for a deeper sense of continuity in the development of science. Such a line of reasoning, which will be taken below, heavily relies on the notions of ontology and of ontological synthesis. These notions, however, will be grounded on a structural approach to underlying entities, and thus the methodological tension between the reductive analysis of science based on the idea of fundamental entities and the holistic stance of structuralism will be reduced.

In this chapter, a structural realist account of the construction of QCD will be given, in which the construction of QCD will be shown to be the result of an ontological synthesis (Section 9.3). This will be preceded by a clarification of the place and roles of fundamental entities in scientific theories and some general discussions about the relationships among ontological categories (entities, properties, relations, processes, etc.), with a conclusion in favor of taking fundamental entities constituted by some individuating factors as the candidate category for the ontology of fundamental physical theories, such as QCD (Section 9.1), and an elaboration of the structural approach to fundamental entities in theoretical sciences (Section 9.2). The chapter will end with remarks on the reality and accessibility of what is structurally constructed (Section 9.4).

9.1 Fundamental entities in theoretical sciences

In the heyday of logical empiricism, when only experience and logic were the legitimate topics in the discussion of science, the assumption of underlying unobservable entities existing at the fundamental level, the so-called ontological commitment of a scientific theory, was regarded as part of metaphysics, and thus had to be expelled from science itself. Rudolf Carnap (1934), for example, took the ontological commitment of a scientific discourse as a question external to the linguistic framework of the scientific discourse, which thus would have no meaningful answer within science. More indicative of the *Zeitgeist* was the fact that Heisenberg's postulate that physicists should try "to establish a quantum-theoretical mechanics based entirely upon relations between observable quantities" (1925) had been hailed as the root of the success of quantum mechanics by many physicists and philosophers who were under the sway of logical empiricism.

With the demise of logical empiricism in the 1950s, however, the discussion of metaphysics reentered philosophical discussion of science. A systematic investigation of the role and function played by metaphysics in science can be found in John Watkins's work (1958, 1975). Thereafter, the role played by ontological commitments in theoretical sciences has attracted increasing attention from historians and philosophers of science (Cao, 1997).

It is an undeniable fact in the history of theoretical sciences that the notion of fundamental entity occupies a central place in their conceptual structure. They are specified by the ontological commitment of a scientific theory. While the notion of ontology generally refers to what exists in the world, an ontological commitment is a commitment to the existence of certain particular entities (or other ontological categories, more discussions on this below) which functions as the irreducible conceptual element in the logical construction of reality in the domain under investigation. In contrasting with appearance, and also opposed to mere heuristic and conventional devices in a theory, the fundamental entities are assumed to represent deep reality, and thus enjoy great explanatory power. That is, all appearances can be derived from them as a result of their behavior. Although the existence of a fundamental entity may be interrelated with other fundamental existence, such as quantum fields with the Minkowskian spacetime, its existence cannot be deduced from anything else within the theoretical discourse.

It is obvious that the notion of fundamental entity has a reductive connotation. Although a strong version of reductionism should be rejected if ontological emergence[1] is to be taken seriously, the fact remains that within a specific domain of scientific investigation, a reductive pursuit is always desirable and productive in terms of simplification, unification, and explanation. In this pursuit, ontological commitments play a crucial role, without which no theoretical science would be possible.

It is interesting to note that the above consideration works also in the reverse direction, and poses a severe constraint on the making of any ontological commitment, in terms of the explanatory function and causal efficacy. Formally, an explanation of an appearance at an upper level in a scientific theory can be defined by the appearance's reduction to, or derivation from, a lower, underlying level, which is described by the fundamental entities of the theory. Physically, however, this explanatory function of fundamental entities requires that only those entities that are causally effective enough to produce the appearance can be a candidate for the fundamental entities. In more physical terms, this requires that the candidate has to be in causal interaction with its like and with other entities; in other words, it has to be dynamical.

Thus, as a mental construction, the ontological commitment provides a bridge from observable phenomena to unobservable hypothetic entities within the framework of the hypothetic-deductive methodology. The risky side of this step is that science is moving away from its empirical warranty. The fertility of this step manifests itself in the stimulations it provides to the further theoretical and experimental investigations on the hypothetical entities, which is one of the major driving forces for the development of science. For this reason, although, according to empiricists, ontological commitments are "metaphysical" in nature, in fact they are not decoupled from conceptual and empirical investigations in science, and thus have to be judged on the merits of their utilities in constructing a workable science.

The origin of ontological commitments

It should be stressed that the ontological commitment of a theory is not a pregiven prejudice, which can be chosen at will. Rather, in normal cases, it has emerged, often quite unexpectedly, from scientific endeavor. We may make an even stronger claim that it is forced upon scientists by the internal logic of scientific investigations. Let us look at quantum field theory (references and more details can be found in Cao, 1997) and the theory of strong interactions as examples for illustration.

As a theory of interactions, which are supposed to be realized through the exchange of quanta, quantum field theory takes the violently fluctuating vacuum field as its basic entity, whose excitations result in real and virtual quanta, subject to various conservation laws and the uncertainty relations. But quantum field theory did not make this ontological commitment until the mid 1930s, a decade after its birth, and the commitment did not take a firm hold in the community until the late 1940s when the successes of renormalization and the implications of the Casimir effect were assimilated into the conceptual structure of quantum field theory.

At the early stage of quantum field theory, in Pascual Jordan's work from 1925 to 1928, or Paul Dirac's work on radiation in 1927, or in Pauli and Heisenberg's work of 1929, the major endeavor was to actively quantize a classical field, either the substantial classical electromagnetic field, or a fermion wave function which was unjustifiably taken as a classical entity. No attention was paid to the vacuum. An exception was Dirac's picture of the vacuum as the infinite sea of negative energy electrons. However, it was introduced in 1928, not in the context of quantum field theory, but in his relativistic theory of the electron. It was not necessary in the conceptual structure of quantum field theory, and was soon abandoned by Robert Oppenheimer and Wendell Furry, Pauli and Victor Weisskopf in 1934, and subsequently by many active players in quantum field theory, from Gregor Wentzel to Julian Schwinger to Steven Weinberg.

An important development was brought out in 1932–33 by Heisenberg in his work on the compound model of neutron, in which he developed the idea that the nuclear force consisted in the exchange of pseudo electrons. The idea was taken over and further developed by Ettore Majorana. Then a crucial step was taken by Enrico Fermi in his work on the beta decay of 1933, in which interactions were conceptualized, not in terms of the exchange of existing particles, but in terms of the couplings of fields, or the creation and annihilation of the relevant quanta at the interacting point. The tacitly assumed conceptual foundation of Fermi's theory was the vacuum fluctuations, which under certain constraints result in the creations and annihilation of real and virtual quanta. A crucial step in laying down this foundation was taken by Niels Bohr and Leon Rosenfeld in 1933 when they investigated the measurability of the field, which brought the idea of field fluctuations into the physics community. This, combined with Heisenberg's uncertainty relations, paved the way for the idea of local fluctuations of the field, whether it is in the vacuum state or in excited states.

But local fluctuations, according to the uncertainty relations, entail the divergence difficulty, which reveals the inconsistency of quantum field theory. Thus in the 1930s and 1940s no firm ontological commitment was made by physicists, and quantum field theory was in a dubious status. However, there was no other framework for dealing with electrodynamics in the atomic world. Thus intensive investigations were still conducted by many physicists, such as Weisskopf, Hendric Kramers, Ernest Stueckelberg, and Sidney Dancoff in the 1930s and Hans Bethe, Sin-itiro Tomonaga, Schwinger, Feynman, and Freeman Dyson in the 1940s, to find a way out of this difficulty. The result was, as Schwinger, among many others, argued convincingly that the concept of a fluctuating vacuum seemed to be a viable one if it was combined with the renormalization scheme developed in the late 1940s. More positively, the Casimir effect discovered in 1948 suggested that the fluctuating vacuum even had an observable effect. Only then, that is only at the end of 1940s, a firm ontological commitment to the fluctuating vacuum field was made, which signaled the maturity of the discipline.

More than six decades have passed since quantum field theory matured, but as far as its ontological commitment to the vacuum field is concerned, nothing has changed. It is true that the algebraic approach posed a challenge against the notion of local fields by arguing that the physical reality of quantum field theory can only be epistemically reconstructed from local observables of the theory, but not from unobservable fields, let alone the vacuum field (Haag, 1992; Schroer, 1997). The challenge, however, cannot be directed against the notion of a vacuum state of a field. In fact, the famous Reeh–Schlieder theorem (1961) convincingly demonstrates that the existence of a cyclic vacuum is the prerequisite for a coherent and consistent theoretical structure of QFT. In general, whatever the representation the algebra of observables chooses to represent physical processes, the representation, as claimed by Rudolf Haag himself (1992), has to satisfy a selection criterion, that is, elementary systems are localized excitations of the vacuum field. Thus the primary existence of the vacuum field is presumed by the algebraic approach itself, just as it is indispensable for, and thus is posited by, other approaches to quantum field theory.

In the case of strong interactions, the previous chapters were devoted to show that a commitment to taking quarks and gluons as the fundamental entities in the realm of strong interactions was reached only after more than two decades of intensive investigations, exploring various possibilities, from hadrons in meson theory and "eightfold way," to sub-hadronic particles in the Sakata model and the constituent quark model as well as the quark model for hadronic currents. The explorations were even extended to ontological categories other than entity, from processes (such as transition amplitudes and currents) to structures of processes (such as Regge pole structures and current algebra). Although Gell-Mann speculated on the idea of gluon as early as 1962, and on the idea of quark in 1964, it was not clear if he took the ideas seriously. He was serious in many senses, but certainly not ontologically. In any case, the gluons and quarks Gell-Mann speculated about in the early 1960s were fundamentally different from what have been committed to in QCD. It was shown in the previous chapters that there was an internal logic in the conceptual development concerning the strong interactions that dictated the ultimate commitment to the existence of quarks and gluons and their underlying explanatory role in our understanding of strong interactions. Similar to the case of quantum field, such an ontological commitment signaled the maturity of the discipline.

Further roles of ontological commitment

The specification by the ontological commitment of what hypothetic entities should be taken as fundamental and investigated has also provided resources for explaining the phenomena in the domain under investigations, dictated the conceptual structure of the theory itself, and provided stimulations and guiding principles for further theoretical and experimental researches, which force the relevant developments into a specific direction. As a result, the evolution and branching of the theory constitute a distinctive coherent research program, whose hard core is the ontological commitment.

In the case of quantum field theory, the specification of the fluctuating vacuum with its local excitations to be the basic entity for investigations requires further study about the unitary representation of the Poincaré group, the implications of virtual quanta, and renormalization for sidestepping the inconsistent consequence of local excitations of the vacuum. In a very strong sense, the theoretical structure of quantum field theory as specified by the axiomatic theorist, is dictated by the basic ontological commitment to a violently fluctuating vacuum, which is constrained only by conservation laws and the uncertainty principle. Various attempts to renormalize models of physical interactions beyond electromagnetic ones represent a clearly defined research program, whose hard core is the commitment to a locally fluctuating vacuum.

In the case of QCD, the situation is even clearer. The studies on asymptotic freedom and confinement were directly dictated by the commitment to the existence of quarks and gluons; and the lattice gauge theory, the quarkonium physics, the jet physics and fragmentation are all originated from the fundamental assumption of quarks and gluons, and thus belong to a distinctive research program.

The notion of ontological commitment is also crucial for our understanding of scientific revolutions. A scientific revolution (or a conceptual revolution in science) is characterized by a change in perspective when scientists look at the same domain of phenomena, and this change is always accompanied by or even resulted from a change of ontological commitment. For example, the change from the mechanical perspective to field-theoretical perspective in looking at physical interactions was accompanied by or resulted from the replacement of a commitment in which force was taken to be an independent entity, as in Newton's case, to that in which force was taken to be only an epiphenomenon of fundamental processes in which fields as fundamental entities are transmitted. If we take the geometrical view, such as what is assumed by John Wheeler, in which force is taken to be only a manifestation of the geometry of spacetime, then this view signals another conceptual revolution in looking at physical interactions. Of course, the view in which interactions are conceptualized as the processes of emission and absorption of quanta by interacting entities through exciting the general background of the vacuum signals is still another conceptual revolution. In the case of strong interactions, the change from the bootstrap perspective to the one of QCD resulted from the replacement of an ontological commitment to equally fundamental hadrons by a commitment to the sub-hadronic colored and fragmentally charged entities.

The notion of ontology is also helpful in conceptualizing progress in science. One might argue that if we accept conceptual revolutions as shifts in ontological commitment, and thus accept that theories before and after a revolution are incommensurable, then how can we claim continuity and progress in science across a revolution? Kuhn claims that discontinuity in science notwithstanding, progress in science can still be granted, although only in an instrumental sense in terms of the increase of our ability in solving puzzles, but not in the sense of the accumulation of objective knowledge of what exists in the world. In denying the second sense of progress, Kuhn firmly claimed that "I can see no coherent direction of ontological development" (1970). But as we will see in Section 9.2, progress in science can be claimed, with the help of the notion of ontological synthesis, in the sense denied by Kuhn.

Justifications for ontological commitments in science

All that has been said about ontological commitments can only be justified by the following assumptions about the physical reality. First, the physical world consists of entities that are all structured and/or involved in larger structures through their causally effective properties. Second, a thus resulted causal-hierarchical structure of a physical domain should be represented by the conceptual structure of a physical theory, whose multi-layered foundational structure has provided a ground for reduction. These assumptions give the ground for realists to read off a metaphysical structure of the world from the conceptual structure of a fundamental theory. Or the realist may go the other way round, trying to construct and readjust the conceptual structure of a theory by consulting the entrenched metaphysical picture of the world.

Various possibilities

Then what should be the exact metaphysical nature of what is ontologically committed in physical theories? The answer depends on specific theoretical context. They are individual objects, such as those in Newtonian mechanics; or non-individual objects, such as those in quantum mechanics that are numerically distinct but indistinguishable by any intrinsic properties or relations; or non-object physical entities, such as fields; they can also be other theoretical entities which themselves are not physical entities, such as processes, e.g. currents or transition amplitudes in particle physics, or some patterns or structures of processes, e.g. current algebra structures and Regge pole structures.

Although a process ontology is not inconceivable, and was in fact once taken seriously by some S-matrix theorists in high energy physics in the 1960s and early 1970s, it requires a stretching of the imagination to conceive a process as ontologically prior and causally effective in generating physical entities that would instantiate and substantiate the process, while it is only natural to conceive a dynamical process as a temporal sequence of a set of interacting physical entities. If physicists take this entity–ontology view of process seriously, then they would not stop and stare at dynamical processes and content themselves with their mathematical description. Rather, they would try to understand them in terms of underlying entities and their interactions, and take their knowledge of dynamical processes only as an epistemic access to the dynamical properties of the underlying entities.

However, entities can be conceived in various ways. An entity can be conceived as a bare particular with its own primitive thisness. A bare entity can be equipped with or deprived of its properties, put into or withdrawn from relations. This way, the entity enjoys ontological priority over its intrinsic properties and relations. But to conceive an entity having its unique individuality but existing alone without being engaged in any relations and being deprived of all possible properties is in contradiction with numerous experiences in science.

A more popular conception of entities is the bundle theory of entity, according to which an entity is nothing but a bundle of intrinsic properties and relations that always appear together, which provides the identity conditions for the entity. Here intrinsic properties and relations enjoy ontological priority over entities and constitute the latter.

It was further argued by some structuralists that in terms of human access no intrinsic properties of an entity can be shown to exist at the fundamental level. The argument is based on a version of the underdetermination thesis: physical properties can only be identified through the relations they are involved in; but identity of relations does not imply identity of intrinsic properties. So we can never know the properties of physical entities in so far as they are intrinsic. Here we are getting closer to a pure relation ontology, in a sense that the relata exist but they are constituted by the relational structure in which they are embedded, which in turn provides a ground for structuralism.

Before diving deep into the structuralist discourse in the next section, some general discussions on the relations among ontological categories (entities, properties and relational structures) seem to be in order.

Entities, properties, and structures

If we deny the existence of any lonely bare entity and assume that entities must be endowed with certain causally effective intrinsic properties and thus are inevitably engaged in various relations which are made possible by particular properties and concrete situations (in which relevant properties of entities are present so that particular relations obtain), we have to acknowledge that, generally speaking, while in a sense entities can be viewed as being constituted by properties they possess and relations they engage in, and properties can be claimed to exist only in relations in which they manifest themselves, there is no ground to argue that relations are ontologically prior to properties and entities, or that properties are ontologically prior to entities. The reason is simple: any relation must presume the existence of relata, and thus is constituted by relata (certain kinds of properties or entities with certain kinds of properties); and free floating properties are equally unimaginable as free floating relations. That is, as far as the ontological categories as categories are concerned, no category can claim priority over others; they are mutually constituted, co-existing, and thus enjoy equal ontological status.

Of course, the above claim about the categories *per se* does not mean that no priority claim can be made in concrete contingent situations. For example, it is perfectly reasonable to investigate if an instance of a particular category is ontologically prior to an instance of another. Surely an electron and a positron can be claimed to be ontologically prior to the structure (the relation in positronium) they formed; and the structural relation in Adler's sum rule can be claimed to be ontologically prior to the particles entering into an experiment in which the validity of the rule obtains.

More specifically, the existence and identity of a concrete property ("first order property") is constituted by its particular nomological (law-like) causal power which gives certain disposition for entities with the property to behave in certain ways and to form certain relations in given contexts. If we take a structure as a stable system of relations among a set of relata, then structural knowledge contains some knowledge of concrete properties (or part of the nature) of entities.[2] Some scholars argue that the nature of an entity cannot be exhausted by its structural relations because an entity always possesses its natures whether or not it is, at any given moment, engaged in all relations it is capable of, which is a contingent matter (Chakravartty, 2004). But if a property never manifests itself in any relation, then how can one be sure that there is such a property? Thus the claim for the structural constitution of properties and, as a consequence, for the structural approach to the nature of entities, is justifiable.

But then one may ask, what ontologically constitute structures? The question is essentially the same question as to which, a structure or its parts, should be taken to be ontologically prior over the other. The answer depends on specific situation. A structure, characterized only by its invariants, may be ontologically prior over its constituents (relata in the relational structure, either as unstructured raw stuff or as place-holders). But if a structure is formed by its constituents through a structuring agent and characterized by structuring laws, which govern the behavior of the constituents and hold them together to be a structure, then the constituents certainly enjoy ontological priority over the structure as a whole.

The difference in the ontological status of structures versus their constituents in the two situations has its root in different allocation of causal power. In the first situation, the constituents (such as spacetime points) are causally idle to the (spacetime) structure (constituted by the metric tensor); what is causally effective is the metric structure, which is thus constitutive of the existence and characteristics of the constituents (spacetime points) (Cao, 2006). In the second situation, however, it is the causally effective properties of the constituents (such as the charges of the electron and proton) that make it possible for a structure (the atomic structure of hydrogen) to be formed through the causal interactions of the constituents (the structuring of elements), while the hydrogen atom has no causal power over the existence and characteristics of the electron and proton.

In the first situation, the existence and characteristics of the constituents are derived from the structure. A holistic conception of structure derived from such a situation suggests that, in deliberations on an ontological commitment, unobservable constituents should be conceived as playing only a heuristic role in allowing the introduction of the structure. It is interesting to note that this metaphysical conception concerning the parts–whole relationship was actually taken by some prominent physicists as the conceptual foundation of their methodology. For example, bootstrappers took mesons as placeholders in the amplitude structures (Chew, 1961), and Gell-Mann, when he first introduced the idea of quarks (1964a), took quarks only as a conceptual device for generating the observable structures, namely the currents in hadronic processes, which would obey the structural relations of current algebra. The epistemological implication of this methodology, as clearly stated by Chew, was that physicists should be content with the investigations about the structure, and should not cherish unreasonable ambition of exploring deeper into the very existence of its constituents and their behavior.

In the second situation, the structure should be taken as being constructed from those conjectured to be ontologically primary ingredients. According to this componential conception of structure, the structure may enjoy some epistemic primacy because it has provided physicists with an epistemic access to its conjectured unobservable constituents. Ontologically, however, it enjoys only a derivative existence. It is interesting to note that this metaphysical conception was also taken by many particle physicists in the 1960s as the conceptual foundation of their methodology. Guided by this understanding of structure, these physicists tried to extract structural knowledge of the unobservable ingredients (quarks) from the explorations of the behaviors of the structures (currents). One prominent example of this methodology was Bjorken's successful inference from the behavior of the electromagnetic currents (namely scaling), to the claim about the structural features of sub-hadronic particles (namely partons): they are point-like and asymptotically free.

It is worth noting that the componential conception of structure remains within the structuralist discourse since the constituents of a structure are embedded in various structural relations and they have their own internal structures. Thus they can be approached from both within and without structurally. Certainly they can never exist by themselves alone. Still, the second case justifies the notion of *atomicity* at each and every level of componential structure. Clearly, a hydrogen atom is a structure of an electron and a proton glued together by electromagnetic forces; but the electron and proton enjoy a status of being the elementary ingredients of the hydrogen structure. The atomicity in this case, and more generally definable at any particular level of discourse, defies an untenable claim that structure enjoys ontological primacy over entities simply because entities can only be reconceptualized in structural terms (French and Ladyman, 2003a).[3]

The typical, also scientifically and philosophically interesting, case in the discussion of structure–constituents relationship is the case in which the structure is observable or epistemically accessible while the constituents are not yet. In this case we still have some knowledge about the "unobservable" constituents. But then, as widely recognized, this knowledge is structural in nature.[4] Some structuralists thus argue that we may have some relational knowledge concerning the places the constituents occupy and the functions they play in the structure, but we can never have precise knowledge about these ingredients and their intrinsic properties. But as early as 1783, Kant had already realized that although "the object in itself [in the sense of thing-in-itself] always remain unknown, but when, by the concepts of the understanding, the connection of the representations of the object, which are given by the object to our sensibility, is determined as universally valid, the object [in the sense of the phenomenal entity] is determined by this relation, and the judgment is objective."

Crucial to the mutual constitution of (fundamental) entity (as relata) and structure (as relation) is a proper understanding of the very nature of entity. As a causal agent an entity is endowed with a certain group of basic properties which dictate its nomological behaviors and thus render it to be embedded into various causal-hierarchical structures. The identities of different kinds of entities and the individuality of each member of a kind of entity are, for this reason, constituted by relevant groups of structural properties, which themselves belong to ontological categories different from the category of entity. Thus a kind of fundamental entity is constituted when a combination of basic factors (the constituting structural features) realized together as a being, while a concrete entity may be considered a nexus consisting of an inner core of tightly co-dependent structural features constituting the entity's essence and a corona of swappable adherent structural features allowing it to vary its features while remaining in existence (Simons, 1994).

Among structural features some structuralists ascribe a special role to symmetry principles and claim that only through symmetry principles as the higher rules that all other structural features would be held together to be an entity. In the context of particle physics, they claim that the basis for this line of thinking was provided by Wigner's association of an elementary particle with an irreducible representation of the spacetime symmetry group, which replaces the loose talk of entities as nodes in structure by a more formal group theoretical understanding in terms of invariants. The invariants or intrinsic and state-independent properties of particles (mass, spin and charges) are further claimed to be just the self-subsistent and permanent relations that Kant thinks would give us determinate objects (French and Krause, 2006).

But invariant properties, though necessary for determining the kind of entity, are not sufficient for distinguishing one member of the kind from other members, although in the quantum realm, the identification of kinds is more important than the individuation of particular entities, which do not have any different intrinsic features. A more important limit of the invariants view of entity, however, is that the reality of some important entities cannot be captured by invariant properties. For example, the Aharonov–Bohm effect (1959) clearly demonstrates that the gauge variant vector potential is a real physical entity. The same can also be said of many entities related to (spontaneous and anomalous) symmetry breakings in particle physics.

Ontological foundations of a fundamental theory

Although sometimes the situation is much simpler, it is better to keep in mind that generally the identity of an underlying entity committed by a fundamental theory is constituted by a group of basic factors (identifying structural features) which indicate the special kind the entity belongs to, combined with a group of qualificators which qualify the entity with variable features. Any change in the theoretical development involving more than a change of qualificator will inevitably lead to a change of the identity of the entity; the result is an ontological shift and a conceptual revolution in the domain.

Even if many important features of a fundamental theory (the major characteristics of its theoretical structure and its further development) can be captured by its ontological commitment to some fundamental entities, the ontological foundation of the theory may not be exhausted by a few kinds of dynamical entity. Rather, it may exhibit a complicated structure in which other components also play an irreducible and indispensable role. Take conventional quantum field theory as an example. In addition to its basic commitment to dynamical global fields (meaning they are defined over a global spacetime manifold), which are ever fluctuating, locally excitable, and quantum in nature (meaning their local excitations obey quantum principles, such as canonical commutation relations and uncertainty relations), it also has another component in its foundational structure, namely, a four dimensional Minkowskian spacetime manifold (or a non-dynamical curved background manifold justified by the equivalent principle). Such a manifold has a fixed chrono-geometrical structure, which underlies the notions of infinite degrees of freedom, localizability, and a global vacuum state, without which no understanding of the conceptual structure of quantum field theory, as formulated, e.g., in Wightman's or Haag's systems, would be possible.

Thus, properly understood, what characterizes the ontological approach to the understanding of theoretical science is not a fixation on a few underlying dynamical entities, but rather a detailed analysis of the ontological structure of the fundamental entities within the context of the ontological structure of a theory's conceptual foundations.

9.2 Structural approach to fundamental entities

Structuralism as an influential intellectual movement of the twentieth century has been advocated by Bertrand Russell, Rudolf Carnap, Nicholas Bourbaki, Noam Chomsky, Talcott Parsons, Claude Leve-Strauss, Jean Piaget, Louis Althusser, and Bas van Fraassen, among many others, and developed in various disciplines such as linguistics, mathematics, psychology, anthropology, sociology, and philosophy. As a method of enquiry, it takes a structure as a whole rather than its elements as the major or even the only legitimate subject for investigations. Here, a structure is defined either as a system of stable relations among a set of elements, or as a self-regulated whole under transformations, depending on the specific subject under consideration. The structuralist maintains that the character or even the reality of a whole is mainly determined by its structuring laws, and cannot be reduced to its parts; rather, the existence and essence of a part in the whole can only be defined through its place in the whole and its relations with other parts.

In the epistemically interesting cases involving unobservable entities, the structuralist usually argues that it is only the structure and the structural relations of its elements, rather than the elements themselves (properties or entities with properties) that are empirically accessible to us. It is obvious that such an anti-reductionist holistic stance has lent some support to phenomenalism. However, as an effort to combat compartmentalization, which urge is particularly strong in mathematics, linguistics, and anthropology, the structuralist also tries to uncover the unity among various appearances, in addition to invariance or stable correlation under transformations, which can help discover the deep reality embodied in deep structures. Furthermore, if we accept the attribution of reality to structures, then the antirealist implications of the underdetermination thesis [which claims that since evidence cannot uniquely determine (or, worse, can even support conflicting) theoretical claims about certain unobservable entities, no theoretical entities should be taken as representation of reality], is somewhat neutralized, because then we can talk about the realism of structures, or the reality of the structural features of unobservable entities exhibited in evidence, although we cannot directly talk about the reality of the entities themselves that are engaged in the structural relations. In fact, this realist implication of structuralism was one of the starting points of current interests in structural realism.

In the philosophy of science, structuralism can be traced back to Henri Poincaré's idea about the physics of the principles (1902). According to Poincaré, different from the physics of central force, which desired to discover the ultimate ingredients of the universe and the hidden mechanisms behind the phenomena, the physics of the principles, such as analytic dynamics and electrodynamics, aimed at formulating mathematical principles. These principles systematized experimental results achieved by more than two rival theories, and expressed the common empirical content and mathematical structure of these rival theories. Thus they were neutral to different theoretical interpretations, but susceptible to any of them. The indifference of the physics of the principles to ontological assumptions about the ultimate ingredients was approved by Poincaré, because it fitted rightly into his conventionalist view of ontology. Based on the history of geometry, Poincaré accepted no *a priori* ontology that was fixed absolutely and had its roots in our mind. For him, ontological assumptions were just metaphors, they were relative to our language or theory, thus would change when the change of language or theory is convenient for our description of nature.

But in the transition from an old theory with its ontology to a new one, some structural relations expressed by mathematical principles and formulations, in addition to empirical laws, may remain valid if they represented the true relations in the physical world. Poincaré's well-known search for invariant forms of physical laws has its roots in the core of his epistemology, namely, that we can have objective knowledge of the physical world, which, however, is structural in nature; we can grasp the structures of the world, but we can never reach the ultimate ingredients of the world.

The inclination for structuralism was reinforced by the rapid development of abstract modern physics (relativity theories and quantum theories) in the first quarter of the twentieth century, and was reflected in the writings of Moritz Schlick and Bertrand Russell. Schlick (1918) argued that we cannot intuit the unobservable entities of mathematical physics since they were not logical constructions of sense data, but we can grasp their structural features by implicit definitions, and this was all scientific knowledge required. In accord with this trend, Russell (1927) introduced objects through structural and implicit definitions because, he argued, the structures of the objective and phenomenal world must be the same although the essence of individuals were different; but these belonged to metaphysics, and had nothing to do with science. Russell thus claimed that in science we can only know structures, but not properties and essences of objects, and structures can be expressed by mathematical logic or set theory.[5]

The Ramsey sentence version

Russell's structuralism was reformulated by Grover Maxwell (1970a, b) in terms of the Ramsey sentence of scientific theory (constructed by replacing all the theoretical terms in a theory by second-order variables and then existentially quantifying over those variables), which is adopted by Worrall when he enters the realism/antirealism debate (1989, 2007). The Ramsey sentence of a theory $T(S_1,...S_n, O_1,...O_r)$, where the S_i are theoretical terms (including intrinsic properties and unobservable entities) and the O_j are observational terms, namely $\exists\Phi_1...\exists\Phi_n\{T(\Phi_1,...,\Phi_n, O_1,...O_r)\}$ "clearly asserts that the 'natural kinds' $\mathbf{S}_1, ..., \mathbf{S}_n$ (the extensions of the theoretical predicates $S_1, ... S_n$ in the initial theory T) exist in reality just as realists want to say" (Worrall, 2007). Here an unobservable entity whose place is held by a bound variable in a Ramsey sentence is whatever it is that satisfies the relations specified by the sentence. Thus we can know about fundamental theoretical notions (properties and entities) only by description, that is, only through their role in the theory in which they appeared. For example, an electron is whatever it is that satisfy our current theory of electron.

This means that in the Ramsey sentence formulation, although we may never know what the natures of theoretical entities are which have no directly observable properties, we still can assert their existence and relations they stand in because these entities are characterized in observable terms and exhibited themselves in the structures of the observable phenomena. According to Maxwell, this feature of the Ramsey sentence formulation has explicated Russell's idea that "our knowledge of the theoretical is limited to its purely structural characteristics and that we are ignorant concerning its intrinsic nature" (Maxwell, 1970b).

In the Ramsey-sentence version of structuralism, all terms in a theory, observable as well unobservable, are theory-laden. This means that it is impossible to establish a connection between a theory and the world in a term-by-term way, which would be possible only if (i) a term can be isolated from its context (the theory), and (ii) one can get a direct access to the world to make possible the comparison between a term in the theory and the world. Here the causal theory of reference is as impotent as the correspondence theory of reference. There is simply no way to ostend an unobservable entity in order to give it a name and then use it through a causal chain. Physicists could not know that they are ostending an unobservable entity (such as a phonon or a quark) without appealing to the theory in which the entity is described.

The aim of current structural realism in the history and philosophy of science, as Worrall (1989) conceives it, against the general background of the Kuhnian view of ontological discontinuity in scientific development, is to address the inter-theoretical relations directly in terms of mathematical structures without fundamental physical entity as a medium. The realist urge of this position is manifested in the attempt to establish a cognitive continuity in scientific development through a referential continuity between mathematical structures (involving law-like statements which go beyond knowledge of empirical regularities) used by physical theories at different historical stages, such as those used by Newton and Einstein. The holistic nature of the position is most clearly exhibited in its replacement of intrinsic properties of entities by relations, its rejection of atomistic metaphysics, according to which objects characterized by their essence and intrinsic properties exist independently of each other, and its taking entities as merely the names of images we substituted for the real objects as Worrall has insisted over years.

Thus although the intrinsic natures of unobservable entities were declared to be unknowable ontological content because all properties and entities have to be phrased in structural terms, which is in sharp contrast with traditional realism in which entities are conceived in non-structural categories (such as haecceity and substance), science is declared to be cumulative and progressive in terms of the discoveries of true relations between real objects and structures in the real world.

But if entities, including those central to a scientific theory (such as the ether for Fresnel's optics or the electromagnetic field for Maxwell's optics) are merely the names of images, then scientific revolutions existing and having played a great role in the evolution of science would be relegated into a status of illusion, and the position is left impotent in giving a convincing account of scientific revolutions. This is inevitable because the mathematical relations without physical interpretation, which as an additional ontological posit is a taboo to structuralism, are neutral to the nature of relata, and thus cannot exhaust the content of the relata. For example, classical mechanics and quantum mechanics share many mathematical relations and structures, and thus these relations and structures can tell us nothing about the classical or quantum nature of the physical entities under investigations. For this reason, concentrating on the shared mathematical structures, though helpful for conceiving the history of physics as a cumulative and progressive process, would render the quantum revolution invisible.

More troublesome is the fact that in the Ramsey-sentence version of structural realism the structure means the structure of observable content, but in order to structure the observable content properly, scientists need fundamental entities central to their theory. Scientists working in any fundamental theory simply cannot formulate the structure of their theory's empirical content (such as the Casimir effect or the three-jets phenomenon) without adopting certain explanatory fundamental entities, such as the vacuum field or quarks and gluons.

To be sure, the Ramsey sentence of a theory does show the way how reference to unobservable entities can be achieved purely by description, but interpreting Ramsifying a theory this way would bring the Kuhnian discontinuity back again: since the Ramsey sentence refers to exactly the same entities as the original theory, Ramsifying a theory does not help address the discontinuity problem at all.

In arguing against the item-by-item referential semantics, Worrall heavily relies on the argument that there is no way to have direct access to the object that exists in the world. That is true. But it is also true that no one can have direct access to the structure in the world either. Ultimately speaking, the only thing that is directly accessible to us would be sense data, not even percepts or concepts, let alone the structures of the world. If, cognitively, structures are accessible to us through our reasoning faculties, then the ingredients of the structure (properties, relations, entities) should also be accessible to us through similar reasoning faculties. The way in which the existence, comparison, and continuity of structures are argued can also be applied in the case of entities although all the reference to properties and entity should be formulated in structural terms.

This restraint in the way entities are conceived, however, should not be advertised as the dissolution of entities into structures, which, however, has been pursued passionately by the ontic structural realists in recent years. I will dispute the move below although at the same time I wish to register my appreciation of its laudable aim of removing a false dichotomy between the structure and the nature of the world: while the former is knowable, the latter remains mysterious in Worrall's version.

The ontic version

Different from the Ramsey-sentence version (or epistemic version) of structural realism, according to which unobservable entities exist but these entities and their metaphysical natures are unknowable, the ontic version denies the very existence of entities and claims that the metaphysical nature and ontological content of the world are nothing but structural relations, which are mathematically expressible and empirically accessible. If there is nothing that is over and above the structural relations, which bear all the ontological weight, then the notion of physical entities, at least those posited in quantum physics, could only be taken as a metaphor of our language and its real content does not go beyond the structural relations.

According to French and Ladyman (2003a), "physical entities are mere placeholders which play only a kind of heuristic role, allowing physicists to apply mathematics and hence getting them up to the group-theoretical structure; but once this is achieved, the entities can be dispensed with." Their most powerful rhetoric is: "if all the 'observable' properties of an object can be represented in structural terms, then what is the nature of the ontological residuum?" Thus they reject the idea of beginning with a "definitely determined entity" which possesses certain properties and then enters into definite relations with other entities, where these relations are expressed as laws of nature. Rather, since entities "cannot be regarded as prior to or ontologically separate from the structure which yields them analytically," we should begin with the laws which express the relations in terms of which the "entities" are constituted. From the ontic structuralist perspective, the entity "constitutes no longer the self-evident starting point but the final goal and end of the considerations."

In their elimination of entity for a pure structure ontology, French and Ladyman heavily rely on an equation of "lack of individuality" with "lack of reality." They first argued that since the metaphysical nature of quantum particles as individuals, which could be shown in some artificially designed formulations (French, 1998), or lacking individuality (as was shown by the quantum statistics in conventional formulation of quantum mechanics) was so important for something to be a physical entity, the situation of underdetermination had effectively deprived quantum particles of the status of being physical entities. The argument was soon extended to all properties and entities involved in holistic structures, which covers all that exists in the world if all structures are interpreted in an exclusively holistic way. Since relata of a holistic structure are only constituted by the structure and lack of any intrinsic nature, the argument for denying the existence of quantum entities was straightforwardly applied to all physical properties and entities in the microscopic world.

Although we have indicated earlier that there are two types of structure, componential and holistic, and the elements (relata) in the componential structure enjoy an ontological priority over the structure formed by the elements (which means that the structure derives its very existence from the existence of its elements), we have to examine the situation in holistic structures closely because what was evoked by the ontic structuralist was mathematical structure and mathematical structure by definition is holistic in nature. However, even in a holistic structure, the ontological priority of a structure over its elements only means that it is only the characteristics of the elements that are constituted by the structure, not that their existence is derived from the structure of which they are elements.[6]

Ontologically, in any holistic structure, the elements are always already embedded in the structure. The embedment has endowed the element with a contextual individuality, which confirms their reality according to the criterion of the ontic structural realism. Thus the existence or reality of the elements and their individuality cannot be separated. The abstract way of talking about placeholders without individuality makes sense only in the realm of epistemology, when we try to approach the individuality and thus the full reality of the elements in a holistic structure through its constituting agent, the structure itself or the structural features of the components dictated by the structure.

The above argument has underlain and justified a constructive version of structural realism in epistemology, according to which reality is structurally constructed step by step, as is vindicated by the construction of a tenable theory of QCD (more discussion on this shortly). It should be stressed, however, the process of construction is only an epistemic process of approaching reality, not a Platonic process of imposing structures upon an otherwise unstructured reality (formless and passive matter) from without.

The ontic structural realist tries to find a way of addressing Kuhn's discontinuity thesis concerning scientific development at the ontological level by rhetorically calling mathematical structures the only existence in the quantum world and dismissing the very possibility of having unobservable entities as the ontological foundation for scientific theories to ground their conceptual structures. If the very idea of entity is taken to be only a metaphor of our language or a heuristic device for the introduction of structures as the only ontology in the world, the ontological discontinuity can be easily rejected because it is not too difficult to argue for the continuity in terms of mathematical structures. But the price for this success is to give up the very idea of fundamental entities, which, as I have argued earlier, occupies a central place in theoretical sciences. Thus this is not a small price to pay.

Once physical entities are removed from the theoretical discourse, physical theories are reduced to their mathematical structures, and structures are understood as "relations without relata" (properties and entities carrying causal powers); no interpretation in terms of physical entities would be possible, or even desirable. In this case, the empirical content can only be smuggled in through a data model, the structure of which is targeted by the mathematical structure of the theory. Then what science can do is only to represent the empirical phenomena solely as embeddable in certain abstract structures (theoretical models), and these abstract structures are describable only up to structural isomorphism. The result is that all deliberations and elaborations on the inter-theoretical relationships, such as the shift from Newton to Einstein, and intra-theoretical relationships, such as those downwards from the theoretical models to the data models and upwards to the mathematical structures, in terms of partial structures and partial isomorphism are self-closed, holding only between the mathematical structures, and not between such structures and "the world" itself, as French and Ladyman (1999) candidly acknowledged.

More disturbing with ontic structural realism is that by removing underlying physical entities from consideration, it takes away an important task from physicists and philosophers of physics, the task of interpreting mathematical structures in terms of physical entities. The phenomenalist move was clearly declared by Howard Stein (1989): "interpretation in terms of 'entities' and 'attributes' can be seen to be highly dubious . . . I think the live problems concern the relation of the forms . . . to phenomena, rather than the relation of (putative) attributes to (putative) entities." Drawing inspiration from such a move, Ladyman concluded that "traditional realism should be replaced by an account that allows for a *global* relation between models and the world, which can support for predictive success of theories, but which does not supervene on the successful reference of theoretical terms to individual entities, or the truth of sentences involving them" (1998).

Such a phenomenalist move underlies a methodological holism and thus has the effect of preventing us from penetrating into the deeper layers of the physical world, and thus is somewhat counter-productive and even detrimental to the development of physics. The decline of S-matrix theory, which focused exclusively on the analytic structure of the overall amplitudes of hadronic physical processes, seems to have exemplified the counter-productive aspect of the holistic methodology.

However, there is also a potentially positive aspect in the ontic version. In denying the ontologically self-sustaining existence of intrinsic properties which are supposed to be possessed by lonely objects, the ontic structuralist rightly rejects the existence of both such properties and entities, and asserts that both can be exhaustively reconceptualized in structural terms. Here the relational conception of entities and properties is justified by appealing to Kant's idea that "all we know in matter is merely relations . . . but among these relations some are self-subsistent and permanent, and through these we are given a determinate object." Thus the ontic structural realist takes electric charge as just such a "self-subsistent and permanent relation" that justifies regarding the electron as a "determinate object" because it, like other intrinsic properties, features in the relevant laws of physics, although conventionally it is understood as an intrinsic or state-independent property of particles.

The ontological representation of the fundamental entities of physics in terms of structure leaves no unknowable mysterious entities lurking in the shadows as in the case of the epistemic version of structural realism.

But the claim by French and his fellow travelers that fundamental entities can be mathematically represented in an exhaustive way by sets of group-theoretical invariants, as was attempted by Weyl, Wigner, Piron, Jauch, and others, is not tenable. At least the case of the vector potential's reality derived from the Aharonov–Bohm effect mentioned earlier offers a clear counter-example. As will be clear shortly in the positive explication of the structural approach to fundamental entities, the reconceptualization of entities is much more complicated than a simple application of sets of group-theoretical invariants.

The constructive version

Different from both the epistemic version of structural realism, according to which only structural relations are knowable but the underlying entities are hidden for ever, and the ontic version, according to which the very existence of physical entities is denied because entities, at least in the realm of quantum physics, are dissolved into mathematical structures, the constructive version, based on the metaphysical understanding that concrete structures of primary scientific interest, which provide the basis for higher-order abstract structures, and their elements are ontologically inseparable and enjoy the same ontological status, has refined traditional realism and takes theoretical entities as referring to real entities existing in the world and knowable through our structural knowledge of the world.

But there is an important difference with traditional realism. Instead of conceiving the physical world as consisting of fixed natural kinds, as the traditional realist usually does, the constructive structural realist conceives the identity (or nature) of physical entities in a structural way, which opens a vast conceptual space for accommodating radical ontological changes during scientific revolutions while maintaining a realist sense of continuity of our knowledge of what exists and happens in the world.

One advantage of the underlying metaphysical assumption about the ontological inseparability of structures and their elements is that it gives some justification to the claim that a discovery of a deep structure, deep in the sense of having unifying power, would help to discover deep entities embedded in the deep structure. One might argue in favor of the advantage claimed that without the discovery of the flavor SU(3) symmetry in hadron physics, it would be unlikely that physicists would ever be able to discover quarks and gluons.

But then the question is how to spell out the way it helps, or the way we move from structures to fundamental entities.

Structures and structural knowledge of entities

In our construction of a theory in mathematical physics such as QCD, we have to use mathematical concepts and constructions, such as Hilbert space, local fields, symmetries, or even ghosts, because this is the only window through which we may have some access to the deep reality. Since typically our knowledge of deep reality takes the form of a set of structural statements (most of them are mathematical) about fundamental though hypothetical unobservable entities and their structuring, it is crucial to the claim (that both the reality of these entities and the objective knowledge of them are warranted by our objective knowledge of structural relations of the world) that the focus of attention should be shifted from mathematical structures as such to the structural knowledge of underlying entities endowed with causally effective and qualitatively distinct properties.

Here are a few remarks to justify the shift.

First, while a mathematical structure (of dynamical processes or other aspects) as such gives us only a holistic knowledge of a relational whole, the reference of structural statements of underlying entities is more than the whole itself, but also points to the hidden constituents of the structure, and thus provides an epistemic access to the underlying entities.

Second, while a mathematical structure is purely relational, and thus is causally inert and qualitatively indifferent, the structural statements of stable physical relations among the underlying entities have to invoke or at least assume something more than relational, namely some qualitatively distinct properties that are causally effective. Thus a door is open for hypothetico-deductive investigations of the intrinsic or even essential properties (such as mass, spin, and charge) of underlying entities. The investigations still have to be couched in relational and structural (mostly mathematical) terms, because no property, not even the intrinsic properties, of an entity can be known or defined without taking into consideration the network the entity is involved in. However, a physical property can only be defined within a physically special network, whose relational structure is interpreted in a physically special way and defined by physically relevant parameters. Thus, no structural statement of fundamental entities could be purely relational, but instead carries with it some special physical interpretation, which is defined in terms of physically specific and causally effective properties.

Third, different from speculations about unobservable entities, the structural knowledge is empirically accessible and thus can be made reliable.

However, since the door to any direct access to unobservable entities is closed, any conception of entities through structural knowledge has to be reconceived time and again with our ever-increasing structural knowledge. In addition, the structural construction of unobservable entities, although reliable, is fallible and subject to revisions. Thus, the attainment of objective knowledge at the level of underlying entities can only be realized through a historical process of negotiations among empirical investigators, theoretical deliberators, and metaphysical interpreters. This character of our approach to fundamental entities of theoretical sciences brought about by the shift is of crucial importance to the realist conceptualization of the history of science, as we will see shortly.

From structural knowledge to entities

But how can we be sure that we can really reach unobservable entities and their intrinsic properties through structural knowledge rather than stay within the confine of such knowledge?

The case for properties is easier to argue. Since concrete structures as specific relations of relata are determined by causally effective properties as their relata, we can make legitimate inference from knowledge of a structure to knowledge of a specific property with specific causal power which is responsible for the formation of the specific relations or structuring.

The case for entities is much more complicated. In order to show that entities of any kind can be approached through their internal and external structural properties and relations that are epistemically accessible, we might begin with the laws expressing the relations in terms of which the entities are constituted, or with invariant properties which are important structural properties (derived from symmetry groups) used widely to characterize entities in the quantum realm. But then the ontic structuralist would argue that since entities are determinable through laws and invariants, which however are ontologically prior to them, entities in fact enjoy only a status of metaphors, which are ontologically dissolved into mathematical relations expressing laws and symmetries.

Surely we can argue that invariant properties such as charge and mass are not sufficient for determining certain kinds of entities. For example, the positron has the same mass as the electron and the same charge as the proton. Thus more is required. That is, a certain relationship that designates the totality of invariant properties associated with a particular kind of entity, and thus is specific to a particular kind of entity, must be found out so that certain kinds of entities can be shown to be objectively constituted and existing in the physical world rather than mere metaphors.

Here a crucial notion is "constraints," which offers a major justification for the emergence of (relatively stable and objective) structural knowledge of the world from (purely subjective and ever fleeting) sensations that are the only things directly accessible to us, according to the sense data foundationalism within the camp of logical empiricism. Equipped with this notion, we can apply it to the new conceptual situation and argue that when certain constraints are satisfied by our structural knowledge, we may have objective knowledge of real unobservable entities[7] in the world. Then the crucial thing is to spell out the constraints necessary for the emergence of objective knowledge of unobservable entities from the structural statements about them, for justifying the introduction of these entities into discourse, and, ultimately, for justifying the reality of the entities constructed from structural knowledge.

When we have a set of empirically adequate and qualitatively distinct structural statements (all of which involve an unobservable entity and describe some of its features), the constraints on the organization of the given set necessary for the emergence and justifications mentioned above can be formulated as follows.

If within the given set there is a subset such that (i) it is stable within a configuration of the set and is reproducible in variations of the configuration; (ii) it occupies a central place (the core) in the configuration; (iii) it describes some physically specific features that can be interpreted as the intrinsic features of the entity, which are different from its accidental features described by those situated on the periphery of the configuration; among these features, (iiia) some of them (e.g., spin) are common to various physical entities, (iiib) others (e.g., fractional charges) are qualitatively specific to the entity (e.g., quark), thus can be taken as its essential features[8] and used as identifying references to characterize the entity and distinguish it from other entities; and (iv) some of its statements describe the causal efficacy of the intrinsic features (essential features in particular), and these causally effective features can be taken as a basis for explanation and prediction[9]; then we are justified to (i) introduce the unobservable entity constructed from the set of structural statements into our discourse; (ii) take it as ontologically inseparable from the structural properties described by each and every statement in the set responsible for the general mechanism (underlying empirical laws) that is resulted from these properties (especially dynamical properties); and (iii) take structural statements in the set as providing us with knowledge of the unobservable entity.

Holistic constitution

But the question remains as the ontic structuralist would ask: "what is the ontological content or metaphysical nature of a real entity that is over and above mathematical structures?"

A proper answer to the question is that the set of empirically adequate and qualitatively distinctive structural statements constituting a physical entity has a new feature that is absent in what is involved in each and every structural statement in the set. Different from an amalgamation of structural statements, which itself is structureless, the constitutive set is hierarchically structured. Most importantly, the set has a stable core subset which provides feature-placing facts about the hypothetical entity, and thus can be used as identifying references to the entity, which render the entity referentially identifiable (Strawson, 1959). As a crystallization of holistic characteristics of a hierarchically structured configuration of the set, of which the holistic characteristics are prescribed by a specific allocation of functions (essential or not) and places (core or periphery) to the statements involved (in addition to the coherent existence of the structured configuration at specifiable spacetime locations), the entity constituted is relatively stable against all changes except for those which have changed the function of core statements, or reallocated core and peripheral statements, and thus have changed the defining features of the configuration as a whole.

Thus the approach to unobservable entity introduced above is structuralist in nature for two reasons. First, epistemically, the ultimate reference of an unobservable entity is made to structural statements, the ontological justification of which is the inseparability of entity with the structural relations it is engaged in. Second, and more importantly, the approach is based on a way of holistic reasoning that is characteristic of the structuralist approach. In order to justify the conception of irreducible unobservable entity on the basis of a set of structural statements, and to use such a conception to give a consistent and coherent account of both ontological changes and ontological continuity (with the help of the notion of ontological synthesis, which will be introduced below) in scientific development, it is unavoidable to appeal to some sort of holistic reasoning as described above. However, different from meaning holism,[10] according to which any change in the configuration of a set of statements which defines the meaning (here the constitution of an entity) entails a radical change of the meaning (the characteristics of an entity),[11] the unobservable entity defined in this approach, in contrast, is relatively stable.

New type of natural kinds

In the structural approach, the entities, such as quarks, gluons, gauge and Higgs bosons, are constituted through structural relations displaying their intrinsic and essential (causally effective) features, and thus can be taken as natural kinds. They are *natural* kinds because, although they are conceived by scientists following the structural approach, the conception has to be approved by nature. A case in point is the Higgs boson, which has been conceived by physicists in structural terms (symmetry-breaking and mass-generation) for a long time, but has not yet been approved by nature.

To be sure, the traditional scientific realist thinking in terms of Aristotle's realist ontology may also use structural terms to formulate their notion of an entity. But a thus conceived entity is a member of a preexisting and fixed natural kind. In contrast, an entity conceived in the structural approach is not a member or an instance of a fixed natural kind, but rather a manifestation of a historically constructed, revisable natural kind, which is subject to reconstitution time and again.

The source of flexibility and openness characteristic of the new type of natural kinds lies in the unavoidable changes of the configuration of the set of structural relations (statements) from which a *kind* is constituted (conceived). With the increase of structural knowledge (statements), the reallocation of some core and peripheral statements, and the change of the function of some core statements (describing essential features or not), the defining features of the configuration change accordingly. As a result, the identifying references, or the content, or the characteristic features, metaphysical or otherwise, of a *kind* also change. That is, what is constituted and thus conceived is a different *kind* from the original one. The completion of such a process of reconfiguration is the substance of a scientific revolution through which theorists have changed their ontological commitment and thus the ontological character of the whole theory.

It is worth stressing that the constitution and reconstitution mentioned above are more than steps in an epistemic process. Rather, they have references to an objective ongoing reconfiguration of the natural world as a hierarchy of entities, which can be accessed by scientists in various ways, depending on contexts and perspectives.

Here the notion of objectivity is not one that is detached from human involvement, which is an illusion. Rather, it is defined in terms of nature's resistance to any arbitrary human construction. Take the case of QCD as an example. The ingredients of hadrons were conceived in various ways. Along one line of conception, they were first conceived through a certain set of structural knowledge to be partons, then partons were reconceived as quarks and gluons (see Sections 5.3 and 6.1). With the importation of the structural constraints posed by the notion of color into the "current quark" picture, they were reconceived again to be colored quarks and gluons (see Sections 7.1 and 7.3), which conception was approved by nature and accepted by the community. Along another line of thought, the ingredients of hadrons were conceived to be integrally charged ones (Section 7.1), which was not approved by nature and thus was forgotten by the community. All these were the result of human construction in terms of structural relations, but the objectivity of some conceptions and constructions rather than others is warranted by nature's approving and disapproving responses. For this reason the construction and reconstruction discussed here cannot be dismissed as purely subjective moves.

Referential continuity and ontological synthesis

The recognition of the variability of structurally constituted natural kinds has provided a theoretical resource for accommodating both the apparent ontological discontinuity and a deeper sense of referential continuity in the development of science, which seems to be the best possible way to address the Kuhnian challenge to scientific realism.

Is it legitimate to talk about the referential continuity of an unobservable entity, an electron for example, across radical changes in a series of theories about the entity? There are various theories of reference trying to establish connections between theoretical terms and what exist in the world. Most popular among them are two: one tries to fix the reference through descriptions, the other appeals to causal roles.

One difficulty with the description line is that theoretical descriptions of an entity change as theory changes, (for example, the description of electrons changed greatly from Thompson's theory to Rutherford's, Bohr's, Heisenberg's, and Dirac's theory), yet we continue to refer to the "same" entity (the electron). Since we don't have direct access to electrons, and the notion electron makes sense only within a particular theoretical context, surely the "sameness" of the electron cannot be justified by radically different descriptions in different theories.

Some scholars argue that the notion of the ether in fact refers to the field since it plays the same causal roles as the field does if its mechanical properties are excluded. This way, the referential continuity between the ether and the field can be assured across the theory change that happened in the mid to late nineteenth century. But in the old conception of natural-kind entities with fixed essence, once the mechanical properties (which are essential properties of the mechanical ether) are excluded, the ether as an *entity* would be abandoned, although certain properties featuring in certain structural relations would be preserved and retained in subsequent theories, where they feature in relevant laws. This means that the notion of causal role is of no help in arguing for the referential continuity of an *entity* (in the traditional sense) across a scientific revolution.

From the perspective of the structural approach to the new type of entities, however, the referential continuity of an entity can be argued by appealing to the notion of reconfiguration discussed above. In the new configuration associated with the new entity constituted thereby that appears after a conceptual revolution, the retained structural features from the old configuration retain their constitutive roles in the new context, although their places (at the core or periphery) and functions (identifying-features-placing or not) would have changed.

The continuity may appear in three different ways. First, if some identifying-features-placing structural statements, such as the lightest mass and smallest negative charge in an atom in the case of an electron, are retained in the new configuration, then no matter how radical a change that happened between theories across conceptual revolutions, such as those between Thompson's theory and Dirac's theory, it is justifiable to say that physicists are basically talking about the same electron. The justification is provided completely within the discourse of structuralism. The notion of the electron as an entity is completely formulated in structural terms because the notions of its essential features (mass and charge) can only be defined in relational terms: no entity would have any mass if it exists lonely in the world in which no other masses exist and thus no gravitational relations with other masses exist; the same can be said of the notion of charge.

Second, it may happen that the expansion and reconfiguration of structural knowledge of an entity and other entities in the domain under investigation results in a change of the ontological status (primary or derivative) of the entity. In the case of strong interactions, for example, the pion in Yakawa's theory was the primary agent for the strong interaction, but later was relegated into a status of epiphenomenon in the quark model and QCD. However, a change of status does not deny its existence, and the referential continuity in this case cannot be denied.

But the referential continuity may also be realized through a mechanism of ontological synthesis, which, different from the two ways mentioned above that can be accepted without too much reflection, is comprehensible only from the perspective of the structural approach to entity. If there are two distinctive configurations of structural statements, each of which is responsible for constituting a distinctive entity, and if an empirically adequate combination of one (or more) constitutive structural statement(s) from the core subset of each configuration constitute a new core subset in an enlarged and/or reconfigured set of structural statements, then the new configuration with a new core subset may be responsible for constituting a new entity, which, if approved by nature, would be a case of ontological synthesis.[12]

Ontological synthesis is by no means a uniquely fixed epistemic process in theoretical construction, nor a uniquely fixed recipe for a natural process happening in the world. Rather, it has various ways to instantiate with two given configurations. For example, with one configuration responsible for constituting the kind of gravitational field, the other for the kind of quantum boson field, the result of ontological synthesis may be a kind of quantum gauge field (Cao, 1997), or a kind of quantum gravitational field (Cao, 2001, 2006). The notion of ontological synthesis will play an important role in our understanding of the construction of QCD, as will be clear in the next section.

9.3 The construction of QCD

The structural approach to fundamental entities offers a realist account for theory construction in fundamental physics. As we noticed in Section 9.1, the central task for constructing a fundamental theory is to lay down its conceptual foundation and clarify what its fundamental entities are. Physicists have no epistemic access to the hypothetical fundamental entities except for some hints from the constraints posed by existing more or less empirically adequate and conceptually well-reasoned theories and reliable observational data. In the case of QCD, these refer to the available empirical data about hadron spectrum, Regge trajectories, the $\pi \rightarrow 2\gamma$ decay, scaling, the ratio of total cross sections for muon and hadron production; and to the theoretical notions of bootstrapping hadrons, eightfold way, current algebra, current quarks, constituent quarks, and partons. How to proceed from hints to a formulation of the conceptual foundation for a new fundamental theory like QCD is a formidable challenge. The challenge in the case of QCD, however, was met satisfactorily by a group of physicists who adopted a method that was in line with the notion of ontological synthesis discussed above. This section is devoted to showing that crucial to the construction of QCD is a synthesis involving structural constraints that constitute the kinds of bootstrapping hadrons, current quarks, constituent quarks, and partons.

According to the bootstrap understanding of hadron physics, hadrons are placeholders in the network of relations without being independently definable relata. Indeed, when Chew questioned the conceptual foundations of the sum rule in current algebra, closure or completeness and locality, his objection aimed directly at the very notion of atomicity, which, with its connotation of indivisibility, grounds the notion of point-like locality, and with that of finiteness, grounds the notion of closure. But atomicity is not a tenable notion, according to Chew, because with impinging energy and momentum transfer sufficiently large, a proton or any other hadron would reveal its internal structure. There is no end to dividing, and thus no fundamental substructure is definable. All are composite, none elementary. So the structural constraint constituting the bootstrapping hadrons is that hadrons are pure structures without any atomicity involved (see Section 4.3).

Chew was wrong, as we indicated in Section 4.4, because at any particular stage of human practice and conception, atomicity is perfectly definable. More importantly, the endless division Chew appealed to is not necessarily the division at the same (hadron) level, but may involve entities at the sub-hadron level. Still, Chew's understanding posed a strong constraint on the construction of QCD, and, in fact, in a transformed version in a new context, it was involved in the ontological synthesis and thus the construction of QCD, as we will see shortly.

Much more important were the developments originated from the research program of exploiting the implications of symmetry to high energy physics. Here the expansion of the empirically vindicated isospin SU(2) symmetry to the hypothetical and hierarchically broken unitary SU(3) symmetry set in motion consequential explorations.

The guiding idea of the symmetry school was to extract useful physical information, namely structural knowledge about the relevant hadrons, from the broken symmetry. For this purpose, Gell-Mann and others derived mass formulae, sum rules, and selection rules – and then submitted them to experimental examinations – from a hierarchy of symmetry breakings that were suggested by phenomenological observations, although they had no idea about the mechanisms responsible for such breakings. Here what were involved, from the structuralist perspective, were hadrons' external structural features that characterize the external relations between hadrons rather than their internal structural constitutions and constraints.

The mathematical structure of the broken SU(3) symmetry suggested a deep unity of hadrons. From the octet classifications for baryons and mesons, Gell-Mann moved forward to explore the properties of the hadronic electroweak currents, while Ne'eman, stimulated by Sakurai, tried an exact SU(3) gauge symmetry of strong interactions. However, at that stage of explorations, the fundamental fermions and boson, including Ne'eman's gauge bosons, were baryon and vector meson octet. With hindsight, it is understandable that similar to Pauli's failure a decade before, Ne'eman's exploration of gauge symmetry at the hadron level would lead to nowhere.

However, Gell-Mann's move, in which the application of the SU(3) unitary symmetry was extended from hadron spectrum to its application to the weak and electromagnetic interactions of hadrons in terms of currents, was more consequential. If a current is conserved, then there is a symmetry. The U(1) symmetry for electromagnetism; the SU(2) symmetry for the electroweak isovector currents (due to CVC); then the hypothetical SU(3) symmetry for the hypothetical unitary currents. Thus currents as mediations for energy exchange in the electroweak processes became representations of the SU(3), and the result was current algebra.

What current algebra described were abstract structural features of a local field theory, which were hoped to be valid in a future true theory. The whole scheme was only a mathematical structure among currents. Since currents expressed the structures of physical processes of some unknown physical entities instead of being physical entities themselves, current algebra can be viewed as a structure of structures.

Then what was the physical relevance of such a mathematical structure? Physically, it was only a *conjecture* to be tested. It is worth noting that although the ingredients of structures at the level of current algebra and at the level of currents were uncertain when the scheme was proposed, the scheme was taken to be the underlying structure of the hadron system, and thus to be an expression of the external structural features of hadrons.

In a sense current algebra is a paradigmatic application of the structuralist methodology. General structural features were derived from the hypothetical symmetry of a field-theoretical model without respect to the details of the model. However, once currents rather than hadrons were assumed to be responsible for hadronic processes, which signaled a subtle shift in ontological commitment from hadrons to currents, a door was open for further investigations of the (external and internal) structural properties of currents. Thus although the approach at first was phenomenological with a seemingly evasive nature (namely avoiding the ontological and dynamical questions in terms of specific entities), the wedge provided by structural knowledge, as we will see, would open a path for exploring deeper layers.

Physicists had extracted many structural properties ("algebraic relations") from current algebra relating various hadronic processes although it was not clear what these currents actually were. The currents could be derived from various models, the sigma model as well as the quark model, for example. Generally these models as heuristic devices were not taken realistically. The quark model, however, did provide an opportunity for exploring the ingredients of hadrons although at first the quarks were not taken to be real entities.

What Gell-Mann originally wanted from the idea of quarks was to use them in the field-theoretical model to abstract presumably valid algebraic relations among hadronic processes, that is, to take quarks as placeholders in currents as structures that satisfy all the structural constraints posed by current algebra. Conceptually, however, the proposal of quarks made it possible for a transition from seeking knowledge of structures to seeking structural knowledge of entities underlying the structures. However, substantial progress had to be made for turning this possibility to reality.

Quarks were introduced not as hadrons' ingredients existing on their own, but were assumed to be enmeshed in a network of relations. Along one line, contrary to Gell-Mann's structural pursuit, quarks were assumed to be the real constituents of hadrons, from which the hadron spectrum can be derived. Along another line initiated by Gell-Mann himself, they were defined by currents (as relational structures) to be their placeholding relata, and their relations to hadrons were not clear at all initially. It was only through the accumulation of the structural features of phenomena derived from current algebra that their relevance to hadrons was gradually revealed. For this reason, this line of research exemplified the structural approach to entities and to theory creation.

A preliminary step in moving from exploring the *external* structural features of hadrons to exploring the *internal* structural features was taken by Adler, suggested by Gell-Mann, in his derivation of the local current algebra sum rule. The Adler sum rule expresses the general structural constraints on the processes where leptons were used to explore the structural features of hadrons. As Bjorken rightly acknowledged, Adler's sum rule was the conceptual foundation on which significant developments were built up. However, the nature of the structural constraints, external or internal, posed by Adler's sum rule itself remained unclear. This was so mainly because the nature of the intermediate states, hadronic states or sub-hadronic states, in the sum rule was not clear.

The crucial step was taken by Bjorken in his scaling hypothesis, guided by his "historical analogy," which clearly transformed Adler's constraints to internal structural constraints: hadrons consisted of point-like and asymptotically free ingredients. This is a move, crucial in the structural approach to entities, which turned the focus of physical research from seeking knowledge of structures to seeking structural knowledge of entities underlying structures. After this step was taken by the high energy physics community in 1969, the structural knowledge of underlying entities accumulated rapidly at SLAC and many other high energy physics laboratories worldwide, and at many institutions of theoretical researches. The most important achievements in this category were closely related to scaling, which led to the idea of partons and a bundle of structural features of partons. In addition to the point-likeness and asymptotic freedom, it was also firmly established that half of the partons in a hadron had quantum numbers of quarks, the other half did not (more in Section 6.1). The knowledge revealed structural features constituting the ingredients of hadrons, and thus was crucial for the conception of the constituents and the construction of QCD.

But quarks and gluons in QCD were not constituted by structural features that constituted partons. In fact, they were jointly constituted by factors that constituted partons and factors that constituted other entities.

More specifically, the notion of color, which carried a huge amount of information about the structural constraints between hadrons and their constituents, and played a crucial role in constituting quarks and gluons as constructed in QCD, had its origin in the SU(6) constituent quark model, but was not a constituting factor for the current quark and partons prior to the advent of QCD. It was transported from the constituent quark model to the current quark model (see Section 7.1), and the transportation was not mechanical but transformative: it was no longer confined within the context of the SU(6) model but could be applied to situations in other contexts, such as the electron–positron processes, which had nothing to do with SU(6) symmetry.

But the color, together with all the structural constraints related to it, imported from the constituent quark model was not really the color in constituting the quarks and gluons in QCD. The latter was the result of another ontological synthesis. That is, it was a combination of the color from the constituent quark model and an additional factor constituting the bootstrapping hadrons, namely the structural constraint that no parts of hadrons can exist in an asymptotic state. In the context in which quarks were conceived to be colored, the color singlet restriction to the asymptotic states of the physical world equaled the conservation of color in asymptotic states of the world. With such a constitutive constraint of color conservation, the path was opened for Fritzsch and Gell-Mann to take the decisive step in the construction of quarks and gluons, namely, to take gluons as forming a color octet and take the gluon octet as the gauge bosons of the gaugeized exact color SU(3) symmetry. Once this step was taken, the construction of the conceptual foundation of QCD was completed.

The notion of ontological synthesis helps us to comprehend the rich structure of scientific developments (both in normal science and during scientific revolutions). In addition to simple discard and retention of fundamental entities across theory changes, the notion offers a mechanism which can accommodate the apparent ontological discontinuity (the new fundamental entities replaced old ones), and at the same time reveal a deep continuity in our knowledge of what *exists* in the world in terms of factors that constitute new types of natural kind entities existing before and after radical theory changes. More interesting in the context of this project, however, is its utility in helping us to understand how a new fundamental theory is created. In addition to the creation of QCD, the combination of the constraints from the factors constituting quarks and those from Reggeism resulted in another complicated set of constraints centered with the notion of duality, which became the constitutive factors of string even before QCD was constructed. It seems that in a deep sense structural realism is now undergoing a phase transition, a transition from the old phase, in which the major concern was with theory changes, to the new phase in which the major concern is with the construction of new fundamental theories, whose methodological guidance and conceptual-historical process can be clearly analyzed in terms of ontological synthesis.

9.4 Remarks

The structural approach to hypothetical underlying entities and to the construction of fundamental theories has brought a few long-lasting questions concerning the accessibility to unobservable reality, the reality of thus-constructed entities, and the nature of knowledge thus produced to the foreground, so some remarks on them seem to be in order.

Accessibility

It is clear that the old constructive empiricist distinction between observables and unobservables is simply impotent in addressing contemporary scientific endeavors, especially in analyzing the construction of fundamental theories, such as QCD, string theory, and quantum gravity. For this reason, this empiricist distinction carries no weight at all in the discussion of human accessibility to the so-called "unobservables." In fact, the whole idea of constructive structural realism originated from the recognition of special features in our access to "unobservable" physical reality.

There are two sources of access. First, the foundational constraints posed by predecessor theories, whose domains wholly or partially overlap with the domain of intended investigations, have provided a major epistemic access to the unobservable reality we intend to investigate and conceptualize in the theory under construction, because these constraints have encoded all reliable structural knowledge humans have acquired in the past that is crystallized in the predecessor theories. The very idea of ontological synthesis is critically based on the recognition of such a source of access.

Second, as Gell-Mann repeatedly reminded the physics community, some structural constraints, he called them "algebraic structure," are in principle knowable. The accessibility of structural knowledge is presumed by structuralists and is widely accepted based on scientific experience.

With these two windows of access, together with other means for cognitive access, such as experimental equipment as well as conceptual, theoretical, and mathematical apparatus, microscopic entities, such as electroweak gauge bosons, quarks, and gluons, are cognitively accessible enough for us to claim the reality of structurally constructed entities.

Of course, some philosophers (French and Ladyman, 2003a) have argued that there are some metaphysical natures of microscopic entities, which can be explicated in terms of such fundamental categories as identity and individuality but are cognitively completely inaccessible in modern physics. Since these metaphysical natures, they further argue, are necessary components in the full constitution of a microscopic entity according to traditional metaphysics, without which no theoretical entity can be taken as a real entity, we cannot really claim that we have full cognitive access to microscopic entities. If these natures are cognitively inaccessible in principle, however, then no matter how fundamental they were in traditional metaphysics, they are irrelevant for modern metaphysics as well as modern physics. Metaphysics has to be modified with the advancement of physics. So these categories have to be removed from contemporary metaphysics. As far as the factors physically (as well as metaphysically[13]) relevant for constituting an entity are concerned, none of them is cognitively inaccessible in principle.

Reality

Are quarks real? For a long time and up to the very moment when the conceptual foundation of QCD was first articulated at the 16th International Conference on High Energy Physics held in Chicago, September 1973, Gell-Mann had maintained that quarks did not have to be real; real quarks were not required by his theory because they did not fill any obviously theoretical need. "Real or not" made no difference to him when he simply wanted to use a mathematical field theory model containing a quark triplet to abstract algebraic relation.

Worse, Gell-Mann even claimed that quarks could not be real. "If quarks are real, then we cannot assign them para-Fermi statistics of rank 3, since that is said to violate the factoring of the S-matrix for distant subsystems" (Fritzsch and Gell-Mann, 1971b). But without assigning such a para-statistics to quarks, quark theorists would be in deep trouble, facing a direct conflict with numerous empirical facts, from the $\pi \rightarrow 2\gamma$ decay to the ratio of total cross sections for muon and hadron production, even if they ignore the troubles in the attempt to derive the hadron spectrum from the SU(6) model.

As we will see in the next chapter, a radical revision of the criteria for being a real particle was under way in the course of the construction and acceptance of QCD by the physics community. For now it would be enough to point out that for Gell-Mann at the time of constructing quarks and gluons, a real particle meant a particle that could exist asymptotically and thus was detectable in the laboratory. Once the criteria have been revised, physicists are fully confident of the physical reality of quarks. For example, Detlev Buchholz (1994, 1996) convincingly argues that the physical reality of quarks, gluons, and colors can be uncovered from the algebra of observables (which is a set of empirically accessible relations describing the structural features of the hypothetical entities), through applying renormalization group transformations to the local observables, as the intrinsic features of the underlying theory based on the algebra of observables. This means, epistemically, we can proceed from our knowledge of structural relationship of observables (the algebra of observables) and reach the physical reality of microscopic entities (quarks and gluons), which can be metaphysically conceived, in line with the Ramsey sentence metaphysics of reference, as the structures of empirical content ("observables").

Since the evidential status of theoretical entities is encoded in theory, the reality of quarks cannot be separated from the validity of QCD within which they are constructed. In case QCD is confirmed, quarks are real, real within the framework of QCD. What if QCD is replaced by a more fundamental theory? This would not deprive quarks and gluons of reality, but only change their ontological status from primary (the underlying reality) to derivative (epiphenomena). This means our objective knowledge of reality is continuously accumulating even if the continuity is punctuated from time to time with the replacement of fundamental entities.

Objectivity and historicity

The structural approach to underlying entities has several interesting features worth commenting on. According to the approach, the underlying entities in the hierarchically organized reality can be approached by theoretical entities, which themselves are constructed by historically acquired knowledge of their structural features. This is the objectivity of historically constructed knowledge of the underlying reality.

On the other hand, the structural approach also attributes a constructive and historical character to the objectivity of concepts or conceptual structure of scientific theories (such as charm, jets, scaling, confinement, and asymptotic freedom) in general, and of the theories' ontological commitments to underlying entities (such as quarks and gluons) in particular. This historicity of objective knowledge prevents the constructive process from ending. As a result, the reality of the underlying entities thus approached can only be partial but not total or complete, extending only to the extent of the information about the structural constraints the entities carry with them confirmed so far, and thus is variable but not fixed. And this has provided an empirical basis for the introduction of a new metaphysical category, the new type of natural kind that is not fixed but variable as we discussed earlier.

Notes

1 Examples of ontological rather then epistemic emergence can be found in the emergence of mind from body, or beauty from the arrangement of colors and shapes.
2 For example, structural knowledge about invariants under symmetry transformations contains knowledge about the intrinsic properties of a physical entity system, such as mass, spin, and various kinds of charge. Although these invariant properties are structurally derived, they are in fact the structural features of the entities, not something external to the relevant entities to be imposed upon the entities.
3 One may protest that the argument here can only be applied to concrete structures, and the situation involving abstract structures would be radically different. But abstract structures are scientifically relevant only when they are physically interpreted. That is, they as higher order properties have to be instantiated by concrete structures. Of course, the multiple instantiations present a further problem. But this is a problem that demands a separate solution. More on this in Section 9.2
4 See, e.g., Schilick, 1918; Russell, 1927; Carnap, 1928; Maxwell, 1970a, b; and Worral, 1989. Please note that there has been an evolution from taking structure as a higher order property of relations to the realization that structures have to be concrete to be scientifically relevant. More on this in Section 9.2.
5 Russell's structuralism was criticized by Newman (1928) that Russell's structure has no empirical content of a domain except the information of its cardinality. Russell himself accepted the criticism and retreated from the position. Worrall (2007) points out that if Russell's structure is partially interpreted in observational terms, as it should be since Russell's "whole structural realist view was based on a sharp distinction between theoretical and observational notions (in his own terms between things known by acquaintance and things known only by description)," then the Newman argument would be irrelevant. To this kind of "diluted" version of structuralism, Michael Redhead (2001) responds that such an interpretation would amount to an additional "ontological posit, but all we have epistemic warrant for is the second order structure."
6 It can be argued that even though quantum particles and relata in a holistic structure are not identifiable (not individuals), they remain as numerically distinct objects of predication.
7 Strictly speaking, what is involved in the discussion of this subtle issue is not unobservable *entities* (e.g., "red" "up"*quarks and electrons*), but unobservable *kinds* of entities (e.g., "red" "up" *quark and electron)* or kinds of kinds of entities (e.g. quark and lepton). This observation applies to all subsequent discussion: whenever an "*entity*" is mentioned, it is just for convenience and should be understood as a "*kind*" of entities or a kind of kinds of entities, depending on specific context.
8 The way in which some intrinsic features are taken to be essential while others are not is highly theory-dependent; or one may say that some important characteristics of a theory as a

whole are dictated by the way some and not other features of its fundamental entity are treated as essential. This complicated intertwining between two levels of theory construction has important bearings on the structuralist approach to ontology and ontological shifts, as we will see shortly.

9 For example, the quark's such causally effective intrinsic features as "point-like structure" (meaning a structure whose internal construction cannot be probed by available experimental energies), "free of interactions at short distances" (a dynamic feature rooted in the quark's other intrinsic feature, carrying an SU(3) color charge) and "fractional electric charge" have provided a basis for predicting and explaining the Bjorken scaling in the deep inelastic scattering experiments observed at SLAC in 1969.

10 Meaning holism is entailed in "reconceptualising objects in structural terms," pursued by French and Ladyman (2003a).

11 "Radical change" means that meanings or entities before and after change are incommensurable.

12 Due to the change of context, the original constituting structural statements have to be partially transformed. For detailed discussion in the case of quantum gravity, see Cao, 2001, 2006.

13 Metaphysical factors in the constitution of an entity refer to the holistic characteristics (constraints) of the set of physical factors (structural constraints) discussed in Section 9.2.

10

Structural realism and the construction of the CA–QCD narrative

Writing a conceptual history of QCD is exciting. The materials of the project are not natural phenomena of nuclear forces, which are inherently meaningless occurrences; not quasi-independent abstract ideas, whose causal effectiveness in moving history forward only idealist historians would believe; but human activities, in which meaning was radically changing with the change of perspective, and whose long-term significance for physics, metaphysics, and culture in general is to be discerned and interpreted by historians of science.

Historians' interpretations, however, are conditioned and constrained by the wide cultural *milieu*. Broadly speaking, at the center of contemporary cultural debate on issues related to science, such as objectivity and progress in science or the nature of scientific knowledge and its historical changes, sit two closely related questions. First, can science provide us with objective knowledge of the world? Second, is the evolution of science progressive in nature in the sense that it involves the accumulation of objective knowledge? The old wisdom that science aims at discovering truths is seriously challenged by such prominent commentators as Richard Lewontin and Arthur Fine. According to Lewontin (1998), taking science as a reality-driven enterprise has missed the essential point of science as a social activity, namely, its ambiguity and complexity caused by the socio-historical setting that has put severe constraints on the thought and action of scientists. In his Presidential Address at the Central Division of the American Philosophical Association, Fine (1998) has skillfully dissolved the notion of objectivity and reduced it to a democratic procedure, which may contribute to enhance our trust in the product of scientific endeavor, but has nothing to do with the objective knowledge the product may provide. More radical claims, such as "science is only an art of persuasion, manipulation, and manufacturing facts and knowledge," "knowledge is only a power move, having nothing to do with objective truth," "objectivity is only an ideology, having nothing to do with how scientific knowledge is socially constructed," are rampant in the profession of science studies.

Philosophically, the reason that objectivity and progress in science have become issues in the debate is closely related to two theses that were made popular mainly by W. V. Quine and Thomas Kuhn. The first thesis claims that scientific theories are underdetermined by empirical evidences (Chapter 1). One interpretation is that our knowledge of the world expressed in scientific theories is by no means an objective representation, supported by reliable evidence, of what the world actually is, but rather as conventions, that are constructed and dictated by our language, culture and other social factors, for conveniently explaining the evidence.

The second thesis is Kuhn's special interpretation of scientific revolutions involving radical change of conceptual framework, according to which concepts and statements across a conceptual revolution are incommensurable. The incommensurability thesis entails that the history of science is discontinuous and that no theory can be taken as true or even survivable. Kuhn (1970) claims that "I can see no coherent direction of ontological development" in the history of science. If Kuhn was right, then there would be no cognitive progress in the history of science, in the sense of the accumulation of our objective knowledge of what exists and happens in the world, although a kind of instrumental progress, in terms of our ability in solving puzzles, would still be imaginable. The antirealist implication of Kuhn's view can be best seen through Hilary Putnam's meta-induction thesis: if no theory in the history can be taken as true from the viewpoint of later theories, then there is no reason to believe that our present theories would enjoy any privilege over their predecessors (Putnam, 1978).

Thus gone with the continuous accumulation of knowledge are the ideas of progress and objectivity, and the result is the rampancy of relativist claims, which gives strong support to the postmodernists in their fight against the idea of progress and even rationality itself.

In the debate against relativism in science studies, or more generally in the attempt to develop a realist conception of science and a cognitively progressive conception of history of science, appealing to formal logic and empiricism is not a great help. According to Carnap (1950, 1956), formal logic is unable to address what he calls the external questions that are related to radical changes of conceptual framework. And it is a truism that empiricism has no theoretical resource to deal with the underdetermination thesis, which challenges the status of empirical evidence as a bridge connecting theoretical entities and physical reality. But as we have shown in the previous chapters, with a structural understanding of ontology (what exists in the world) and a realist understanding of structural knowledge, we would be able to, first, make a realist claim that we could have objective knowledge of even the hidden-for-ever underlying entities such as quarks and gluons; second, understand the radical changes of ontological commitment in the development of science in terms of the accumulation and reconfiguration of recognizable, cumulative, and modifiable structural statements that are constitutive of our conception of ontology; and third, recognize a pattern in some cluster of scientific revolutions in terms of ontological synthesis.

This means that structural realism is highly relevant to the history of science. This chapter is devoted to a closer look at this relevance in the context of composing the CA–QCD narrative.

10.1 Objectivity and progress in science

Major controversies about progress in science are phrased in cognitive rather than instrumental terms; and cognitive progress is defined in terms of the accumulation of objective knowledge of the world; thus different views on the objectivity of knowledge ground the contending views on progress.

Objectivity: from Kant to Kuhn

Most, if not all, modern and postmodern conceptions of objectivity, including the structural conception explored in Chapter 9, have their roots in, or are critical responses to, Kant's philosophy. According to Kant, objective experience is possible only when subjective perception is submitted to a priori concepts of the understanding (such as, substance as the permanent substratum underlying all changes, causality dictating the changes of phenomena, etc.), and reordered according to the universal laws of nature prescribed by the understanding; the resultant schematization of the pure concepts of the understanding when they are applied to the spatio-temporal forms of sensibility provides an objective metaphysical foundation for natural sciences or our knowledge about nature in general. Since the origin of the universal order of nature is the understanding and the unity of objects is entirely determined by the understanding, it is possible, with the help of the concepts of the understanding, to have objective knowledge of the world. Kant himself remarked that although "the object in itself always remain unknown, but when, by the concepts of the understanding, the connection of the representations of the object, which are given by the object to our sensibility, is determined as universally valid, the object is determined by this relation, and the judgment is objective" (1783).

Two transcendental notions constituted Kant's conception of objectivity, the notion of a priori concepts of the understanding and the notion of spatio-temporal forms of intuition, were seriously challenged by the developments in science since the early nineteenth century and soon collapsed. The responses to their collapse constituted the major part of the intellectual history of the conception of objectivity (Friedman, 1992, 1999).

Kant realized that when a pure concept, such as substance, was applied to supersensible objects, such as souls, the result would not be objective knowledge but dialectical illusion. Such pure use of reason, transcending all possible sense experience, according to Kant, must be criticized because in this case the reality of concepts can never be determined. Then what about the applicability of pure concepts to microscopic entities and processes which are inaccessible to intuition? If in classical physics the visualization of microscopic entities and processes was still imaginable, then the conceptual developments in quantum physics had completely destroyed any possibility of spatio-temporal visualization of the microscopic entities and processes.

A more serious challenge to Kant's conception of objectivity targeted his apriorism. With the discovery of non-Euclidean geometries and the rise of classical field theory in the nineteenth century, many *a priori* principles (such as the axioms of Euclidean geometry and those about an absolute reference framework and action at a distance) that Kant took as indispensable for the construction of objective experience were discarded. With the collapse of apriorism, the role played by a priori concepts of the understanding in constituting our experience and its meaning has been taken over by historically constructed contingent conceptual frameworks. As a result, the universality and necessity of our experience and knowledge entailed by Kant's conception of objectivity, or even the very idea of objectivity, become questionable.

The rejection of Kant's apriorism, however, was not immediately followed by the rejection of objectivity. Rather, objectivity was retained, but Kant's schematism was replaced by a modified notion of symbolism first developed by Kant in his *Critique of Judgment*, in which Kant distinguished Schematism (a demonstrative, direct presentation of the concept in intuition responsible for scientific knowledge) from Symbolism[1] (an analogous and indirect presentation without corresponding sensible intuition, unsuitable for scientific knowledge, and can only be used by judgment). As a historical fact, following Goethe, Humboldt, and Helmholtz, who took all forms of knowledge (sensation, language, art, and science) as symbolic in nature rather than representations of things, Hertz, Cassirer, Husserl, Schlick, and many other thinkers in their attempt to modify Kant's conception of science, frequently appealed to symbolic rather than representative interpretation of science. This symbolism movement provided a background for the rise of Hilbert's formalism and Poincaré's conventionalism, both of which proposed new and consequential conceptions of objectivity.

Hilbert's formalist program was motivated by a special conception of objectivity. For him, a complete formalization required only axioms, implicit definitions, and internal consistency, leaving no room for intuition or experience. One of its entailments was a notion of structural objectivity divorced from experience. This was the legacy inherited and further elaborated by Schlick, Carnap, and others, with all of its ramifications, including the temptation of using verification to reestablish the connection between thought and experience, which was provided by Kant's pure intuition, but was removed by Hilbert's complete formalization.

Poincaré took experience more seriously than Hilbert. But for him, our objective (in the sense of neutral to different interpretations) experience is possible only for the structures of the world, but can never reach the nature and content of objects (the ontological basis of the structures).

Facing the failure of spatial-temporal intuition in quantum theory, Bohr (1925), heavily drawing on the symbolic tradition of German thought, appealed to a symbolic interpretation of its theoretical structure for grasping its objective character, in the sense of unambiguous communication of experience.

In *Aufbau* (1928) Carnap declared that "science wants to speak about what is objective," but, following Hilbert and Poincaré, the objectivity of science can only be captured by restricting it to purely formal and structural statements. Under the influence of Wittgenstein's *Tractatus*, Carnap in his *Logical Syntax of Language* of 1934 took scientific theories as languages. Equipped with new tools of modern logic and set theory, he distinguished meta-language from object-language. While each scientific theory discusses its subject within an object-language (with its existential or ontological commitment), philosophy uses logic as a meta-language to formulate and investigate the syntactical rules (formation and transformation) and structure (axioms and rules of inference) of all object-languages or linguistic frameworks, thus is neutral, in terms of ontological commitment, towards them.

For Carnap at that time, the objective meaning of an expression, no matter whether it was a logical expression or a descriptive (empirical) one, was not derived from any direct experience. Rather, its meaning was solely determined by its purely syntactical behavior in a language. Closely related with this holistic conception of meaning was a rejection of an absolute notion of a fixed reality, independent of any linguistic structure: the questions of the reality of entities or the truth of statements, or even the empirical facts, the so-called internal questions, can only be answered within a linguistic framework; but there was no way to discuss the reality of new entities outside of a certain framework, or the reality of a linguistic framework itself (the so-called external questions). According to Carnap, the external questions are pseudo-questions of metaphysical nature (and had to be expelled from theoretical discussion) because the discussion of which used both the object-language and the meta-language at the same time, thus no meaningful answer would be possible.

Within the framework of *Logical Syntax*, in terms of object-language and meta-language, a transition from one object-language to a new one is possible, but the decision for taking a new language is made purely on pragmatic considerations. With the change of objective-language would come radical changes in our interpretations of statements and our understanding of the phenomenal domain under investigation, so radical that they would not be translatable from or into the old ones. These changes notwithstanding, logical syntax as the meta-language (or a universal requirement) and the core of scientific methodology, would remain the same and serve as the basis of objectivity and rationality in science.

The problems with Carnap's system are many, and I only mention two here. First, since no conceptual system can be completely formalized without introducing any non-logical terms, Carnap's structure-based conception of objectivity that is divorced from experience and his quasi-Kantian notion of syntax that is immune from revision are not tenable. Second, his separation of internal questions (rational and cognitively significant) from external questions (practically relevant yet without any cognitive significance) makes it impossible to take radical changes in science as cognitively significant and having revealed the progressive nature of scientific development.

Contrary to common wisdom, Kuhn's notion of objectivity is quite similar to Carnap's and is closely related with his notion of conceptual frameworks (his terminology evolved from paradigm and disciplinary matrix to lexical structure). Different from Kant's notion of *a priori* concepts of the understanding, Kuhn's conceptual framework refers to historically constructed structures, in which facts and concepts are interwoven into a hermeneutical circle by which nature is discussed and understood. Each of these frameworks has its own ontological commitment, own problems and rules, own meaning and reference for concepts, own standards of what is to be explained and the forms of explanation. Thus they are holistic in nature and incommensurable in the sense that they cannot be translated into each other item by item faithfully. The change of allegiance from one framework to another is the core of scientific revolutions.

In clarifying the meaning of the external world we are dealing with, Kuhn (2000) denies that he is a nominalist who divides individuals into kinds at will. Kuhn acknowledges the existence of a stable and permanent world which, however, similar to Kant's thing-in-itself, is ineffable, indescribable, and undiscussible. For Kuhn, a phenomenal world is constituted, in a quasi-Kantian sense, by a given lexical structure consisting of patterns of similarity/difference relations shared by members of a linguistic community, which makes communication possible and binds members of the community together. Kuhn maintains that a structured lexicon is an embodiment of the stable part of human knowledge about the world, which underlies all the processes of differentiation and change, thus is a precondition for describing the world and cognitively evaluating truth claims, and serves as an Archimedean platform for human knowledge.

Insofar as the structure of the world can be experienced and the experience communicated, it is constituted by the structure of the lexicon of the community inhabited in the world. Yet the world is not constructed by the inhabitants, because, first, it is experientially given. That is, people born into a world constituted by a lexical structure must take it as they find it: it is entirely solid, not in the least respectful of an observer's wishes and desires, quite capable of providing decisive evidence against invented hypotheses which fail to match its behavior. Second, although people are able to interact with the world and change both it and themselves in the process, what they can change is not the whole world, but some aspects of it, with the balance remaining as before.

Thus for Kuhn, both the world with its structure and the subject with his ability to act are historically situated: they are empirically given and practically changeable. Kuhn's notion of objectivity is entirely constrained by his notions of the world and subject, and contains two components: intersubjectivity in terms of community consensus resulting from interactions (communications) among members of the community, and the constraints posed by the empirically given world structured by a given lexical structure. From our explorations in Chapter 9, it is not difficult to see that Kuhn's notion of objectivity was a historical and conceptual background, both positively and negatively, against which the structuralist conception of objectivity has been developed.

The structuralist conception of objectivity and progress

The structuralist conception of objectivity, as a response to Kuhn's conception, was based on two general positions. Negatively, it agrees with Kuhn and argues that objectivity in the sense of mirroring nature should be rejected for two reasons. First, science is not a subjective representation of an independently existing object. Rather, it is a result of interactions between the plastic, changeable, and transformable object and a similarly plastic and adaptable subject. In addition to intervention (experiment), our understanding of an object also involves interpretation (theory) and application (industry, technology, or social project). All these make mirroring impossible. Second, rooted in the nature of the knowing subject appears the mediative character of science. Simply put, the knowing subject is neither a tabula rasa, nor a disengaged rational being. Rather, biologically he has inherited a set of cognitive faculties which, as a result of adaptation, enable him to internalize and interpret the outside world in an increasingly appropriate way. Socially, the root of his identity is in a community which is delimited by a set of shared factors, including language, tradition, culture, and a network of economic and political structures. Thus knowledge is always mediated by these pregiven specificities of the epistemic subject, and this mediation mechanism makes mirroring impossible.

Positively, the notion of objectivity in science can be defined in the following way. First, science is knowledge of an outside world. As a reflective construction from internalizing the environment, which as an essential part of adaptation for survival is a precondition for the subject's existence, science is empirically increasing its adequacy. Second, although the object is not immune to change and transformation by the action of the subject, different from absolute fluidity, this plasticity of the object (which is defined in terms of the relationship between a subject and an object) has its limit at any given moment, which is historically set jointly by the pregiven state of the object and by the subject's ability to penetrate into the object and change its state. Thus the intervention of the subject based on structural knowledge will meet resistance if it reaches the limit of the object's plasticity.

The ontological limit to the plasticity of object has its epistemological counterpart. That is, our knowledge is inevitably constrained by an outside world. In some cases, the constraints can be very rigid. In general, the limits of the plasticity of the outside world explain the continuity of the scientific enterprise: even at the most dramatic revolutionary moments in science, most of our knowledge always remains unchanged, although nothing is unchangeable in principle or in the long run. Thus both foundationalism and social constructivism have to be rejected. Foundationalism has to be rejected because it is incompatible with the historicity of objectivity, which is constituted by a historically defined epistemic subject (rather than a transcendental subject) with his objective structural knowledge; social constructivism has to be rejected because it is incompatible with the idea of nature's resistance in terms of its structural responses to human conception and action, which has provided the ontological basis for objectivity.

Nature's structural responses will sooner or later be internalized by scientists who will readjust their conception and action accordingly. The ongoing feedback process manifests itself in the historically constructed increasingly objective knowledge, as we elaborated in the last chapter.

With such a structural realist understanding of the dynamics between nature and human conception and action, the cognitive progress in science can be defined in terms of the accumulation of objective knowledge of what exists in the world; and with the help of a structural understanding of ontology and ontological synthesis, the seemingly difficult question about the cognitive nature of conceptual revolutions in science can be answered in terms of the accumulation of structural knowledge and its reconfigurations, as we have discussed in the last chapter. This answer clearly confirms the progressive nature of conceptual revolutions in science.

10.2 Historical enquiries of science

History of science as a special kind of historical enquiry should pay attention to the characteristic feature of science that distinguishes science from other human knowledge. Our understanding of science, however, has undergone a radical change in the last half century (Hesse, 1980).

Scientific enquiries

Traditionally, it was claimed that the truth status of a scientific statement (an explanatory hypothesis or a universal law), in terms of Tarski's correspondence between the statement and what happens in the world, can be fixed objectively and uniquely by empirical testing, either through observation or by experiment. Since the 1950s, however, it was realized that observational and experimental data have to be selected and interpreted by theory or hypothesis. This theory-ladenness of data does not mean that theory is not constrained by data obtained in empirical testing, and thus no arbitrariness is entailed. Still, the data/evidence/experience foundationalism could not be rescued, and a hypothesis–data hermeneutical circle has forcefully presented itself as a reigning paradigm in the understanding of science as a system of meaning production.

Another time-honored observation that was brought to the forefront in the understanding of science in the 1950s is that no hypothesis or theory can be uniquely determined by empirical data. This underdetermination thesis entails the existence of a multitude of hypotheses, which are empirically equally adequate. How can we objectively determine which in the multitude is true in the sense that it describes the reality while others are not? Or, if there is no effective means to differentiate the true theory from untrue but empirically equally adequate theories, are we going to be forced to accept the bizarre notion of a multitude of realities?

Some philosophers tried to appeal to other criteria, the so-called coherence conditions to establish the uniqueness of the true theory. Coherence conditions, such as rational postulates about consistency and simplicity, or certain categorical properties of space, time, matter, and causality, are not directly derived from empirical testing, but constrain the type of theory that could be permitted. The trouble with this strategy is that coherence conditions themselves are, *pace* Kant, not uniquely fixed, but changing with the development of science and mathematics.

The worst for the traditional view of science came from historical study, which has ruthlessly established that all well-established true scientific theories, in the course of time, have been, sooner or later, replaced by others, and no one can survive from being somewhat discredited. It was Kuhn who had pushed the notion of scientific revolutions (paradigm shifts) to the farthest and tried hard to establish an incommensurability thesis, which claims that the worlds described by theories before and after a scientific revolution are incommensurable, each of them has its own true theory for its objective description. Philosophers and historians, let alone scientists, have felt threatened that such a radical version of relativism has completely undermined the very notion of objectivity in science. Numerous efforts have been made to evade the logical force of Kuhn's argument and other allied arguments mentioned above, but the old empiricist and naïve realist views of science are dead, and a new post-empiricist image of science has emerged. The new conception of science, in terms of its internal operations and external relations, can be summarized as follows.

First, as a system of knowledge production, science is a pragmatic learning device in dealing with the empirical world, with effective self-correcting mechanisms. Internally, experience is organized by hypothesis and empirical data are selected and interpreted by hypothesis, while the hypothesis is suggested and constrained by the data available. Data and hypothesis are thus mutually constitutive and mutually corrective, forming a hermeneutical circle in interpreting the meaning of experience and producing the understanding of the world. But the circle is not closed: data are forever expanding due to continuing human praxis; and hypothesis, as an expression of the myth or metaphysics of a society (e.g., bootstrap for democracy, symmetry for equality, unification for a nation state, or divisibility for class struggle, etc.) and thus as part of the internal communication system of that society, is in constant feedback with other cultural expressions of the society.

Secondly, as a social institution, science functions through various exclusively organized communities that are embedded into a society, from which they get problems to solve, the means for solving problems (funding, organization, equipments, ideas, metaphors, hypotheses, and theories for conceptualizing the world), and for which they produce acceptable solutions. To a great extent, a scientific community and a Kuhnian paradigm coincide. A scientific community, which is taken by the social constructivist to be the major determining factor in the production of scientific knowledge, can be used as the institutional incarnation of a paradigm to support Kuhn's cognitive relativism; while a Kuhnian paradigm is the major intellectual resource for the internal cohesion of a community.[2]

Finally, in contrast with other systems of knowledge production, science is characterized by its ultimate subjection to a pragmatic criterion, namely its instrumental effectiveness in dealing with the empirical world. It is required that any knowledge, in order to be scientific, has to be able to explain, predict, or even control the empirical world. This criterion is a very effective mechanism for filtering out all subjective elements in science, such as the subjective interests, values, and biases that are involved in scientific hypotheses and judgments, and the built-in conceptual relativism in its organization of knowledge within a conceptual framework. In terms of the filtering-out mechanisms, and only in relation to these mechanisms, we can talk about reality, truth, and objectivity of scientific knowledge. This implies that, first, the notions of truth and ontology (what exists in the world) are internal to science. Second, the notion of truth is not defined in terms of correspondence between knowledge and reality, but by being effective in terms of explanation, prediction, and control of the empirical world; the notion of reality is not characterized and classified by natural kinds with fixed essence, but rather by variable natural kinds as we discussed in the last chapter; and the notion of objectivity is not taken to be absolute, its opposition to arbitrariness can only be defined partially and historically.

The impact of philosophical understanding of science on the practice of history of science is beyond doubt. Sometimes, historians could go so far as to take history of science as a means to illustrate philosophical points. For example, given the underdetermination thesis, it is epistemologically interesting to see how any scientific conclusion can be reached with a limited amount of empirical information, or when experiments should end and a hypothesis should be accepted (Galison, 1987).

Historical enquiry

History of science is not science, but a historical enquiry with its own special features.

No historical enquiry can start with "what happened in the past," because "what happened in the past" is not in historians' immediate experience, and can only be known as a result of historical enquiry. Rather, it has to start from what has survived to historians' reach, that is, from those data that have continued to exist in the present as documents or as ideas, conventions, and institutions.

But data for historical enquiry are almost always both too many and too few. Too many so that we have to select relevant, interesting, and informative ones from numerous noises; too few for a meaningful picture of what actually or even plausibly happened in the past so that we have to fill the gap with our reconstructive efforts. How can a historian select from what is available to him and reconstruct a narrative of what actually or plausibly happened in the past for understanding and communication without some guiding ideas, namely some presuppositions about what happened in the past? Thus there is no way for a historian to escape from taking some working hypotheses in historical enquiry, just as physical hypotheses are necessary for theoretical physicists to reconstruct an unobservable layer of the physical world from the observable data.

Data selected by hypotheses for historical enquiry must be historically relevant. That is, they are neither merely philosophically relevant, relevant only to some abstract concepts or ideas that were not historically effective; nor merely socially relevant, relevant only to some social problems and their possible solutions without a clear connection between these problems and solutions of the present to those in the past. The historical relevance of data means that these data are relevant to the understanding of the connection between the past and the present.

If both data and hypotheses are necessary in historical enquiry, what we have said about scientific enquiry, namely underdetermination and a multitude of realities as its entailment and the data–hypothesis hermeneutic circle in meaning fixing, are all applicable to historical enquiry, and the old empiricist, positivist, and naïve realist views of historiography cannot demand serious respect anymore.

Surely, historical and scientific enquiries cannot be all the same in terms of their hypothetico-reconstructivist and interpretive methodology. The most significant differences have their roots in the difference of the subject matter of two disciplines. Science deals with inherently meaningless natural occurrences, which are more or less causally related to each other and thus have relatively stable patterns. In contrast, historiography deals with ever-changing human actions guided by actors' understanding of the meaning of events in the environment and by their idiosyncratic strategies. The feedback loops in human actions, which are characterized by strategic considerations of an agent about the perceived strategic considerations of other agents, have rendered causal analysis, which presumes a clear cut separation between a cause and an effect, useful only to a very limited extent. With causality relegated into a secondary place in historical enquiry, causal explanation and prediction follow suit. As a result, no meaningful appeal to the pragmatic criterion of instrumental effectiveness, that characterizes scientific enquiry, can be made for filtering out subjective elements in the enquiry so that the notion of objective truth about reality can be properly defined.

Then what is the specific content of historical knowledge that is produced by historical enquiry and what is the nature of the content? Certainly, all kinds of historical knowledge involve information about facts and evidences of whether certain events occurred, involve general laws and causal explanations of certain events and processes. But this kind of information is only secondary to historical enquiry, because the real aim of historical enquiry is to understand the character of events, the meaning of information, and in particular the reason of human action. Under certain conditions within a social structure, certain human actions and social events are followed by others. Some of them may become understandable by causal explanation, others through structural analysis. But not all desirable understanding of what happened in the past can be obtained this way if we take the unavoidable contingent human agency into consideration. For the same reason, both teleology and idealism have to be rejected, because human reasoning, that is the ultimate driving force for what actually happened in the past, is quite different from atemporal rationality, no matter whether it is divine wisdom or otherwise.

Any real understanding of the character of events, meaning of information and reason for human action a good piece of historical knowledge may supply can only be found in the narrative structure of scattered events. That is, the real content of historical knowledge lies in the significant whole of a narrative, or the configurational dimension of the whole, whose meaning is over and above the meaning of its parts. While the latter sometimes can be fixed by scientific method, the former in general cannot, but is obtainable only by a proper description that would expose the inner working of certain historical forces, or by telling a plausible or convincing story that is structured by a narrator, who discerns the significance of the whole episode from a later perspective.

The hypothetic, interpretive, and holistic nature of historical knowledge (not the information about the events, but the understanding of the meaning of a whole episode) has raised a serious question: is it possible for any historical enquiry to supply any true knowledge of objective reality? The holism challenge can be somewhat mitigated by appealing to the so-called decoupling argument, which says that a quasi-atomistic sense of meaning can be practically defined when, in some circumstances, the link between a part and the rest of the whole is so weak that the impact on the part from the rest is ignorable. But the interpretive nature of historical knowledge has posed a more serious threat of skepticism that deserves careful analysis.

If we follow the empiricist or naïve realist view and take truth as defined by correspondence to reality, which is described by fixed natural kinds and explained by causal laws, and objectivity as interest-independent and value-independent universality, that is, take objective truth as something historians may be able to choose, but not to change or create, then, historical knowledge, as a narrative of what happened in the past, which is mediated by interests, problems, and purposes and hypothetically constructed under the constraints of survived data, is not in this category of true knowledge of objective reality. But this does not mean that historical knowledge gives only subjective fictions or conventions, having nothing to do with reality. On the contrary, as pragmatic knowledge that aims at understanding rather than explanation, at communication rather than prediction, at consensus rather than control, historical knowledge is objective in the sense that it has a reference to reality, and thus is not arbitrary.

Then what is reality in historical enquiry? It is not what occurred in the past, but what has a place in the narrative. That is, historical reality is not naked experience encoded in data, but a particular organization of data by a historian's hypothesis, which has its roots in the praxis of the society, from within which a historian's picture of reality is conceptualized. Then, due to the underdetermination thesis of hypothesis by data, historical reality is by no means fixed and unique, but exists in a multitude, each of which is irreducible and thus autonomous, with certain valid reference to certain praxis of a given age. This means that objectivity of historical knowledge, in the sense of having a reference to historical reality, can only be appreciated in a local and perspectival way. But once locality and perspective are fixed, no arbitrariness can escape critical assessments.

Historical knowledge has an important function to serve. Its importance mainly comes from the historicity of the present, that is, from the very existence of the present as the praxis connecting the past and the future. In an important sense, historical enquiry is a critical engagement with the past and the present. It could be very critical of currently reigning ideology of a society by interrogating its accepted categories and historicizing them in order to challenge their aura of naturalness or inevitability. But frequently it serves more apologetic purposes for the confirmation of self-understanding and self-identification of current engagements and projects.

The way historical knowledge functions has deep roots in human nature. A human has no nature other than being aware of belonging to a present of understanding, achievements, and relationships which stretch back into a past, thus to evoke a past that is recognized as one's own is the precondition of all self-consciousness. From time to time, in a fluctuating balance of not always compatible past dispositions, sympathies, attachments, loyalties, wants, purposes, devotions, some of them may temporarily become the dominant consideration in the self-understanding of a social group, and in terms of which the group may respond to the emergent situations, which responses constitute their critical existence at the present. Such a past in which people seek a reflection of itself and in terms of which it may come to understand itself, is an *evoked past*. To be aware of such a past and to cultivate this awareness constitutes the education which endows the educated with an identity. That is, human self-recognition as a relation between the self and others, and between the present and the past has to be created, learned, and cultivated. And the most effective way of doing so is to be engaged in historical enquiry.

In historical enquiry, the past and the present are constitutive of each other. The resultant self-identity is an assertion of a relationship to past events. In particular, a present intellectual adventure usually would seek to confirm its self-understanding in terms of corresponding awareness in the past and to link the present intellectual experience with a *matching past*. In any sustained intellectual engagement, such as theoretical physics with its imagined magnificence, the *matching past* is always evoked to give shape to, to identify, and to sustain this new style of self-understanding. For example, the contemporary theoretical physics community engaged in pursuing symmetry and unification has frequently evoked a past of similar character to celebrate and to confirm its own identity, mainly through having constructed an intellectual legend of Einstein. The Einstein legend serves the community very well in having fortified its current engagement in seeking symmetry and unification, and in having authorized its pursuit of certain directions of enquiry, sanctioned the exploration of some experiences, mainly in mathematical manipulations, and restrained that of others, which, however, is not without having met the resistance from those who are eager to pursue complexity rather than simplicity, layered effective theories based on the idea of diversity rather than a fundamental theory based on the alleged ultimate unity of the universe.

With these general features of historical enquiry in mind, we are ready to move to history of science itself.

History of science

What differentiates history of science from other historical enquiries is the special features of its subject matter, namely science as a distinctive system of knowledge production has posed severe constraints on the practice of historical study of science.

If we accept that science is a learning device and scientific activities are intellectual endeavor for knowledge production, then we cannot have a history of science with little scientific knowledge in it. But a collection or a chronicle of scientific results in itself is not history of science. For lack of historical treatment, these results may be useful for knowing science, but not for history of science. But what is historical treatment? Can we have a history of science in which one scientific idea leads to another in a historically effective way? Although the importance of the conceptual aspect of science should be properly appreciated as we will do in Section 10.4, scientific ideas by themselves simply could not be historically effective.

What are historically effective in the production of one idea followed by another, in my view, are scientists' activities. Scientists, organized in a community, faced with certain specific sets of problems, utilize available conceptual resources, together with financial, organizational, and technical resources, to propose solutions to the problems and to test these proposals in various ways, including the way of competing with the scientists of other communities with different research programs. Scientific knowledge produced this way is problem-specific, resources-specific, environment-specific and culture-specific, and its meaning has to be assessed in the historically specific cultural environment in which it is produced

As a historical enquiry, the aim of history of science is to understand the historical meaning of scientists' activities, not the cognitive meaning of a scientific endeavor. Of course, a qualified historian should be able to give a correct explication of a scientific endeavor. If he gets it wrong, this means that he did not even get his data right. It requires great effort for a historian of science to get his data right if he is working on the history of frontier science. But the cognitive meaning of a scientific endeavor is not the historical meaning a historical enquiry of science is supposed to deliver.

Then what is the historical meaning the present narrative about the genesis of QCD intends to deliver?

What is intended in the CA–QCD narrative?

The appreciation by the US government of physicists' contributions to national security during World War II led to its lavish support of researches in nuclear physics and particle physics, with newly designed accelerators built one after another, ever since the war was over and in the whole period in which QCD was conceived, formulated and accepted (Schweber, 1997).

The postwar accelerators provided enough energy to probe the nucleus deeply. New particles were discovered, wrenched out of the strong force field of the nucleus. The proliferation of the new particles and resonances soon led to the recognition of the organization of the properties of old and new particles, which provided clues to their underlying structure that was soon to be encoded in the idea of a hierarchy of symmetries, which was developed by Pais, Gell-Mann, and others (Section 2.1).

But the real actors in the story were not those individuals but a community in high energy physics, the hadron physics community. Originally, the community was identified by its major concern with the strong nuclear forces, although the weak and electromagnetic forces had also to be taken into consideration in some hadron processes. In the late 1940s and early 1950s, after the success of quantum electrodynamics, physicists tried to use a similar theory, the meson theory, to understand the strong force. Understanding here, as was conceived by a leading practitioner in this pursuit then, means, first, the theoretical calculations should match the measurements of hadron processes; second, the theory should have a clear physical picture of what is going on, and third, it has to have a precise and consistent mathematical formalism. Although some agreements could be cooked out by manipulations, the meson theory definitely failed on the other two terms (Dyson, 2004).

Then with the growing suspicions, since the mid 1950s, against QFT in general, physicists, out of desperation, turned to other ways as substitutes for dynamical understanding. Two strategies were pursued, and the results were two research programs that dominated practice in hadron physics for two decades. One, the S-matrix theory (with many other names which stress different aspects of the theory, such as dispersion relations, nuclear democracy, bootstrap, Regge pole, dual resonance model, and finally the old string theory) appeals to general principles such as analyticity, unitarity, and crossing symmetry for the calculations of hadron processes. The other, current algebra, explores the mathematical implications of assumed broken symmetries of hadron physics. The two programs share one feature in common, that is that no reference was made to the physical picture of dynamical processes as quantum field theory would suggest.

At the end of the story, however, several communities had speciated from the original one, and each of them has its own concerns and specialties. Interesting enough, the QCD community, one among those speciated, has quite different concerns and specialties, and thus has a quite different identity indeed.

The rise of QCD in a brief period of time was a major event in the history of science. Together with a shift in ontological commitment from hadrons to quarks and gluons, the fundamental dynamics was also reconceived. The strong force among hadrons was no longer regarded as a fundamental force of nature; rather, it was taken to be merely the un-cancelled residue of far stronger long-range forces mediated by gluons between individual quarks within hadrons.

How could such profound changes have happened during such a short period of time? The question is particularly challenging if we remember that first, the old ideas about elementary hadrons and fundamental forces of nature were well-entrenched; secondly, the evidential support for the new ideas was highly unusual in character, remotely indirect at best; and finally, the short period in which these changes occurred was marked by deep intellectual turmoil: confusions and conflicts among hadron physicists were rife. In Section 4.3 we reviewed a major conflict between the bootstrap idea (and community indeed) and the idea of atomicity then implicit in the local current algebra sum rule (and thus also a community).

In Section 7.2 we also reviewed profound confusions related with the idea of gauge symmetry. For example, Ne'eman tried to pursue the idea, similar to what Pauli, Yang and Mills, and Sakurai had tried, at a wrong level, the level of hadrons. The level was wrong in the sense that nature did not provide an ontological basis (proper entities) for the gauge symmetry at that level. Nambu also tried to pursue the idea, and the level he chose (the sub-hadronic entities) to pursue was right. But Nambu did not really get the ontological basis for the gauge symmetry right. His assumed integrally charged fundamental triplets was not a right ontological basis for the color gauge symmetry because on that basis, the color symmetry would have to be mysteriously mixed with the electromagnetic U(1) symmetry, and this was not a scheme that nature would approve.

When physicists, who contributed to the genesis of QCD, expressed their curiosity, sentiments, conjectures, conflicting ideas, and confused beliefs, they were acutely aware of deserting a well-trodden track (ordinary quantum field theory without confinement and asymptotic freedom) and embarking upon intellectual adventures that were invited by new technical engagements (such as the deep inelastic lepton–proton scattering and the electron–positron collision that became available then) and by the prospect of novel experiences. What this narrative intends to offer is to do full justice to the historical experience physicists acquired in the making of QCD.

For example, in order to justify any claim to reality when something is unobservable in principle (in the traditional sense that is not accessible asymptotically), the constructive structural realist approach adopted in this project tries to explain, in physical terms, how physicists were led to probe the unobservable particles, if such a probe is possible, with the help of structural knowledge suggested by theoretical deliberations in the framework of local current algebra and acquirable from experiments available then, rather than appealing to abstract analysis as to what makes a reality claim to theoretical entities justifiable. Here, all depends on a careful examination of interactions between observations and interpretations (such as scaling and partons), between theoretical claims and experimental constraints on these claims (such as quark-like ingredients and the momentum distribution data, see Section 6.1). I hope, with this kind of conceptual analysis, the historical experience offered by the genesis of QCD could be properly appreciated, and could provide further materials for philosophers to reflect on.

Since the historical meaning can only be found in a particular narrative structure, this narrative intends to offer a historical-critical structure in which conceptual steps crucial to the rise of QCD are presented in the historical context in which they were taken, with the intellectual justification for each such step being critically assessed in the context of the great debates then current among different schools of thought; most important among the debates were those between field theorists and anti-field theorists (Landau and his followers), between bootstrap advocates and composite modelers (Sections 2.2 and 4.3), between a purely mathematical conception (Gell-Mann and his followers) and a naïve realistic conception of the sub-hadronic entities (Nambu and advocates of constituent quarks).

Broadly speaking, a narrative on the rise of QCD can be constructed as an important step in a linear progress for understanding the strong forces, or a step in exploring larger and larger symmetries for understanding the physical world, or a step for the ultimate unification in the final theory, or as a case for illustrating how unobservable entities could be socially constructed; it can also be structured as a narrative of experiments and equipment. This project, however, is structured in different ways. Starting from a post-empiricist conception of science, the narrative intends to derive some lessons for the understanding of high energy physics, most important among them is the possibility of constructing entities through the structural approach, by describing the epical discoveries, dramatic confrontations, and the constitution and remolding of the identity of a high energy physics community during the course in which QCD was taking shape.

How could this actually happen is an interesting story full of activities and experiences by physicists, full of changes of the meaning of experience, changes of the direction of events, and changes of aims and identity of the community. But the meaning of all that happened understood by contributing scientists when the history was in the making is most likely to be quite different from what a critical historian would understand due to the difference in perspectives, and thus scientists' judgments, even those concerning their own contributions, cannot be unreflectively taken for granted, but have to be critically assessed and properly interpreted. For this reason, what this narrative intends to do is to critically interpret the meaning of scientists' theoretical and experimental explorations helped by hindsight without distorting the actual historical process, in which scientists who contributed to the rise of QCD had their own perceptions and understanding of what was going on and gave consequential responses to the events perceived and understood that way.

10.3 Major concerns in the CA–QCD narrative

In delivering the messages within the narrative structure mentioned above, there are three major concerns which guided the organization of the narrative: how the narrative end was reached in the actual movement of scientific endeavors that happened in the past; what the criterion for the theoretical discovery of quarks and gluons is, and what the cultural implications of scientific discoveries are in the case of QCD. This section is devoted to highlight a few results out of these concerns.

Historically effective contingent activities

Structural realism can help physicists to understand and appreciate what they have achieved in the construction of QCD, namely the discovery of a new type of natural kinds at the sub-hadronic level. It can also help historians to understand and appreciate the way this result was achieved.

The movement from Gell-Mann's global approach to Adler's local approach to Bjorken's penetrating explorations into protons may give readers an impression that it was a linear movement dictated by an internal logic. In fact, each step was only one of many contingent responses, which were not recounted in this narrative except for the selected one,[3] to the conceptual situation created by the previous step. The contingent nature of crucial steps in the movement can be seen much more clearly in the case of what Kenneth Wilson contributed to the whole zigzag movement toward the creation of QCD.

In his non-Lagrangian approach to current algebra (1969), which was a response to the situation created by the existing models of current algebra and by scaling and anomalies, Wilson developed his views on the broken scale invariance and incorporated them into his formulation of operator product expansions. This step set in motion a series of developments. Its immediate consequence was the formulation of the Callan–Symanzik equation, which was an application of Wilson's ideas (Section 6.2). But then the application of the Callan–Symanzik equation to non-abelian gauge theories resulted in the discovery of asymptotic freedom under certain constraints. This had opened the door for further explorations in the high energy sector of QCD which is characterized by the super-strong quark–gluon interaction that has no direct relevance to the original concern of the strong forces. Thus, a new community has emerged with perturbative QCD as its special concern, which requires skills in theoretical deliberations and experimental testing that are quite different from those that were required by the traditional hadron physics community, and thus has endowed the community with a new identity that is quite different from that of the old community.

As a response to the discovery of asymptotic freedom, Wilson, originally not interested in gauge theory and thus lagging behind, got excited and was eager to catch up. By utilizing his expertise in statistical mechanics, he soon invented a new formulation of QCD defined on a lattice, and after some deliberations he published it a year after (Section 8.3). Wilson's new lattice formulation of QCD was able to approach the low energy sector of QCD and had implication for the understanding of confinement, and thus was potentially able to address the traditional concerns of the hadron physics, which were still addressed by two sub-communities, one was identified with the Regge theory, and the other with the constituent quark model. But the expertise and theoretical and experimental resources that are needed by the lattice gauge theory are so different from those needed by the Regge theory and the constituent quark model, that a speciation of yet another sub-community with a different identity is in display.

However, the recalcitrance of confinement to physicists' understanding, lattice formulation included, has forced them to adopt the idea of supersymmetry for a decisive attack against this last bastion of resistance to their understanding of QCD (Veneziano and Yankielowicz, 1982; Affleck, Dine, and Seiberg, 1983, 1984). The introduction of supersymmetry has opened still another new direction of explorations, with its own problems, expertise and techniques of addressing these problems.

The above brief clearly suggests that the direction of actors' activities frequently change under the pressure of each previous step of progress in scientific exploration. The change of actors' direction is characterized by the contingent circumstances and idiosyncratic strategic considerations of the agents; and the direction of the narrative is realized only by following the changes of actors' direction, and thus has nothing to do with the direction of events themselves that happened in the past. There is no meaningful way to talk about the coherent direction of events, and thus nothing is predictable for future developments. But this view of the history of physics also suggests that even if QCD turned to be dismissed in the future, and as a result the narrative direction in future enquiry may be quite different from the one chosen in this project, in terms of the genesis of QCD, however, nothing essential would be changed by future developments and future enquiries.

Another interesting question in this regard was raised recently by Wilson's observation (2004) that "the lattice gauge theory was a discovery waiting to happen, once asymptotic freedom was established." If each step in the conceptual development is prepared by previous developments, as it is always the case in the history of physics, and thus has some inevitability, what is the meaning of originality with which we can credit each breakthrough made by individual scientists?

In the case of QCD examined by this project, it seems clear that the originality of a step in scientific development can only be defined in terms of its consequences rather than its novelty or the sophistication of the ideas involved, or the lack of predecessor for these ideas. In fact, many consequential breakthroughs originated from very simple ideas that have clear inheritance from their predecessor. For example, Pauli's non-abelian gauge theory can be viewed only as an extension of QED, and the quark model as an extension of the Sakata model, although current algebra and renormalization group are more complicated and lack clear predecessor lineage. What is crucial in historian's assessment of a conceptual step is whether it has provided a new vision, opened a new direction for explorations, and thus initiated a new research program, all of which can only be known by historians with the help of hindsight. The justification for such a criterion of originality is provided by the following fact: history is a succession of events, one followed by others, and only those consequential steps are historically effective in moving science ahead.

When was the theoretical discovery made?

The experimental "observations" of quarks and gluons in the events of naked charmed particles and three jets cannot be viewed as real discoveries. Rather, they were only confirmations of the theoretical discoveries made in QCD. But scholars have argued all the way down to present that the quarks and gluons were discovered much earlier.

For example, Lipkin (1997) argued, as early as 1966, if not earlier, there were reasons to believe that quarks existed due to a set of structural knowledge about them. Great progress was made in 1966 in understanding the hadron spectrum through a symmetrical quark model which classifies the spectrum in terms of the group $SU(6) \times O(3)$. It described all baryons as three quark states with wave functions satisfying Bose statistics, which entailed and in fact was followed by the introduction of parastatistics or color, and having orbital and radial excitations with quantum numbers qualitatively described by a harmonic-oscillator shell model.

Andrei Sakharov, Lipkin further argued, was a pioneer in hadron physics who took quarks seriously and "anticipated QCD by assuming a quark model for hadron with a flavor-dependent linear mass term and a two-body interaction whose flavor dependence was all in a hyperfine interaction." Nambu was another physicist who "derived just such a universal mass formula for meson and baryons from a model in which colored quarks were bound into color singlet hadrons by an interaction generated by coupling the quarks to a non-abelian SU(3) color gauge field with spin effects neglected".

By hindsight, however, we know that there were many false ideas about quarks and gluons (concerning their charges and asymptotic states, for example) up to 1966 and beyond, and physicists' understanding of them had undergone substantial changes from 1964 when the idea was first proposed to 1972 when QCD was first formulated. In this case can we really claim that the discovery was made in 1966?

Surely the implication of the evolution of scientists' understanding of an entity on what was discovered and when the discovery was made is a complicated issue. Priestley is not regarded as the discoverer of oxygen although he observed and described many of the crucial structural effects of oxygen, which was labeled as "deflogisticated air"; but Thomson is still regarded as the discoverer of the electron even though his conception of electron has been radically revised by quantum physicists. The different treatment is not without justification, which can be understood in terms of the core subset of structural statements, which serves as the identifying reference to the hypothetical entities under investigations as discussed in Section 9.2. If the core subset remains the same through radical changes in understanding the underlying entity, then the initial discovery would be respected, and this is the case for Thomson. Otherwise, the initial observation or conception would not be regarded as the discovery of the entity conceived by later scientists, and this is the case for Priestley.

How about quarks? Quarks were conceived to be the ingredients of hadrons with a holistic set of structural properties, such as discrete charges, spin, and parity, which always cohere together, although they do not fit into Wigner's definition for being elementary particles or being set-theoretical structures fixed within the framework of group theory, that is as irreducible representations of the Poincaré group with some invariants such as mass, spin, and parity. For our concern, however, the core subset of structural statements can be viewed as consisting of three assertions: being ingredients of hadrons, being fractionally charged, and being colored and color-confined.

One answer in this regard was given by Lipkin: "The Nobel prize for QCD as a description of strong interactions might have been awarded to Sakharov, Zel'dovich, and Nambu. They had it all figured out in 1966: the Balmer formula, the Bohr atom, and the Schrodinger equation of strong interactions. All subsequent developments leading to QCD were just mathematics and public relations with no new physics" (1997).

But if one takes the core subset as the identifying reference seriously, then uncolored quarks, from which all those achievements were derived, were essentially different entities from the quarks of QCD. As to Nambu's more sophisticated discovery of integral charged and non-confined quarks and his gluons carrying both color and electromagnetic charges, if we compare it with the discovery made in QCD, then it is clear that his quarks and gluons were essentially different from those discovered in QCD.

What about Gell-Mann? Gell-Mann's quarks and gluons before QCD were all uncolored. Worse, he did not even take quarks and gluons as real entities, real ingredients of hadrons. For this reason, his 1964 proposal of the quark model, although contributing a crucial discovery of quarks' structural property, namely they are fractionally charged, cannot be credited as a step in which quarks were discovered.

Science and culture (physics and metaphysics)

Another major concern of the project is cultural in character. As stated earlier, what were historically effective in moving physics ahead were not ideas *per se*, but scientists' activities. Since scientific knowledge produced is problem-specific, resources-specific, environment-specific, and culture-specific, as we argued earlier, the meaning of scientists' activities has to be assessed in the specific cultural environment in which it was produced.

More specifically, since scientists' experiences obtained from their activities were encoded in ideas, concepts, and formulations, and thus always appeared as linguistic events, their meaning can only be comprehended within an established meaning structure, or a conceptual framework or a paradigm. A paradigm within which scientists encoded their experiences would dictate the meaning of scientists' experience. But then how can we understand the radical changes of the meaning of such terms as "constituents" and "reality" once they had to be comprehended from the perspective of color confinement in QCD?

In the case of "constituents," the debate over the divisibility of hadrons between field theorists and bootstrap theorists, whose arguments against exploring more deeply into the hadrons were both obstacles and stimulus for field-theorists' explorative efforts, had resulted in a consensus among physicists that the notion of constituents simply cannot be understood purely in a naïve reductionist way (the constituent quark model),[4] but had to accommodate it to the idea of emergence in a reconstructive sense (the confinement thesis in QCD). More specifically, once the notion of a running coupling and the consequent confinement were accepted, it was realized that the idea that constituents had to be capable of existing in isolation was valid only for weakly coupled field theories, but had to be revised when the coupling was strong. That is, they may exist permanently in the confined rather than asymptotic states.

Since the change of meaning which accompanied the profound development in science occurred within an existing cultural environment with a prevailing metaphysical structure, the task for physicists to digest the change and absorb it into a meaning structure was very difficult. Thus Gell-Mann's painful struggle with the question of whether a permanently unobservable individual quark is real cannot be frivolously dismissed as mathematical instrumentalism, but has to be put in the context that when the idea of quark was proposed, the very idea of reality, according to the prevailing metaphysics of the time, was inseparably tied with observability, which was defined in terms of asymptotic states. For this reason, Gell-Mann understandably did not know how to handle the complicated situation when, on the one hand, the progress in understanding the hadron spectrum gave him good reasons to believe that hadrons consist of quark constituents; and on the other hand, he had deep reasons to believe that quarks could not be observed individually.

Were quarks real? This was a question that had frequently tortured him. When he said that he used quarks to get useful algebraic relations although quarks themselves "may or may not have anything to do with reality," and thus the function of quarks was similar to veal "in French cuisine: a piece of pheasant meat is cooked between two slices of veal, which are then discarded" (1964b), he was frequently accused of taking quarks only as a fiction. Later Gell-Mann (1997) defended himself by saying that he could not take quarks seriously because the use of quarks in his original theory with scalar gluon was wrong; if he had a theory with vector gluons he would take them seriously. This implies that he took the reality of quarks as depending on the validity of a quark–gluon theory. However, quarks and gluons in a correct theory remained unobservable, were they real?

Constrained by various logical inferences, Gell-Mann and the whole community soon realized that when the notion real applies to any physical hidden entity, in the light of confinement, it refers to more than merely the epistemic accessibility of the entity as a separate individual. Rather, it had widened its reference to a combination of some structural features without any assignment of individual observability to the entity.

Conceived this way, it was easy for them to acknowledge the reality of quarks and gluons. But then the meaning of reality has already been radically changed: now it is more complicated than being observable individually. Thus with the advent of QCD, physicists had convincingly revised the metaphysical criterion of what counts as a real existence (Section 9.4).

In my view, Gell-Mann's painful experience in his personal struggle with the reality of quarks in fact was only a human expression of the difficult adjustment of metaphysics under the pressure of the advancement of physics.

An important point here is that metaphysics, as reflections on physics rather than a prescription for physics, cannot be detached from physics. It can help us to make physics intelligible by providing well-entrenched categories distilled from everyday life and previous scientific experiences. But with the advancement of physics, it has to move forward and revise itself for new situations: old categories have to be discarded or transformed, new categories have to be introduced to accommodate new facts and new situations.

Another point here is that paradigm is not a completely rigid structure assigning fixed meaning to each term involved in it. Rather, the meaning of each individual term (co-dictated jointly by a paradigm and ever changing living experiences) and a paradigm as a whole are mutually adjusting to each other, which have rendered paradigm a dynamic entity in a long duration.

The above discussion seems to have vindicated a broad view articulated by Julian Schwinger: "The scientific level of any period is epitomized by the current attitude toward the fundamental properties of matter. The world view of the physicist sets the style of the technology and the culture of the society, and gives direction to future progress" (1965). Thus the neglect of fundamental physics, in terms of the formation and change of physicists' worldview, would leave a significant lacuna in our understanding of an essential element of our culture. One implication of this view for the history of science is that the frontiers of physics, particle physics and cosmology, have to be brought into the forefront of history and philosophy of science rather than left in the backwater.

10.4 In defense of conceptual history of science

Different understandings of the nature of scientific enquiry suggest different subject matter for historical treatment. For example, the resurgence of positivism, under the guise of stressing the importance of experiments, and the flourishing of social constructivism in the conception of science, have gravely affected the practice of history of science in the last three decades.

It is often stressed that equipment is necessary for experiments and experiments are crucial for the suggestion, testing, confirmation, and revision of theory. These are undeniable facts. But if one further claims, from the positivist position, that the basis for scientific progress is mainly the expansion of observations and experiments, and all other components of scientific enquiry, (theoretical and metaphysical), are secondary, then the instruments (detectors and accelerators) necessary for observations and experiments arguably should be the focus of attention.

Sometimes the positivist influence on the selection of subject matter in history of science is indirect and subtler. For example, according to the logical empiricist Hans Reichenbach (1951), scientific discoveries escape the rational analysis, and thus the concern should be only with the context of justification; or as another logical empiricist Herbert Feigl (1974) puts it, an account of the origin and development of scientific knowledge is not as important as a logical reconstruction of the knowledge. This conception of scientific endeavor leads people to pay more attention to positive results rather than the processes with which the results were obtained.

Similarly, based on the fact that social aspects of science simply cannot be ignored, the social constructivist claims that the social and institutional histories of science should be taken as the major subject matter for the historical treatment.

Both positions share an opinion of dismissing the importance of traditional concerns with the truth and objectivity of theoretical concepts, which in fact are crucial to the long-term effectiveness of science in explanation and prediction. As a result, the conceptual history, which is pejoratively labeled as an "internal history" and accused of dealing only with ideas, is relegated to a disreputable place in the history of science.

But if we take science as an intellectual endeavor for explaining, predicting, and controlling the world, we cannot ignore the basic fact that theoretical deliberations are the driving force for the advancement of science, although the deliberations are constrained by the possibility of experimenting and the latter is constrained by the equipment available to the community. The cognitive content of science is highly intellectual in its nature and thus simply cannot be socially constructed; neither can it be dictated by experiments nor by the availability of experimental equipment. The role of experiments and equipment is only secondary in the conceptual development of science because they have to be subject to theoretical deliberations. Thus it is safe to claim that scientific knowledge is crystallized in theory, not in any other places.

Furthermore, there are deeper implications that can be derived from the special features of scientific knowledge for the history of science. First, the existence of the filtering-out mechanisms implies that truth, ontological commitment (what actually exists in the world from which all phenomena can be derived) and objectivity are internal to scientific knowledge. In particular, scientific knowledge is always organized as a causal hierarchical structure, with a few fundamental theories (dealing with underlying entities) as its conceptual foundations, at any given time. Cases in point can be found in Newtonian mechanics during the period from the seventeenth to the nineteenth century, relativity theories, quantum physics and the standard model, and molecular biology about DNA in the twentieth century and up till now. The fundamental theories have in fact provided the basic parameters in defining the worldview of the time, as Schwinger indicated (quoted in Section 10.3). How can we envision the world at the present time without deploying the notions of spacetime curvature, quantum and DNA? This special feature in the way scientific knowledge organizes itself has demanded that historians of science have to pay primary attention to these fundamental theories. Without proper understanding of fundamental theories, our historical knowledge of science will be very limited and fragmented, no matter how much is done on smaller and relatively insignificant events that are more or less related with science.

Secondly, according to the structural realist view of science, the world is classified and described, not by natural kinds with fixed essence, but by natural kind entities whose identities are constituted and characterized by stable but variable sets of constitutive structural constraints, which are variable enough to be revised and reconstructed, after its initial construction, through a hermeneutical process (a feedback process between data and hypothesis). Thus, the ontological foundation of any fundamental theory has to be historicized, and a conceptual history dealing with the historical process, in which the ontological foundation of a fundamental theory is formed, revised, and reconstructed, is an imperative task for the profession of history of science to undertake if the profession aspires to provide any serious historical understanding of science.

It should be stressed that a conceptual history is by no means a narrowly defined "internal history" dealing only with ideas. Rather, a good piece of conceptual history should integrate theoretical deliberations and experimental testing and probing in science into a cognitive system of explorations, and further integrate cognitive and social aspects of science into a system of knowledge production.

In the conceptual history, the historical meaning of a scientific endeavor is delivered not only by a plausible description of the historical process through which a theory had taken shape, but also by critical assessments on scientists' statements and judgments about the history and logic of the theory. Of course, this requirement does not mean that a conceptual historian should endorse rational reconstruction as the right way of doing history of science, which was discredited by the disastrous distortions in Lakatos's notorious work of this kind (1971). Rather, it only means that conceptual historians, helped by hindsight, have to get the data or the scientific meaning of the theory right.

In addition, a critical conceptual historian should pay special attention to the special features that were displayed in the making of the theory (such as the ways to address the accessibility status of quarks in the making of QCD), as well as the special role played by the presumed metaphysical presupposition (such as the one that assumes that the reality of an entity can only be claimed if it is accessible in its asymptotic states) and the special methodological approach (such as Gell-Mann's structural approach to current algebra).

Instead of narrowly dealing with scientific ideas, the conceptual history of science has a wide range of complicated issues to address: philosophical issues about reconceptualization of ontology and reality and about a mutually constitutive relationship between theory and experiment; cultural issues concerning the mutual readjustment of newly acquired experiences (such as permanently unobservable particles) and existent cultural setting, as well as the successes and failures of reductionism;[5] historiographical issues about the direction of events and the possible reassessment of the past from a changed perspective at present or in the future; and also sociological issues about a scientific community (its hierarchical structure and its intellectual leadership, the shaping and remolding of its identity). What is distinctive of conceptual history is that all these issues should be addressed in a way that is not detached from a close examination of the conceptual development of the scientific endeavor under investigations.

Arguably the conceptual history of science should be taken as the foundation for other sub-branches of history of science, such as the histories of experiments and equipment, or the social and cultural histories of science. Surely, these historical enquiries of science are valuable in their own right, and are valuable additions to the conceptual history of science. However, these branches of history of science certainly cannot be substitutes for the conceptual history. In fact, without a proper conceptual history as their foundation, these historical enquiries would unconsciously take some non-reflective quasi-conceptual histories or legends or folklores for granted, and reinforce the misleading implications of these non-reflective legends by unconsciously using them. This means that no proper historical understanding of science as a system of knowledge production could be obtained without a conceptual history as a starting point.

Let us look at an example. The highly indirect character of the evidence for quarks made this an appealing subject for a social constructivist account. But such an account given by Andrew Pickering (1984) without a conceptual history as its foundation was external to science as a system of knowledge production. That is, his account failed to show that social and institutional aspects have penetrating effects on the content and nature of scientific knowledge that was produced in that social and institutional environment. Rather, in terms of content, Pickering heavily relied on accounts offered by participating scientists who were unreflective and often unaware of their own intellectual history, without critically assessing the conceptual underpinnings of these accounts.

For example, crucial to Pickering's project is an articulation of "the transformation between the old (physics: old quark model and bootstrap) and the new (physics: current quark and gauge theory)" (p. 16). But the transformation was possible only because the gauge theory was proven to be renormalizable in the early 1970s. Pickering knew this and tried to clarify how this was realized by devoting Section 6.3 ("The Renormalization of Gauge Theory") to this end. But the section is only a scant seven pages based almost entirely on interviews – a story told by two players without any critical explication. No predecessors, no history, no context, no crucial theoretical arguments, just a story of two great men, two theorists, and no critical assessment of their claims.

Or, consider the hostile conceptual environment against which QCD emerged: there was an active community of baryon resonance bootstrappers and S-matrix advocates who provided a critical background for the line of conceptual developments which ultimately led to the discovery of QCD. But in Pickering's account, only the term of bootstrap was mentioned, but no serious arguments against the notion of deeper ingredients within hadrons and, on that basis, against the whole field-theoretical framework, that were held and advocated by giant figures like Heisenberg, Landau, and Chew, were presented, let alone analyzed. Likewise the issue of confinement, crucial to the consistency of QCD as a scientific theory, was not even attempted. Thus Pickering's achievement in collecting a huge amount of material related to the construction of QCD should be duly acknowledged, but his failure in delivering a proper historical understanding of the relevant history should also be registered.

Of course, a social history of science can avoid this kind of externality if it is based on a proper conceptual history of the subject. For example, both the Sakata model proposed in the 1950s and the straton model, directly pressed by Mao to be invented in China in the mid 1960s, were the product of Marxist ideology. According to that ideology, as Lenin asserted, everything, even small as electrons, can be endlessly divided. Pursuing a social history along this line, the penetration of the ideology into the content of the models, which asserted the existence of sub-hadronic particles and their underlying roles for the understanding of hadronic phenomena, or the scientific knowledge thus produced, would be clear. And this certainly would help us to understand a certain tendency or preference among some of the Japanese and Chinese physicists in their conception of hadron physics. For example, from Nambu's pursuit of integrally charged and thus real triplets, we can certainly detect some impact from the Marxists Sakata and Taketani.

A more fitting example to illustrate the claim, that with a proper conceptual history as a firm foundation certain issues in the social study of science can be properly addressed, is the issue of intellectual leadership within a scientific community. The intellectual leadership mentioned here refers to the ability in setting the agenda for the community, providing key conceptual resource and technical device for pursuing it, and the ability to mobilize and persuade the community to actually pursue it. In all these terms, Gell-Mann was the leader of the particle physics community from the late 1950s to the early 1970s, and his leadership was consequential for the discovery of QCD.

First, his proposition of current algebra, through the work of Adler (who was directly mobilized by him), Bjorken (who was stimulated by Adler) and many others, led to the scaling hypothesis, which was confirmed by the SLAC deep inelastic scattering experiments. The profound implications of scaling (asymptotic freedom) had imposed severe constraints on the theory construction in hadron physics.

Second, his proposition of the quark model was highly consequential. Taking quarks as constituents of hadrons, physicists were able to explain the low lying hadron states successfully. But quarks can also be understood as entities in model field theories for extracting information about various hadron currents that can be used in current algebra.

Third, his own work on the S-matrix theory and his strong support and promotion of it resulted in the dual resonance model and the old string theory for the strong interactions, which played heuristic roles in the deliberations of confinement (see Section 8.3).

Fourth, his work on high energy behavior of QED and the resulting Gell-Mann–Low equation, together with works by others (Stueckelberg and Petermann, Bogoliubov and Shirkov) pursuing the same idea of renormalization group, through the work of his disciple Wilson, led to a new version of renormalization group equation, the Callan–Symanzik equation. One sensational result came from an application of the Callan–Symanzik equation to certain non-abelian gauge theories, whose renormalizability was recently demonstrated by Veltman and 't Hooft, and this was asymptotic freedom.

Fifth, after having integrated various developments related to the idea of color and the idea of an associated gauge group that would be responsible for the dynamics of a field-theoretical system of hadrons' ingredients, Gell-Mann together with Fritzsch proposed a formulation of QCD. The high energy sector of QCD, the one that deals with the super-strong interactions between quarks and gluons, dictated by the color gauge group, is very successful, while its low energy sector is still struggling with a proper understanding of confinement even now.

Gell-Mann's leadership has been widely acknowledged: "Gell-Mann is a great physicist whose work and ideas had a tremendous impact on the work and thinking of practically everyone attending this history conference, including myself" (Lipkin, 1997).[6] Without a firm foundation of a proper conceptual history of particle physics from the 1950s to the 1970s, however, it would be impossible to properly understand and appreciate his intellectual leadership.

Notes

1 Notice that Kant's notion of symbolism does not refer to the conventional designation of concepts by signs or words, but signifies a similarity of relations, or the transfer of a structure, between two quite dissimilar things.

2 But the relationship between the two is not completely symmetrical: the notion of paradigm does not logically entail the skeptical claim, made by the social constructivist, that scientific

knowledge does not have any cognitive content that is not a reflection of social interests. The quarrel between Kuhn himself and the self-claimed Kuhnians, the social constructivists, in the 1990s testifies to this asymmetrical relationship.

3 They will be reviewed in the full-scale historical treatment of the episode (forthcoming).

4 For example, the successful SU(6) prediction of $-3/2$ for the ratio of the proton and neutron magnetic moments was taken to be striking evidence for compositeness, since only a composite model gave a simple ratio for total moments.

5 Reductionism is a dominant theme in scientific enquiry. In the context of QCD, it concerns the sense in which a hadron can or cannot be dissolved into its constituents.

6 It referred to the 1992 SLAC conference on the rise of the standard model.

References

AJP	*American Journal of Physics*
AP	*Annals of Physics (NY)*
CMP	*Communications in Mathematical Physics*
DAN	*Doklady Akademie Nauka, USSR*
JETP	*Soviet physics, Journal of experimental and theoretical physics*
JMP	*Journal of Mathematical Physics*
NC	*Il Nuovo Cimento*
NP	*Nuclear Physics*
PL	*Physics Letters*
PR	*Physical Review*
PRL	*Physical Review Letters*
PRS	*Proceedings of Royal Society (London)*
PTP	*Progress of Theoretical Physics*
RMP	*Review of Modern Physics*
ZP	*Zeitschrift für Physics*

Adler, S. (1964) "Tests of the conserved vector current and partially conserved axial-vector current hypotheses in high energy neutrino reactions" *PR*, **135**: B963–966.

Adler, S. (1965a) "Consistency conditions on the strong interactions implied by a partially conserved axial vector current, I and II" *PR*, **137**: B1022–B1033; *PR*, **139**: B1638–1643.

Adler, S. (1965b) "Calculation of the axial-vector coupling constant renormalization in beta decay" *PRL*, **14**: 1051–1055.

Adler, S. (1965c) "Sum rules for the axial-vector coupling constant renormalization in beta decay" *PR*, **140**: B736–B747.

Adler, S. (1966) "Sum rules giving tests of local current commutation relations in high energy neutrino reactions" *PR*, **143**: 1144–1155.

Adler, S. (1967) "Current algebra" in *Proceedings of International School of Physics, "Enrico Fermi" Course 41, Selected Topics in Particle Physics* (July 17–29, 1967) (ed. Steinberger, J.; Academic Press) 1–54.

Adler, S. (1969) "Axial-vector vertex in spinor electrodynamics" *PR*, **177**: 2426–2438.
Adler, S. (1970): "π^0 decay" in *High-Energy Physics and Nuclear Structure* (ed. Devors, S.; Plenum Press) 647–655.
Adler, S. (2006) "Commentaries" in *Adventures in Theoretical Physics* (World Scientific) 19.
Adler, S. and Bardeen, W. A. (1969) "Absence of higher-order corrections in the anomalous axial-vector divergence equation" *PR*, **182**: 1517–1536.
Adler, S. and Gilman, F. J. (1967) "Neutrino or electron energy needed for testing current commutation relations" *PR*, **156**: 1598–1602.
Adler, S. and Tung, W. K. (1969) "Breakdown of asymptotic sum rules in perturbation theory" *PRL*, **22**: 978–981.
Adler, S. and Tung, W. K. (1970) "Bjorken limit in perturbation theory" *PR*, **D1**: 2846–2859.
Affleck, I., Dine, M., and Seiberg, S. (1983) "Supersymmetry breaking by instantons" *PRL*, **51**: 1026–1029.
Affleck, I., Dine, M., and Seiberg, S. (1984) "Calculable nonperturbative supersymmetry breaking" *PRL*, **52**: 1677–1680.
Aharonov, Y. and Bohm, D. (1959) "Significance of electromagnetic potentials in the quantum theory" *PR*, **115**: 485–491.
Anderson, P. (1963) "Plasmons, gauge invariance, and mass" *PR*, **130**: 439–442.
Appelquist, T. and Politzer, D. (1975) "Heavy quarks and long-lived hadrons" *PR*, **D12**: 1404–1414.
Appelquist, T., Barnett, R., and Lane, K. (1978) "Charm and beyond" *Annu. Rev. Nucl. and Particle Sci.*, **28**: 387.
Augustin, J.-E. *et al.* (1974) "Discovery of a narrow resonance in e^+e^- annihilation" *PRL*, **33**: 1406–1408.
Bardeen, W., Fritzsch, H., and Gell-Mann, M. (1972) "Light cone current algebra, π^0 decay and e^+e^- annihilation" (first circulated as a CERN Preprint, Ref. TH.1538-CERN; 21 July 1972), in *Scale and Conformal Symmetry in Hadron Physics* (ed. Gatto, R.; Wiley) 139–153.
Bartoli, B. *et al.* (1970) "Multiple particle production from e^+e^- interactions at C.M. energies between 1.6 and 2 Gev" *NC*, **70A**: 615–631.
Belavin, A., Polyakov, A., Schwarz, A., and Tyupkin, Y. (1975) "Pseudoparticle solutions of the Yang-Mills equations" *PL*, **59B**: 85–87.
Bell, J. (1967) "Current algebra and gauge invariance" *NC*, **50A**: 129–134.
Bell, J. and Jackiw, R. (1969) "A PCAC puzzle: $\pi^0 \to \gamma\gamma$ in the σ-model" *NC*, **60A**: 47–61.
Bethe, H. A. (1947) "The electromagnetic shift of energy levels" *PR*, **72**: 339–341.
Bjorken, J. (1966a) "Inequality for electron and muon scattering from nucleons" *PRL*, **16**: 408.
Bjorken, J. (1966b) "Applications of the chiral $U(6)\otimes U(6)$ algebra of current densities" *PR*, **148**: 1467–1478.
Bjorken, J. (1967a [2003]) "Inelastic lepton scattering and nucleon structure." An incomplete unpublished manuscript written in March 1967, later published in *In Conclusion* (World Scientific), 27–39.
Bjorken, J. (1967b) "Current algebra in small distances" in *Proceedings of International School of Physics, "Enrico Fermi" Course 41, "Selected Topics in Particle Physics"* (July 17–19, 1967), (ed. Steinberger, J.; Academic Press) 55–81.
Bjorken, J. (1967c) "Theoretical ideas on inelastic electron and muon scattering" in *Proceedings of the 1967 International Symposium on Electron and Photon*

Interactions at High Energies (Stanford, September 5–9, 1967) (International Union of Pure and Applied Physics, US Atomic Energy Commission with a Foreword by S. M. Berman) 109–127.
Bjorken, J. (1968) "Asymptotic sum rules at infinite momentum" SLAC preprint: SLAC-PUB-510.
Bjorken, J. (1969) "Asymptotic sum rules at infinite momentum" *PR*, **179**: 1547–1553.
Bjorken, J. (1973a) "A theorist's view of e^+e^- annihilation" in *Proceedings of the 6th International Symposium on Electron and Photon Interaction at High Energies, 1973, Boun* (eds. Rollnik, H. and Pteil, W.; North-Holland), reprinted in Bjorken, 2003, 199–224.
Bjorken, J. (1973b) "High transverse momentum processes" *Le Journal de Physique, Colloques, Supplement*, **34** (C1): 385–400.
Bjorken, J. (1979) "The new orthodoxy: How can it fail?" A talk presented at *Neutrino 79 – International Conference on Neutrinos, Weak Interactions and Cosmology* (Bergen, Norway, June 18–22, 1979), printed in Bjorken, 2003, 291–303.
Bjorken, J. (1997) "Deep-inelastic scattering: from current algebra to partons" in *The Rise of The Standard Model* (eds. Hoddeson, L., Brown, L., Riordan, M., and Dressden, M.; Cambridge University Press) 589–599.
Bjorken, J. (2003) *In Conclusion* (World Scientific).
Bjorken, J. and Glashow, S. (1964) "Elementary particles and SU (4)" *PL*, **11**: 255–257.
Bjorken, J. and Paschos, E. (1969) "Inelastic electron-proton and γ-proton scattering and the structure of the nucleon" *PR*, **185**: 1975–1982.
Bjorken, J. and Tuan, S. F. (1972) "Is the Adler sum rule for inelastic lepton-hadron processes correct?" *Comm. of Nucl. Part Phys.* **5**: 71–78.
Bjorken, J. and Walecka, J. (1966) "Electroproduction of nucleon resonances" *AP*, **38**: 35–62.
Blankenbecler, R., Cook, L. F., and Goldberger, M. L. (1962) "Is the photon an elementary particle?" *PRL*, **8**: 463–465.
Bloom, E., and Gilman, F. (1970) "Scaling, duality, and the behavior of resonances in inelastic electron-proton scattering" *PRL*, **25**: 1140–1143.
Bloom, E. D. and Taylor, R. E. (1968) "To: Program Coordinator" (May 10, 1968) Supplement No. 1 to SLAC Proposal 4–b.
Bloom, E. D. *et al.* (1969) "High energy inelastic e-p scattering at 6^0 and 10^0" *PRL*, **23**: 930–934.
Bloom, E. D. *et al.* (1970) MIT-SLAC Report No. SLAC-PUB-796 (unpublished report presented to the 15th International Conference on High Energy Physics, Kiev, USSR, 1970.
Bludman, S. A. (1958) "On the universal Fermi interaction" *NC*, **9**: 433–444.
Bodek, A. (1973) "Comment on the extraction of nucleon cross sections from deuterium data" *PR*, **D8**: 2331–2334.
Bogoliubov, N. (1958) "A new method in the theory of superconductivity" *JETP*, **34** (7): 41–46, 51–55.
Bollini, D., Buhler-Broglin, A., Dalpiaz, P. *et al.* (1968) "An evidence for a new decay mode of the X°-meson: X°→2γ" *NC*, **58A**: 289–296.
Bohr, N. (1925) "Atomic theory and mechanics" *Nature*, **116**: 845–853.
Brandt, R. A. (1967) "Derivation of renormalized relativistic perturbation theory from finite local fiend equations" *AP*, **44**: 221–265.
Brown, L. M. (1978) "The idea of the neutrino" *Physics Today*, **31**(9): 23–27.

Buchholz, D. (1994) "On the manifestation of particles" in *Mathematical Physics Toward the 21st Century* (eds. Sen, R. N and Gersten, A.; Ben Gurion University Press).
Buchholz, D. (1996) "Quarks, gluons, color: facts or fiction?" *NP*, **469**: 333–353.
Cabibbo, N. (1963) "Unitary symmetry and leptonic decays" *PRL*, **10**: 531–533.
Cabibbo, N., Parisi, G., and Testa, M. (1970) "Hadron production in e^+e^- collisions" *Lettere NC*, **4**: 35–39.
Callan, C. (1970) "Bjorken scale invariance in scalar field theory" *PR*, **D2**: 1541–1547.
Callan, C. and Gross, D. (1968) "Crucial test of a theory of currents" *PRL*, **21**: 311–313.
Callan, C. and Gross, D. (1969) "High-energy electroproduction and the constitution of the electric current" *PRL*, **22**: 156–159.
Callan, C. and Gross, D. (1973) "Bjorken scaling in quantum field theory" *PR*, **D8**: 4383–4394.
Callan, C. and Treiman, S. (1966) "Equal-time commutators and *K*-meson decays" *PRL*, **16**: 153–157.
Callan, C., Coleman, S., and Jackiw, R. (1970) "A new improved energy-momentum tensor" *AP*, **59**: 42–73.
Callan, C., Dashen, R., and Gross, D. (1976) "The structure of the vacuum" *PL*, **63B**: 334–340.
Callan, C., Dashen, R., and Gross, D. (1978) "Toward a theory of the strong interactions" *PR*, **D17**: 2717–2763.
Cao, T. Y. (1997) *Conceptual Developments of 20th Century Field Theories* (Cambridge University Press).
Cao, T. Y. (2001) "Prerequisites for a consistent framework of quantum gravity," *Stud. Hist. Philos Mod. Phys.*, **32**(2): 181–204.
Cao, T. Y. (2003a) "Structural realism and the interpretation of quantum field theory" *Synthese*, **136**(1): 3–24.
Cao, T. Y. (2003b) "Can we dissolve physical entities into mathematical structures?" *Synthese*, **136**(1): 57–71.
Cao, T. Y. (2003c) "What is ontological synthesis? A reply to Simon Saunders" *Synthese*, **136**(1): 107–126.
Cao, T. Y. (2006) "Structural realism and quantum gravity" in *Structural Foundation of Quantum Gravity* (eds. Rickles, D., French, S., and Saatsi, J.; Oxford University Press) 42–55.
Cao, T. Y. (forthcoming) *The Making of QCD* (Cambridge University Press).
Capra, F. (1979) "Quark physics without quarks: a review of recent developments in S-matrix theory," *AJP*, **47**: 11–23.
Capra, F. (1985) "Bootstrap physics: a conversation with Geoffrey Chew" in *A Passion for Physics: Essays in Honour of Geoffrey Chew* (eds. De Tar, C., Finkelstein, J. and Tan, C. I.; Taylor and Francis) 247–286.
Carmeli, M. (1976) "Modified gravitational Lagrangian" *PR*, **D14**: 1727.
Carmeli, M. (1977) "SL_{2c} conservation laws of general relativity" *NC*, **18**: 17–20.
Carnap, R. (1928) *Der Logische Aufbau der Welt*. (Felix Meiner Verlag; English translation by Rolf A. George, 1967, *The Logical Structure of the World. Pseudoproblems in Philosophy*. University of California Press).
Carnap, R. (1934) *Logische Syntax der Sprache*. (English translation 1937, *The Logical Syntax of Language*. Kegan Paul).
Carnap, R. (1950) "Empiricism, semantics and ontology" *Revue Internationale de Philosophie* **4**: 20–40; also in Carnap (1956), 205–221.

Carnap, R. (1956) *Meaning and Necessity* (University of Chicago Press).
Casher, A., Kogut, J., and Susskind, L. (1973) "Vacuum polarization and the quark-parton puzzle" *PRL*, **31**: 792.
Cassirer, E. (1936) *Determinism and Indeterminism in Modern Physics* (Yale University Press).
Cassirer, E. (1944) "Group concept and perception theory," *Philos. Phenom. Res.*, **5**: 1–35.
Chakravartty, A. (2004) "Structuralism as a form of scientific realism" *Int. Stud. Philos. of Sci.*, **18**: 151–171.
Chandler, C. (1968) "Causality in S-matrix theory" *PR*, **174**: 1749–1758.
Chandler, C. and Stapp, H. P. (1969) "Macroscopic causality and properties of scattering amplitudes" *JMP*, **10**: 826–859.
Chang, C. *et al.* (1975) "Observed deviations from scale invariance in high-energy muon scattering" *PRL*, **35**: 901–904.
Charap, J. M. and Fubini, S. (1959) "The field theoretic definition of the nuclear potential-I," *NC*, **14**: 540–559.
Chew, G. F. (1953a) "Pion-nucleon scattering when the coupling is weak and extended" *PR*, **89**: 591–593.
Chew, G. F. (1953b) "A new theoretical approach to the pion-nucleon interaction" *PR*, **89**: 904.
Chew, G. F. (1961) *S-Matrix Theory of Strong Interactions* (W. A. Benjamin).
Chew, G. F. (1962) "S-matrix theory of strong interactions without elementary particles" *RMP*, **34**: 394–401.
Chew, G. F. (1967) "Closure, locality, and the bootstrap" *PRL*, **19**: 1492–1495.
Chew, G. F. (1989) "Particles as S-matrix poles: hadron democracy" in *Pions to Quarks: Particle Physics in the 1950s* (eds. Brown, L. M., Dresden, M. and Hoddeson, L.; Cambridge University Press) 600–607.
Chew, G. F. and Frautschi, S. C. (1960) "Unified approach to high- and low-energy strong interactions on the basis of the Mandelstam representation" *PRL*, **5**: 580–583.
Chew, G. F. and Frautschi, S. C. (1961a) "Potential scattering as opposed to scattering associated with independent particles in the S-matrix theory of strong interactions" *PR*, **124**: 264–268.
Chew, G. F. and Frautschi, S. C. (1961b) "Principle of equivalence for all strongly interacting particles within the S-matrix framework" *PRL*, **8**: 41–44.
Chew, G. F. and Frautschi, S. C. (1961c) "Principle of equivalence for all strongly interacting particles within the S-matrix framework" *PRL*, **7**: 394–397.
Chew, G. F. and Low, F. E. (1956) "Effective-range approach to the low-energy p-wave pion-nucleon interaction" *PR*, **101**: 1570–1579.
Chew, G. F. and Mandelstam, S. (1960) "Theory of low energy pion-pion interaction" *PR*, **119**: 467–477.
Chew, G. F., Goldberger, M. L., Low, F. E., and Nambu, Y. (1957a) "Application of dispersion relations to low-energy meson-nucleon scattering" *PR*, **106**: 1337–1344.
Chew, G. F., Goldberger, M. L., Low, F. E., and Nambu, Y. (1957b) "Relativistic dispersion relation approach to photomeson production" *PR*, **106**: 1345–1355.
Chew, G., Karplus, R., Gasiorowicz, S., and Zachariasen, F. (1958) "Electromagnetic structure of the nucleon in local field theory" *PR*, **110**: 265–276.
Cho, Y. M. (1975) "Higher-dimensional unifications of gravitation and gauge Theories" *JMP*, **16**: 2029–2035.

Cho, Y. M. (1976) "Einstein Lagrangian as the translational Yang-Mills Lagrangian" *PR*, **D14**: 2521–2525.
Coleman, S. and Jackiw, R. (1971) "Why dilatation generators do not generate dilatations" *AP*, **67**: 552–598.
Coleman, S. and Gross, D. (1973) "Price of asymptotic freedom" *PRL*, **31**: 851–854.
Coleman, S. and Weinberg, E. (1973) "Radiative corrections as the origin of spontaneous symmetry breaking" *PR*, **D7**: 1888–1910.
Cornwall, J. M. and Jackiw, R. (1971) "Canonical light-cone commutators" *PR*, **D4**: 367–378.
Coster, J. and Stapp, H. P. (1969) "Physical-region discontinuity equations for many-particle scattering amplitudes. I" *JMP*, **10**: 371–396.
Coster, J. and Stapp, H. P. (1970a) "Physical-region discontinuity equations for many-particle scattering amplitudes. II" *JMP*, **11**: 1441–1463.
Coster, J. and Stapp, H. P. (1970b) "Physical-region discontinuity equations" *JMP*, **11**: 2743–2763.
Creutz, M., Jacobs, L., and Rebbi, C. (1979) "Monte Carlo study of Abelian lattice gauge theories" *PR*, **D20**: 1915–1922.
Crewther, R. (1972) "Nonperturbative evaluation of the anomalies in low-energy theorems" *PRL*, **28**: 1421–1424.
Cushing, J. T. (1990) *Theory Construction and Selection in Modern Physics: The S-Matrix Theory* (Cambridge University Press).
Cutkosky, R. E. (1960) "Singularities and discontinuities of Feynman amplitudes" *JMP*, **1**: 429–433.
Dalitz (1966) "Quark models for the 'elementary particles'" in *High Energy Physics* (eds. DeWitt, C. and Jacob, M.; Gordon and Breach) 251–324.
Dalitz (1967) "Symmetries and the strong interactions" in *Proceedings of the 13th International Conference on High Energy Physics* (University of California Press) 215–234.
de Forest, T., and Walecka, J. D. (1966) "Electron scattering and nuclear structure" *Adv. Phys.*, **15**: 1–109.
De Rujula, A. (1976) "Plenary report on theoretical basis of new particles" in *Proceedings of the 18th International Conference on High Energy Physics* (Tbilisi, USSR) II: N111–127.
De Rujula (1979) "Quantum Chromo Dynamite" *Proceedings of the International Conference on High Energy Physics* (European Physical Society, Geneva, June 27–July 4, 1979), 418–441.
Deutsch, M. (1951) "Evidence for the formation of positronium in gases" *PR*, **82**: 455–456.
DeWitt, B. S. (1964) "Theory of radiative corrections for non-Abelian gauge Fields" *PRL*, **12**: 742–746.
DeWitt, B. S. (1967a) "Quantum theory of gravity I. The canonical theory" *PR*, **160**: 1113–1148.
DeWitt, B. S. (1967b) "Quantum theory of gravity II. The manifestly covariant theory" *PR*, **162**: 1195–1239.
DeWitt, B. S. (1967c) "Quantum theory of gravity III. Applications of the covariant theory" *PR*, **162**: 1239–1256.
Dolen, R., Horn, O., and Schmid, C. (1967) "Prediction of Regge parameters of ρ poles from low-energy πN data" *PRL*, **19**: 402–407.
Dolen, R., Horn, O., and Schmid, C. (1968) "Finite-energy sum rules and their applications to πN charge exchange" *PR*, **166**: 1768–1781.

Doncel, M. G., Hermann, A., Michel, L, Pais, A. (eds.) (1987) *Symmetries in Physics (1600–1980)* (Servei de Publicacions, UAB, Barcelona).
Drell, S. and Lee, T. D. (1972) "Scaling properties and the bound-state nature of the physical nucleon" *PR*, **D5**: 1738–1763.
Drell, S. and Schwartz, C. L. (1958) "Sum rules for inelastic electron scattering" *PR*, **112**: 568–579.
Drell, S. and Walecka, J. D. (1964) "Electrodynamic processes with nuclear targets" *AP*, **28**: 18–33.
Drell, S., Levy, D., and Yan, T. M. (1969) "Theory of deep-inelastic lepton-nucleon scattering and lepton-pair annihilation processes. I" *PR*, **187**: 2159–2171.
Duhem, P. (1906) *The Aim and Structure of Physical Theory* (Trans. Philip P. Wiener; Princeton University Press, 1954).
Dyson, F. J. (1951) "The renormalization method in quantum electrodynamics" *PRS*, **A207**: 395–401.
Dyson, F. (2004) "A meeting with Enrico Fermi" *Nature*, **427**: 297.
Eichten, E., Gottfried, K., Kinoshita, T. *et al.* (1975): "Spectrum of charmed quark-antiquark bound states" *PRL*, **34**: 369–372.
Ellis, J. (1974) "Theoretical ideas about $e^+e^- \rightarrow$ hadrons at high energies" in *Proceedings of the 17th International Conference on High Energy Physics*, London, July 1974 (ed. Smith, J. R.; Rutherford Laboratory, Science Research Council), IV: 20–35.
Ellis, J., Gaillard, M. K., and Ross, G. (1976) "Search for gluons in e^+e^- annihilation" *NP*, **B111**: 253–271.
Faddeev, L. D. and Popov, V. N. (1967) "Feynman diagrams for the Yang-Mills field" *PL*, **25B**: 29–30.
Federbush, P., Goldberger, M. L., and Treiman, S. (1958) "Electromagnetic structure of the nucleon" *PR*, **112**: 642–665.
Feigl, H. (1974) "Empiricism at bay?" in *Boston Studies in the Philosophy of Science* (eds. Cohen, R. and Wartofsky, D.; Reidel) XIV: 8.
Fermi, E. and Yang, C. N. (1949) "Are mesons elementary particles?" *PR*, **76**: 1739–1743.
Feynman, R. P. (1963) "Quantum theory of gravity" *Acta Phys. Polonica*, **24**: 697–722.
Feynman, R. P. (1969a) "Very high-energy collisions of hadrons" *PRL*, **23**: 1415–1417.
Feynman, R. P. (1969b) "The behavior of hadron collisions at extreme energies" in *High Energy Collisions* (eds. Yang, C. N., Cole, J. A., Good, M., Hwa, R., Lee-Franziui, J.; Gordon and Breach) 237–258.
Feynman, R. P. and Gell-Mann, M. (1958) "Theory of the Fermi interaction" *PR*, **109**: 193–198.
Feynman, R. P., Gell-Mann, M., and Zweig, G. (1964) "Group U(6) $\otimes$ U(6) Generated by Current Components" *PRL*, **13**: 678–680.
Field, R. D. (1979) "Dynamics of high energy reactions" in *Proceedings of the 19th International Conference on High Energy Physics* (August 23–30, 1978, Tokyo; eds. Homma, S., Kawaguchi, M., and Miyazawa, H.; Physical Society of Japan) 743–773.
Fine, A. (1998) "The viewpoint of no-one in particular" in *Proceedings and Addresses of The American Philosophical Association*, **72**: 9–20.
Fisher, M. E. (1964) "Correlation functions and the critical region of simple fluids" *JMP*, **5**: 944–962.

Frazer, W. R and Fulco, J. R. (1959) "Effect of a pion-pion scattering resonance on nucleon structure" *PRL*, **2**: 365–368.
Frazer, W. R and Fulco, J. R. (1960) "Effect of a pion-pion scattering resonance on nucleon structure. II" *PR*, **117**: 1609–1615.
Fregeau, R. and Hofstadter, R. (1955) "High-energy electron scattering and nuclear structure determination. III. Carbon-12 nucleus" *PR*, **99**: 1503–1509.
French, S. (1998) "On the withering away of physical objects" in *Interpreting Bodies: Classical and Quantum Objects in Modern Physics* (ed. Castellani, E.; Princeton University Press) 93–113.
French, S. and Krause, D. (2006) *Identity in Physics: A Historical, Philosophical, and Formal Analysis* (Oxford University Press).
French, S. and Ladyman, J. (1999) "Reinflating the semantic approach" *Int. Stud. Philos. Sci.*, **13**: 103–121.
French, S. and Ladyman, J. (2003a) "Remodelling structural realism: Quantum physics and the metaphysics of structure" *Synthese*, **136**: 31–56.
French, S. and Ladyman, J. (2003b) "Between platonism and phenomenalism: Reply to Cao" *Synthese*, **136**: 73–78.
Friedman, J. (1968) "Preliminary data from a SLAC–MIT inelastic electron–proton scattering experiment," presented at the conference, not published, but listed as paper 563 on page 456 in the *Proceedings of 14th International Conference on High Energy Physics* (Geneva, CERN, 1968), with the truncated title "Inelastic scattering from protons."
Friedman, J. (1991) "Deep inelastic scattering: Comparison with the quark model" *RMP*, **63**: 615–627.
Friedman, J. (1997) "Deep inelastic scattering and the discovery of quarks" in *The Rise of the Standard Model* (eds. Hoddeson, L., Brown, L., Riordan, M., and Dresden, M.; Cambridge University Press) 566–588.
Friedman, J. and Kendall, H. (1972) "Deep inelastic electron scattering" *Annu. Rev. of Nucl.*, **22**: 203–254.
Friedman, M. (1992) *Kant and the Exact Sciences* (Harvard University Press).
Friedman, M. (1999) *Reconsidering Logical Positivism* (Cambridge University Press).
Frishman Y. (1970) "Scale invariance and current commutators near the light cone" *PRL*, **25**: 966–969.
Fritzsch, H. and Gell-Mann, M. (1971a) "Scale invariance and the light cone" in *Proceedings of the 1971 Coral Gables Conference on Fundamental Interactions at High Energy* (eds. Dal Cin, M., *et al.*; Gorfon and Breach) 1–53.
Fritzsch, H. and Gell-Mann, M. (1971b) "Light cone current algebra" in *Proceedings of the International Conference on Duality and Symmetry in Hadron Physics"* (ed. Gotsman, E.; Weizmann Science Press) 317–374.
Fritzsch, H. and Gell-Mann, M. (1972) "Current algebra, quarks and what else?" in *Proceedings of the XVI International Conference on High Energy Physics, September 6–13, 1972* (eds. Jackson, J. D. and Roberts, A.; National Accelerator Laboratory, Batavia), vol. 2: 135–165.
Fritzsch, H., Gell-Mann, M., and Leutwyler, H. (1973) "Advantages of the color octet gluon picture" *PL*, **47B**: 267–271.
Fubini, S. and Furlan, G. (1965) "Renormalization effects for partially conserved currents" *Physics*, **1**: 229–247.
Gaillard, M. (1979) "QCD phenomenology" in *Proceedings of the 19th International Conference on High Energy Physics* (23–30 August, 1978, Tokyo; eds. Homma, S., Kawaguchi, M., and Miyazawa, H.; Physical Society of Japan) 390–417.

Gaillard, M., Lee, B. and Rosner, J. (1975) "Search for charm" *RMP*, **47**: 277–310.
Galison, P. (1987) *How Experiments End* (University of Chicago Press).
Garwin, R. L., Lederman, L. M., and Weinrich, M. (1957) "Observations of the failure of conservation of parity and charge conjugation in meson decays: the magnetic moment of the free muon" *PR*, **105**: 1415–1417.
Gell-Mann, M. (1953) "Isotopic spin and new unstable particles" *PR*, **92**: 833–834.
Gell-Mann, M. (1956a) "The interpretation of the new particles as displaced charged multiplets," *NC Suppl.*, **4**: 848–866.
Gell-Mann, M. (1956b) "Dispersion relations in pion-pion and photon-nucleon scattering" in *High Energy Nuclear Physics, Proceedings of the Sixth Annual Rochester Conference* (Interscience), sec. III: 30–36.
Gell-Mann, M. (1957) "Model of the strong couplings" *PR*, **106**: 1296–1300.
Gell-Mann, M. (1958) "Test of the nature of the vector interaction in b-decay" *PR*, **111**: 362–365.
Gell-Mann, M. (1961) "The eightfold way: a theory of strong interaction symmetry" California Institute of Technology Synchrotron Laboratory Report No. CTSL-20; reprinted in Gell-Mann and Ne'eman (1964), 11–57.
Gell-Mann, M. (1962) "Symmetries of baryons and mesons" *PR*, **125**: 1067–1084.
Gell-Mann, M. (1964a) "A schematic model of baryons and mesons" *PL*, **8**: 214–215.
Gell-Mann, M. (1964b) "The symmetry group of vector and axial vector currents" *Physics*, **1**: 63–75.
Gell-Mann, M. (1969) "Symmetry violation in hadron physics" University of Hawaii Summer School lectures, later published in Hawaii Topical Conference in Particle Physics, **1**: 168–188.
Gell-Mann, M. (1972) "Quarks: developments in the quark theory of hadrons" *Acta Phys. Austriaca*, **9**: 733–761.
Gell-Mann, M. (1987) "Particle theory from S-matrix to quarks" in *Symmetries in Physics (1600–1980), Proceedings of the First International Meeting on the History of Scientific Ideas* (Catalonia, Spain, 1983; eds. Doncel, M. G., Hermann, A., Michel, L. and Pais, A.; Bella Terra) 474–497.
Gell-Mann, M. (1997) "Quarks, color, and QCD" *The Rise of the Standard Model* (eds. Hoddeson, L., Brown, L., Riordan, M., and Dresden, M.; Cambridge University Press) 625–633.
Gell-Mann, M. and Goldberger, M. L. (1954) "The scattering of low energy photons by particles of spin ½" *PR*, **96**: 1433–1438.
Gell-Mann, M. and Levy, M. (1960) "The axial vector current in beta decay" *NC*, **16**: 705.
Gell-Mann, M. and Low, F. E. (1954) "Quantum electrodynamics at small distances" *PR*, **95**: 1300–1312.
Gell-Mann, M. and Ne'eman, Y. (eds.) (1964) *The Eightfold Way* (W. Benjamin).
Gell-Mann, M. and Pais, A. (1954) "Theoretical views on the new particles" in *Proceedings of the Glasgow International Conference on Nuclear Physics* (eds. Bellamy, E. H. and Moorhouse, R. G.; Pergamon) 342–350.
Gell-Mann, M., Goldberger, M. L., and Thirring, W. (1954) "Use of causality conditions in quantum theory" *PR*, **95**: 1612–1627.
Gershstein, S. and Zeldovich, J. (1955) "On corrections from mesons to the theory of β-decay", *Zh. Eksp. Teor. Fiz.*, **29**: 698–699. (English trans. *JETP*, **2**: 576).
Glashow, S. L. (1959) "The renormalizability of vector meson interactions" *NP*, **10**: 107–117.
Glashow, S. L. (1961) "Partial symmetries of weak interactions" *NP*, **22**: 579–588.

Glashow, S. L., Iliopoulos, J., and Maiani, L. (1970) "Weak interactions with lepton-hadron symmetry" *PR*, **D2**: 1285–1292.
Goldberger, M. L. (1955a) "Use of causality conditions in quantum theory" *PR*, **97**: 508–510.
Goldberger, M. L. (1955b) "Causality conditions and dispersion relations I. Boson fields," *PR*, **99**: 979–985.
Goldberger, M. L. and Treiman, S. B. (1958) "Decay of the pi meson" *PR*, **110**: 1178–1184.
Goldhaber, M. (1956) "Compound hypothesis for the heavy unstable particles. II" *PR*, **101**: 433–438.
Goldhaber, G. *et al.* (1976): "Observation in e^+e^- annihilation of a narrow state at 1865 Mev/c^2 decaying to $K\pi$ and $K\pi\pi\pi$" *PRL*, **37**: 255–259.
Goldstone, J. (1961) "Field theories with 'superconductor' solutions" *NC*, **19**: 154–164.
Goto, T. and Imamura, T. (1955) "Note on the non-perturbation-approach to quantum field theory" *PTP*, **14**: 396–397.
Goto, T (1971): "Relativistic quantum mechanics of one-dimensional mechanical continuum and subsidiary condition of dual resonance model" *PTP*, **46**: 1560–1569.
Greenberg, O. (1964) "Spin and unitary-spin independence in a paraquark model of baryons and mesons" *PRL*, **13**: 598–602.
Greenberg, O. and Zwanziger, D. (1966) "Saturation in triplet models of hadrons" *PR*, **150**: 1177–1180.
Gross, D. (1997) "Asymptotic freedom and the emergence of QCD" in *The Rise of the Standard Model* (eds. Hoddeson, L., Brown, L., Riordan, M., and Dresden, M.; Cambridge University Press) 199–232.
Gross, D. and Treiman, S. (1971) "Light cone structure of current commutators in the gluon-quark model" *PR*, **D4**: 1059–1072.
Gross, D. and Wilczek, F. (1973a) Ultraviolet behavior of non-abelian gauge theories" *PRL*, **30**: 1343–1346.
Gross, D. and Wilczek, F. (1973b) "Asymptotic free gauge theories: I" *PR*, **D8**: 3633–3652.
Gross, D. and Wilczek, F. (1973c) "Asymptotic free gauge theories: II" *PR*, **D9**: 980–983.
Gulmanelli, P. (1954) *Su una Teoria dello Spin Isotropico* (Pubblicazioni della Sezione di Milano dell'istituto Nazionale di Fisica Nucleare, Casa Editrice Pleion, Milano).
Gürsey, F. and Radicati, L. A. (1964) "Spin and unitary spin independence" *PRL*, **13**: 173–175.
Haag, R. (1958) "Quantum field theories with composite particles and asymptotic conditions" *PR*, **112**: 669–673.
Haag, R. (1992) *Local Quantum Physics* (Springer).
Han, M. Y. and Nambu, Y. (1965) "Three-triplet model with double SU(3) symmetry" *PR*, **139B**: 1006–1010.
Hand, L. (1963) "Experimental investigation of pion electroproduction" *PR*, **129**: 1834–1846.
Hara, Y. (1964) "Unitary triplets and the eightfold way" *PR*, **B134**: 701–704.
Harrington, B. J., Park, S. Y., and Yildiz, A. (1975) "Scalar-meson dominance model in the decay $\psi' \to J+2\pi$" *PR*, **D12**: 2765–2767.
Hehl, F. W. *et al.* (1976) "General relativity with spin and torsion: foundations and prospects" *RMP*, **48**: 393–416.

Heisenberg, W. (1925) "Über quantentheoretische Umdeutung kinematischer und mechanischer Beziehung" *ZP*, **33**: 879–883.
Heisenberg, W. (1943a) "Die 'beobachtbaren Grössen' in der Theorie der Elementarteilchen" *ZP*, **120**: 513–538.
Heisenberg, W. (1943b) "Die 'beobachtbaren Grössen' in der Theorie der Elementarteilchen. II" *ZP*, **120**: 673–702.
Heisenberg, W. (1944) "Die 'beobachtbaren Grössen' in der Theorie der Elementarteilchen. III" *ZP*, **123**: 93–112.
Heisenberg, W. (1958) "Research on the non-linear spinor theory with indefinite metric in Hilbert Space" in *1958 Annual International Conference on High Energy Physics at CERN* (CERN, Geneva), 119–126.
Heisenberg, W. (1960) "Recent research on the nonlinear spinor theory of elementary particles" *Proceedings of the 1960 Annual International Conference on High Energy Physics at Rochester* (Interscience), 851–857.
Hesse, M. B. (1980) *Revolutions and Reconstructions in the Philosophy of Science* (Harvester).
Higgs, P. W. (1964a) "Broken symmetries, massless particles and gauge fields" *PL*, **12**: 132–133.
Higgs, P. W. (1964b) "Broken symmetries and the masses of gauge bosons" *PRL*, **13**: 508–509.
Hill, C. (2005) "Conjecture on the physical implications of the scale anomaly" *Fermilab-Conf-05/482-T*.
Hofstadter, R. (1956) "Electron scattering and nuclear structure" *RMP*, **28**: 214–254.
Hofstadter, R. and McAllister, R. W. (1955) "Electron scattering from the proton" *PR*, **98**: 217–218.
Hofstadter, R., Fechter, H. R., and McIntyre, J. A. (1953) "Scattering of high-energy electrons and the method of nuclear recoil" *PR*, **a2**: 978–987.
Houard, J. C. and Jouvet, B. (1960) "Etude d'un modèle de champ à constante de renormalisation nulle" *NC*, **18**: 466–481.
Iagolnitzer, D. and Starpp, H. P. (1969) "Macroscopic causality and physical region analyticity in the S-matrix theory" *CMP*, **14**: 15–55.
Ikeda, M., Ogawa, S. and Ohnuki, Y. (1959) "A possible symmetry in Sakata's model for bosons-baryons system. II" *PTP*, **22**: 715–724.
Jackiw, R. and Preparata, G. (1969) "Probe for the constituents of the electromagnetic current and anomalous commutators" *PRL*, **22**: 975–981.
Jackiw, R. and Rebbi, C. (1976) "Vacuum periodicity in a Yang-Mills quantum theory" *PRL*, **37**: 172–175.
Jackiw, R., Royen, R. V., and West, G. B (1970) "Measuring light-cone singularities" *PR*, **D12**: 2765–2767.
Johnson, K. (1961) "Solution of the equations for the Green's functions of a two dimensional relativistic field theory" *NC*, **20**: 773–790.
Johnson, K. and Low, F. (1966) "Current algebra in a simple model" *PTP*, **37–38**: 74–93.
Jost, R. (1947) "Über die falschen Nullstellen der Eigenwerte der S-matrix" *Helv. Phys. Acta.*, **20**: 256–266.
Kadanoff, L. P. (1966) "Scaling laws for Ising models near T_c" *Physics*, **2**: 263–272.
Kamefuchi, S. (1960) "On Salam's equivalence theorem in vector meson theory" *NC*, **18**: 691–696.

Kant, I. (1783). *Prolegomena zu einer jeden künftigen Metaphysik die als Wissenschaft wird auftreten können* (J. F. Hartknoch, Riga; English translation: Hackett, 1977).
Kendall, H. (1991) "Deep inelastic scattering: experiment on the proton and the observation of scaling" *RMP*, **63**: 597–614.
Kibble, T. W. B. (1961) "Lorentz invariance and the gravitational field" *JMP*, **2**: 212–221.
Klein, O. (1959) "On the systematics of elementary particles" *Ark. Fys. Swedish Academy of Sciences*, **16**: 191–196.
Kogut, J. and Susskind. L. (1974) "Vacuum polarization and the absence of free quarks in four dimensions" *PR*, **D9**: 3501–3512.
Kogut, J. and Susskind, L. (1975) "Hamiltonian formulation of Wilson's lattice gauge theories" *PR*, **D11**: 395–408.
Komar, A. and Salam, A. (1960) "Renormalization problem for vector meson theories" *NP*, **21**: 624–630.
Kramers, H. A. (1944) "Fundamental difficulties of a theory of particles" *Ned. Tijdschrift Voor Natuurkunde*, **11**: 134–147.
Kronig, R. (1946) "A supplementary condition in Heisenberg's theory of elementary particles" *Physica*, **12**: 543–544.
Kuhn, T. S. (1970) *The Structure of Scientific Revolutions* (Second enlarged edition, University of Chicago Press).
Kuhn, T. S. (2000) *The Road Since Structure* (University of Chicago Press).
Ladyman, J. (1998) What is structural realism? *Stud. Hist. Philos. Sci.*, **29**: 409–424.
Lakatos, I. (1970) "Falsification and the methodology of scientific research programmes" in *Criticism and the Growth of Knowledge* (eds. Lakatos, I. and Musgrave, A.; Cambridge University Press) 91–195.
Lakatos, I. (1971) "History of science and its rational reconstructions" in *Boston Studies in the Philosophy of Science* (eds. Cohen, R. and Wartofsky, D.; Reidel) VIII: 91–136.
Landau, L. D. (1955) "On the quantum theory of fields" in *Niels Bohr and the Development of Physics* (ed. Pauli, W.; Pergamon) 52–69.
Landau, L. D. (1959) "On analytic properties of vertex parts in quantum fined theory" *NP*, **13**: 181–192.
Landau, L. D. and Pomeranchuck, I. (1955) "On point interactions in quantum electrodynamics" *DAN*, **102**: 489–491.
Landau, L. D., Abrikosov, A. A., and Khalatnikov, I. M. (1954a) "The removal of infinities in quantum electrodynamics" *DAN*, **95**: 497–499.
Landau, L. D., Abrikosov, A. A., and Khalatnikov, I. M. (1954b) "An asymptotic expression for the electron Green function in quantum electrodynamics" *DAN*, **95**: 773–776.
Landau, L. D., Abrikosov, A. A. and Khalatnikov, I. M. (1954c) "An asymptotic expression for the photon Green function in quantum electrodynamics" *DAN*, **95**: 1117–1120.
Landau, L. D., Abrikosov, A. A., and Khalatnikov, I. M. (1954d) "The electron mass in quantum electrodynamics" *DAN*, **96**: 261–263.
Landau, L. D., Abrikosov, A. A., and Kalatnikov, I. M. (1956) "On the quantum theory of fields" *NC (Suppl.)*, **3**: 80–104.
Lattes, C. M. G., Muirhead, H., Occhialini, G. P. S., and Powell, C. F. (1947) "Processes involving charged mesons" *Nature*, **159**: 694–697.

Lattes, C. M. G., Occhialini, G. P. S., and Powell, C. F. (1947) "Observations on the tracks of slow Mesons in photographic emulsions, part I, II," *Nature*, **160**: 453–456; 486–492.

Lee, B. W. (1972) "Perspectives on theory of weak interactions" in *Proceedings of the XVI International Conference on High Energy Physics, September 6–13, 1972* (eds. Jackson, J. D. and Roberts, A.; National Accelerator Laboratory, Batavia) vol. 4: 249–305.

Lee, T. D. (1954) "Some special examples in renormalizable field theory" *PR*, **95**: 1329–1334.

Lee, T. D. and Yang, C. N. (1956) "Question of parity conservation in weak interactions" *PR*, **104**: 254–258.

Lee, T. D. and Yang, C. N. (1962) "Theory of charge vector mesons interacting with the electromagnetic field" *PR*, **128**: 885–898.

Lee, T. D., Weinberg, S., and Zumino, B. (1967) "Algebra of fields" *PRL*, **18**: 1029–1032.

Lee, Y. K., Mo, L. W., and Wu, C. S. (1963) "Experimental test of the conserved vector current theory on the beta spectra of B^{12} and N^{12}" *PRL*, **10**: 253–258.

Lehmann, H., Symanzik, K., and Zimmerman, W. (1955) "Zur Formalierung quantisierter Feldtheorie," *NC*, **10**(1): 205–225.

Leutwyler, H. (1968) "Is the group $SU(3)_l$ an exact symmetry of one-particle-states?" Preprint (University of Bern).

Leutwyler, H. and Stern, J. (1970) "Singularities of current commutators on the light cone" *NP*, **B20**: 77–101.

Levin, E. M. and Frankfurt, L. L. (1965) "The quark hypothesis and relations between cross sections at high energies" *JETP Letters*, **2**(3): 65–67.

Lewontin, R. (1998) An exchange on higher superstition with Paul R. Gross and Norman Levitt, *New York Review of Books*, **45**(19): 59–60.

Lipkin, H. (1997) "Quark models and quark phenomenology" in *The Rise of the Standard Model* (eds. Hoddeson, L., Brown, L., Riordan, M., and Dresden, M.; Cambridge University Press) 542–560.

Llewellyn-Smith, C. H. (1971) "Inelastic lepton scattering in gluon models" *PR*, **D4**: 2392–2397.

Low, F. E. (1954) "Scattering of light of very low frequency by system of spin ½" *PR*, **96**: 1428–1432.

Low, F. E. (1962) "Bound states and elementary particles" *NC*, **25**: 678–684.

Mack, G. (1968) "Partially conserved dilatation current" *NP*, **B5**: 499–507.

Maglic, B. C., Alvarez, L. W., Rosenfeld, A. H., and Stevenson, M. L. (1961) "Evidence for a T=0 three-pion resonance" *PRL*, **7**(1): 78–182.

Maki, Z., Ohnuki, Y., and Sakata, S. (1965) "Remarks on a new concept of elementary particles and the method of the composite model" *PTP (Supplement:* A Commemoration Issue for the 30th Anniversary of the Meson Theory by Dr. H. Yukawa) 406–415.

Mandelstam, S. (1958) "Determination of the pion-nucleon scattering amplitude from dispersion relations and unitarity. General theory" *PR*, **112**: 1344–1360.

Mandelstam, S. (1968a) "Feynman rules for electromagnetic and Yang-Mills fields from the gauge-independent field-theoretic formalism" *PR*, **175**: 1580–1603.

Mandelstam, S. (1968b) "Feynman rules for the gravitational field from the coordinate-independent field-theoretic formalism" *PR*, **175**: 1604–1623.

Maxwell, G. (1970a) "Structural realism and the meaning of theoretical terms" in *Analyses of Theories and Methods of Physics and Psychology*, Minnesota Studies in the Philosophy of Science, IV (eds. Winokur, S. and Radner, M.; University of Minnesota Press) 181–192.

Maxwell, G. (1970b) "Theories, perception and structural realism" in *Nature and Function of Scientific Theories*, University of Pittsburgh Series in the Philosophy of Science, IV (ed. Colodny, R., University of Pittsburgh Press) 3–34.

Muta, T. (1979a) "Deep inelastic scattering beyond the leading order in asymptotically free gauge theories" in *Proceedings of the 19th International Conference on High Energy Physics* (August 23–30, 1978, Tokyo; eds. Homma, S., Kawaguchi, M., and Miyazawa, H.; Physical Society of Japan) 234–236.

Muta, T. (1979b) "Note on the renormalization-prescription dependence of the quantum-chromodynamic calculation of structure-function moments" *PR*, **D20**: 1232–1234.

Nakano, T. and Nishijima, K. (1953) "Charge independence for V-particles" *PTP*, **10**: 581–582.

Nambu, Y. (1956) "Structure of Green's functions in quantum field theory. II" *PR*, **101**: 459–467.

Nambu, Y. (1957) "Possible existence of a heavy neutral meson" *PR*, **106**: 1366–1367.

Nambu, Y. (1965a) "Dynamical symmetries and fundamental fields" in *Symmetry Principles at High Energies* (eds. Kursunoglu, B., Perlmutter, A. and Sakmar, A.; Freeman) 274–285.

Nambu, Y. (1965b) "A systematics of hadrons in subnuclear physics" in *Preludes in Theoretical Physicse* (eds. de Shalit, A., Feshbach, H., and Van Hove, L.; North-Holland) 133–142.

Nambu, Y. (1969) "Quark model and the factorization of the Veneziano amplitude" in *Proceedings of International Conference on Symmetries and Quark Models, Wayne State University June 1969* (ed. Chaud, R.; Gordon and Breach) 269–278.

Nambu, Y. (1970) "Duality and hadrodynamics" (notes prepared for the Copenhagen high energy symposium, August 1970, unpublished. Texts now available from *Broken Symmetry*, eds. Eguchi, T. and Nishijima, K.; World Scientific, 1995) 280–301.

Nambu, Y. and Jona-Lasinio, G. (1961a) "A dynamical model of elementary particles based on an analogy with superconductivity. I" *PR*, **122**: 345–358.

Nambu, Y. and Jona-Lasinio, G. (1961b) "A dynamical model of elementary particles based on an analogy with superconductivity. II" *PR*, **124**: 246–254.

Ne'eman, Y. (1961) "Derivation of strong interactions from a gauge invariance" *NP*, **26**: 222–229.

Newman, M. H. A. (1928) "Mr Russell's causal theory of perception" *Mind*, **37**: 137–148.

Nielsen, H. and Olesen, P. (1973) "Vortex-line models for dual strings" *NP*, **B61**: 45–61.

Nielsen, H. and Susskind, L. (1970) *TH 1230-CERN* (Preprint, September 1970).

Nielsen, H. (1971) "Connection between the Regge trajectory universal slope α' and the transverse momentum distribution of partons in planar Feynman diagram model" *PL*, **35B**: 515–518.

Nishijima, K. (1955) "Charge independence theory of V particles" *PTP*, **13**: 285–304.

Nishijima, K. (1957) "On the asymptotic conditions in quantum field theory" *PTP*, **17**: 765–802.

Nishijima, K. (1958) "Formulation of field theories of composite particles" *PR*, **111**: 995–1011.

Nissani, N. (1984) "SL (2, C) gauge theory of gravitation: conservation laws" *PR*, **109**: 95–130.

Ogawa, S. (1959) "A possible symmetry in Sakata's composite model" *PTP*, **21**: 209–211.
Ohnuki, Y. (1960) "Composite model of elementary particles" in *Proceedings of the 1960 Annual International Conference on High Energy Physics at Rochester* (eds. Sudarshan, E. C. G., Tinlot, J. H., and Melissinos, A. C.; University of Rochester) 843–850.
Olive, D. I. (1964) "Exploration of S-matrix theory," *PR*, **135B**: 745–760.
Pais, A. (1953) "Isotopic spin and mass quantization" *Physica*, **19**: 869–887.
Pais, A. (1962) "Theorem on local action on lepton currents" *PRL*, **9**: 117–120.
Pais, A. (1964) "Implications of spin-unitary spin independence" *PRL*, **13**: 175–177.
Pais, A. (1965) "Comments" in *Recent Developments in Particle Symmetries* (ed. Zichichi, A.; Academic Press) 406.
Pais, A. (1986) *Inward Bound* (Oxford University Press).
Panofsky, W. (1968) "Low electrodynamics, elastic and inelastic electron (and muon) scattering" in *Proceedings of the 14th International Conference on High Energy Physics, Vienna, August 28–September 5, 1968* (eds. Prentki, J. and Steinberger, J.; Geneva, CERN) 23–39.
Panofsky, W. and Allton, E. A. (1968) "Form factor of the photopion matrix element at resonance" *PR*, **110**: 1155–1165.
Parisi, G. (1973) "Deep inelastic scattering in a field theory with computable large-momenta behavior" *NC*, **7**: 84–87.
Pauli, W. (1953) Two letters to Abraham Pais, see Pauli (1999); two seminar lectures, reported in Paolo Gulmanelli (1954).
Pauli, W. (1999) *Wissenschaftlicher Briefwechsel*, Vol. IV, Part II (Springer-Verlag, edited by K. V. Meyenn). Letter [1614]: to A. Pais (July 25, 1953); letter [1682]: to A. Pais (December 6, 1953); letter [1727]: to C. N. Yang (February 1954).
Pecceei, R. D. and Quinn, H. R. (1977) "CP conservation in the presence of pseudoparticles" *PRL*, **38**: 1440–1443.
Perkins, D. H. (1972) "Neutrino interactions" in *Proceedings of the XVI International Conference on High Energy Physics, September 6–13, 1972* (eds. Jackson, J. D. and Roberts, A.; National Accelerator Laboratory, Batavia), vol. 4: 189–247.
Peruzzi, I. *et al.* (1976) "Observation of a narrow charged state at 1876 MeV/c^2 decaying to an exotic combination of Kππ" *PRL*, **37**: 569–571.
Peterman, A. and Stueckelberg, E. C. G. (1951) "Restriction of possible interactions in quantum electrodynamics" *PR*, **82**: 548–549.
Pevsner, A. *et al.* (1961) "Evidence for a three-pion resonance near 550 Mev," *PRL*, **7**: 421–423.
Pickering, A. (1984) *Constructing Quarks* (Edinburgh University Press).
Poincaré, H. (1902) *La science et l'hypothese* (Flammarion).
Politzer, H. D. (1973) "Reliable perturbative results for strong interactions?" *PRL*, **30**: 1346–1349.
Politzer, H. D. (1974) "Asymptotic freedom: An approach to strong interactions" *Phys. Rept.*, **14**: 129–180.
Politzer, D. (2004) "The dilemma of attribution" Nobel lecture delivered in December 2004; for the text, see *RMP*, **77**(3): 851–856.
Polyakov, A. M. (1974) "Particle spectrum in quantum field theory" *JETPL*, **20**: 194–195.
Putnam, H. (1978) *Meaning and the Moral Sciences* (Routledge and Kegan Paul).
Quigg, and Rosner, J. (1979) "Quantum mechanics with applications to quarkonium" *PR*, **56**: 167–235.

Quine, W. V. O. (1951) "Two Dogmas of Empiricism" *Philoso. Rev.*, **60**: 20–43.
Redhead, M. (2001) "The intelligibility of the universe" in *Philosophy at the New Millennium* (ed. O'Hear, A.; Cambridge University Press) 73–90.
Rech, H. and Schlieder, S. (1961) "Bemerkungen zur unitaraquivalenz von Lorentzinvaranten Feldern" *NC*, **22**: 1051–1068.
Regge, T. (1958a) "Analytic properties of the scattering matrix" *NC*, **8**: 671–679.
Regge, T. (1958b) "On the analytic behavior of the eigenvalue of the S-matrix in the complex plane of the energy" *NC*, **9**: 295–302.
Regge, T. (1959) "Introduction to complex orbital momenta" *NC*, **14**: 951–976.
Regge, T. (1960) "Bound states, shadow states and Mandelstam representation," *NC*, **18**: 947–956.
Reichenbach, H. (1951) *The Rise of Scientific Philosophy* (University of California Press).
Rosenbluth, M. (1950) "High energy elastic scattering of electrons on protons" *PR*, **79**: 615–619.
Rosner, J. (1969) "Graphical Form of duality" *PRL*, **22**: 689–692.
Russell, B. (1927) *The Analysis of Matter* (Kegan Paul).
Sakata, S. (1956) "On the composite model for the new particles" *PTP*, **16**: 686–688.
Sakata, S. (1963) "Concluding remarks at the meeting on Models and Structures of Elementary Particles, Hiroshima University, March 1963" Soryushiron Kenkyu (mimeographed circular in Japanese), 28, 110; English translation, *PTP* (*Supplement*) **50** (1971): 208–209.
Sakita, B. (1964) "Supermultiplets of elementary particles" *PR*, **B136**: 1756–1760.
Sakurai, J. J. (1960) "Theory of strong interactions" *AP*, **11**: 1–48.
Sakurai, J. (1969) "Vector-meson dominance and high-energy electron-proton inelastic scattering" *PRL*, **22**: 981–984.
Salam, A. (1960) "An equivalence theorem for partially gauge-invariant vector meson interactions," *NP*, **18**: 681–690.
Salam, A. (1962) "Lagrangian theory of composite particles" *NC*, **25**: 224–227.
Salam, A. (1968) "Weak and electromagnetic interactions" in *Elementary Particle Theory: Relativistic Group and Analyticity, Proceedings of Nobel Conference 8* (ed. Svartholm, N.; Almqvist & Wiksell) 367–377.
Salam, A. and Ward, J. C. (1959) "Weak and electromagnetic interactions" *NC*, **11**: 568–577.
Schiff, L. I. (1949) *"Microwave Laboratory technical report no. 102"* Stanford University, November 1949.
Schlick, M. (1918) *General Theory of Knowledge* (trans. by Blumberg, A. E. and Feigl, H.; Springer-Verlag).
Schroer, B (1997) "Motivations and physical aims of algebraic QFT" www.esi.ac.at/preprints/esi371.pdf
Schweber, S. (1997) "A historical perspective on the rise of the standard model" in *The Rise of the Standard Model* (eds. Hoddeson, L., Brown, L., Riordan, M., and Dresden, M.; Cambridge University Press) 645–684.
Schwinger, J. (1957) "A theory of the fundamental interactions" *AP*, **2**: 407–434.
Schwinger, J. (1959) "Field theory commutators" *PRL*, **3**: 296–297.
Schwinger, J. (1962a) "Gauge invariance and mass" *PR*, **125**: 397–398.
Schwinger, J. (1962b) "Gauge invariance and mass. II" *PR*, **128**: 2425–2429.
Schwinger, J. (1965) "The future of fundamental physics" in *Nature of Matter: Purposes of High Energy Physics* (ed. Yuan, L. C. L.; Brookhaven National Laboratory) 23.

Silvestrini, V. (1972) "Electron-positron interactions" in *Proceedings of the XVI International Conference on High Energy Physics, September 6–13, 1972* (eds. Jackson, J. D. and Roberts, A.; National Accelerator Laboratory, Batavia), vol. 4: 1–40.

Simons, P. (1994) "Particulars in particular clothing: three trope theories of substance" *Philos. Phenom. Res.*, **54**: 553–574.

Slavnov, A. (1972) "Ward identities in gauge theories" *Theor. Math. Phys.*, **19**: 99–104.

Stapp, H. P. (1962a) "Derivation of the CPT theorem and the connection between spin and statistics from postulates of the S-matrix theory," *PR*, **125**: 2139–2162.

Stapp, H. P. (1962b) "Axiomatic S-matrix theory," *RMP*, **34**: 390–394.

Stapp, H. P. (1965) "Space and time in S-matrix theory," *PR*, **B139**: 257–270.

Stapp, H. P. (1968) "Crossing, hermitian analyticity, and the connection between spin and statistics," *JMP*, **9**: 1548–1592.

Stein, H. (1989) "Yes, but some skeptical remarks on realism and anti-realism" *Dialectica*, **43**: 47–65.

Straumann, N. (2000) "On Pauli's invention of non-abelian Kaluza-Klein theory in 1953" in *The Ninth Marcel Grossmann Meeting. Proceedings of the MGIXMM Meeting held at The University of Rome "La Sapienza", July 2–8 2000* (eds.: Gurzadyan, V. G., Jantzen, R. T. and Remo Ruffini, R.; World Scientific) Part B: 1063–1066.

Strawson (1959) *Individuals* (Anchor Books, Doubleday & Company).

Stueckelberg, E. C. G. and Peterman, A. (1953) "La normalisation des constances dans la théorie des quanta" *Helv. Phys. Acta.*, **26**: 499–520.

Sudarshan, E. C. G. and Marshak, R. E. (1958) "Chirality invariance and the universal Fermi interaction" *PR*, **109**: 1860–1862.

Susskind, L. (1968) "Models of self-induced strong interactions" *PR*, **165**: 1535–1546.

Susskind, L. (1969) "Harmonic-oscillator analogy for the Veneziano model" *PRL*, **23**: 545–547.

Susskind, L. (1970) "Structure of hadrons implied by duality" *PR*, **D1**: 1182–1186.

Susskind, L. (1997) "Quark confinement" in *The Rise of the Standard Model* (eds. Hoddeson, L., Brown, L., Riordan, M., and Dresden, M.; Cambridge University Press) 233–242.

Sutherland, D. G. (1967) "Current algebra and some non-strong mesonic decays" *NP*, **B2**: 433–440.

Symanzik, K. (1970) "Small distance behavior in field theory and power counting" *CMP*, **18**: 227–246.

Symanzik, K. (1971) "Small-distance behavior analysis and Wilson Expansions" *CMP*, **23**: 49–86.

Symanzik, K. (1972) "On theories with massless particles" in *Renormalization of Yang-Mills Fields and Applications to Particle Physics, Proceedings of Marseille Conference* (June 19–23 1972, ed. Korthals-Altes, C. P.; Centre de Physique Theorique).

Symanzik, K. (1973) "Infrared singularities and small-distance-behavior analysis" *CMP*, **34**: 7–36.

Taylor, J. C. (1958) "Beta decay of the pion" *PR*, **110**: 1216.

Taylor, J. C. (1971) "Ward identities and the Yang-Mills field" *NP*, **B33**: 436–444.

Taylor, R. E (1967) "Nucleon form factors above 6 GeV" in *Proceedings of the 1967 International Symposium on Electron and Photon Interactions at High Energies* (Stanford, September 5–9, 1967) (Printed by International Union of Pure and

Applied Physics, US Atomic Energy Commission with a Foreword by S. M. Berman) 78–101.
Taylor, R. E. (1968) "To: Program Coordinator" (July 26, 1968) Supplement No. 2 to SLAC Proposal 4-b.
Taylor, R. E (1975) "Inelastic electron-nucleon scattering" in *Proceedings of the International Symposium on Lepton and Photon Interactions at High Energies*, Stanford, 21–27 August 1975 (ed. Kirk, W. T.; SLAC) 679–708.
Taylor, R. (1991) "Deep inelastic scattering: the early years" *RMP*, **63**: 573–595.
't Hooft, G. (1971a) "Renormalization of massless Yang-Mills fields" *NP*, **B33**: 173–199.
't Hooft, G. (1971b) "Renormalizable Lagrangians for massive Yang-Mills fields" *NP*, **B35**: 167–188.
't Hooft, M. (1972) "Remarks after Symansik's presentation" in *Renormalization of Yang-Mills Fields and Applications to Particle Physics, Proceedings of Marseille Conference* (19–23 June 1972, ed. Korthals-Altes, C. P.; Marseille, Centre de Physique Theorique), CERN preprint Th. 1666, May, 1973.
't Hooft, G. (1974) "Magnetic monopoles in unified gauge theories" *NP*, **B79**: 276–284.
't Hooft, G. (1975) "Gauge theories with unified weak electromagnetic and strong interactions" in *E. P. S. International Conference on High Energy Physics, Palermo, 23–28 June, 1975* (ed. Zichichi, A.; Editrice Compositori).
't Hooft, G. (1976a) "Symmetry breaking through Bell-Jackiw anomalies" *PRL*, **37**: 8–11.
't Hooft, G. (1976b) "Computation of the quantum effects due to a four-dimensional pseudoparticle" *PR*, **D14**: 3432–3450.
't Hooft, G. (1998) "The creation of quantum chromodynamics" in *The Creation of Quantum Chromodynamics and the Effective Energy* (ed. Lipatov, L. N.; World Scientific) 11–50.
't Hooft and Veltman, M. (1972a) "Renormalization and regularization of gauge fields" *NP*, **B44**: 189–213.
't Hooft and Veltman, M. (1972b) "Combinatorics of gauge fields" *NP*, **B50**: 318–353.
Toll, J. (1952) *The dispersion relation for light and its application to problem involving electron pairs*. Unpublished Ph. D. dissertation, Princeton University.
Toll, J. (1956) "Causality and the dispersion relation: logical foundations," *PR*, **104**: 1760–1770.
Tomozawa, Y. (1966) "Axial-vector coupling constant renormalization and the meson-baryon scattering lengths" *NC*, **A46**: 707–717.
Tryon, E. P. (1972) "Dynamical parton model for hadrons" *PRL*, **28**: 1605–1608.
Umezawa, H. and Kamefuchi, S. (1961) "Equivalence theorems and renormalization problem in vector field theory (the Yang-Mills field with non-vanishing masses)" *NP*, **23**: 399–429.
Utiyama, R. (1956) "Invariant theoretical interpretation of interaction" *PR*, **101**: 1597–1607.
Vainshtein, A. I. and Ioffe, B. L. (1967) "Test of Bjorken's asymptotic formula in perturbation theory" *JETP Letter*, **6**: 341–343.
Valatin, J. G. (1954a) "Singularities of electron kernel functions in an external electromagnetic field" *PRS*, **A222**: 93–108.
Valatin, J. G. (1954b) "On the Dirac-Heisenberg theory of vacuum polarization" *PRS*, **A222**: 228–239.

Valatin, J. G. (1954c) "On the propagation functions of quantum electrodynamics" *PRS*, **A225**: 535–548.
Valatin, J. G. (1954d) "On the definition of finite operator quantities in quantum electrodynamics" *PRS*, **A226**: 254–265.
Van Fraassen, B. (1997) "Structure and perspective: philosophical perplexity and paradox" in *Logic and Scientific Methods* (eds. Dalla Chiara M. L. *et al.*; Kluwer) 511–530.
Vaughn, M. T., Aaron, R., and Amado, R. D. (1961) "Elementary and composite particles," *PR*, **124**: 1258–1268.
Veltman, M. (1966) "Divergence conditions and sum rules" *PRL*, **17**: 553–556.
Veltman, M. (1967) "Theoretical aspects of high energy neutrino interactions" *PRS*, **A301**: 107–112.
Veltman, M. (1968) "Perturbation theory of massive Yang-Mills fields" *NP*, **B7**: 637–650.
Veltman, M. (1970) "Generalized Ward identities and Yang-Mills fields" *NP*, **B21**: 288–302.
Veneziano, G. (1968) "Construction of a crossing-symmetric, Regge-behaved amplitude for linearly rising trajectories" *NC*, **57A**: 190–197.
Veneziano, G. (1979) "U(1) without instantons" *NP*, **B159**: 213–224.
Veneziano, G. and Yankielowicz, S. (1982) "An effective Lagrangian for the pure N = 1 supersymmetric Yang-Mills theory" *PL*, **113B**: 231.
Watkins, J. (1958) "Confirmable and influential metaphysics" *Mind*, **67**: 345–365.
Watkins, J. (1975) "Metaphysics and the advancement of science" *Brit. J. Philos. of Sci.*, **26**: 91–121.
Weinberg, S. (1966a) "Current commutator theory of multiple pion production" *PRL*, **16**: 879–883.
Weinberg, S. (1966b) "Current commutator calculation of the K_{l4} form factors" *PRL*, **17**: 336–340.
Weinberg, S. (1966c) "Pion scattering lengths" *PRL*, **17**: 616–621.
Weinberg, S. (1967) "A model of leptons" *PRL*, **19**: 1264–1266.
Weinberg, S. (1978) "A new light boson?" *PRL*, **40**: 223–226.
Weisberger, W. (1965) "Renormalization of the weak axial-vector coupling Constant" *PRL*, **14**: 1047–1055.
Wess, J. (1960) "The conformal invariance in quantum field theory" *NC*, **13**: 1086–1107.
Widom, B. (1965a) "Surface tension and molecular correlations near the critical point", *J. Chem. Phys.*, **43**: 3892–3897.
Widom, B. (1965b) "Equation of state in the neighborhood of the critical point" *J. Chem. Phys.*, **43**: 3898–3905.
Wiik, B. (1976) "Plenary report on new particle production in e^+e^- colliding beams" in *Proceedings of the 18th International Conference on High Energy Physics, Tbilisi, USSR, July 1976* (eds. Bogoliubov *et al.* USSR) **II**: N75–86.
Wiik, B. (1979) "First results from PETRA" in *Proceedings of Neutrino 79* (Bergen, June 18–22, 1979; eds. Haatuft, A. and Jarlskog, C.) **I**: 113–154.
Wilczek, F. (1978) "Problem of strong P and T invariance in the presence of instantons" *PRL*, **40**: 279–282.
Wilson, K. G. (1969) "Non-Lagrangian models of current algebra" *PR*, **179**: 1499–1512.
Wilson, K. G. (1970a) "Model of coupling-constant renormalization" *PR*, **D2**: 1438–1472.

Wilson, K. G. (1970b) "Operator-product expansions and anomalous dimensions in the Thirring model" *PR*, **D2**: 1473–1477.

Wilson, K. G. (1970c) "Anomalous dimensions and the breakdown of scale invariance in perturbation theory" *PR*, **D2**: 1478–1493.

Wilson, K. G. (1971a) "The renormalization group and strong interactions" *PR*, **D3**: 1818–1846.

Wilson, K. G. (1971b) "Renormalization group and critical phenomena. I. Renormalization group and the Kadanof scaling picture" *PR*, **B4**: 3174–3183.

Wilson, K. G. (1971c) "Renormalization group and critical phenomena. II. Phase-space cell analysis of critical behavior" *PR*, **B4**: 3184–3205.

Wilson, K. G. (1972) "Renormalization of a scalar field in strong coupling" *PR*, **D6**: 419–426.

Wilson, K. G. (1974) "Confinement of quarks" *PR*, **D10**: 2445–2459.

Wilson, K. (2004) "The origins of lattice gauge theory" arXiv:hep-lat/0412043v2 *NP, Proceedings Supplements* B**140**: 3–19.

Witten, E (1979a) "Theta vacua in two-dimensional quantum chromodynamics" *NC*, **A51**: 325.

Witten, E (1979b) "Current algebra theorems for the U(1) 'Goldstone boson'" *NP*, **B156**: 269.

Witten, E (1979c) "Dyons of charge e theta/2 pi" *PL*, **B86**: 283–287.

Worrall, J. (1989) "Structural realism: the best of both worlds?" *Dialectica*, **43**: 99–124.

Worral, J. (2007) *Reason in Revolution: a Study of Theory-Change in Science* (Oxford University Press).

Wu, C. S., Ambler, E., Hayward, R. W., Hoppes, D. D., and Hudson, R. P. (1957) "Experimental test of parity conservation in beta decay" *PR*, **105**: 1413–1415.

Wu, S. L. (1997) "Hadron jets and the discovery of the gluon" in *The Rise of the Standard Model* (eds. Hoddeson, L., Brown, L., Riordan, M., and Dresden, M.; Cambridge University Press) 600–621.

Wu, S. L. and Zobernig, G. (1979) "A three-jet candidate (run 447, event 13177)" TASSO Note No. 84 (June **26**, 1979).

Yang, C. N. (1983) *Selected Papers, 1945–1980, with Commentary* (Freeman).

Yang, C. N. and Mills, R. L. (1954a) "Isotopic spin conservation and a generalized gauge invariance" *PR*, **95**: 631.

Yang, C. N. and Mills, R. L. (1954b) "Conservation of isotopic spin and isotopic gauge invariance" *PR*, **96**: 191–195.

Yoh, J. (1997) "The discovery of the b quark at Fermilab in 1977: The experiment Coordinator's story" Fermilab preprint: Fermilab-Conf. 97/432-E (E288).

Yukawa, H. (1935) "On the interaction of elementary particles, I." *Proc. Phys.-Math. Soc. (Japan)*, **17**: 48–57.

Zachariasen, F. and Zemech, C. (1962): "Pion resonances" *PR*, **128**: 849–858.

Zahar, E. (1996) "'Poincaré's structural Realism and his logic of Discovery" in *Henri Poincaré: Science and Philosophy* (eds. Greffe, J., Heinzmann, G., and Lorenz, K.; Academie Verlag) 45–68.

Zahar, E. (2001) *Poincare's Philosophy: From Conventionalism to Phenomenology* (Open Court).

Zee, A. (1973) "Study of renormalization group for small coupling constants" *PR*, **D7**: 3630–3636.

Zichichi, A. *et al.* (1977) "Search for fractionally charged particles produced in proton-proton collisions at the highest ISR energy" *NC*, **40A**: 41.
Zichichi, A. *et al.* (1978) "Search for quarks in proton-proton interactions at $\sqrt{s} = 52.5$ Gev" *NC*, **45A**: 171.
Zimmermann, W. (1958) "On the bound state problem in quantum field theory," *NC*, **10**: 597–614.
Zimmermann, W. (1967) "Local field equation for A^4-coupling in renormalized perturbation theory" *CMP*, **6**: 161–188.
Zumino, B. (1973) "Relativistic strings and supergauges" CERN Preprint TH-1779.
Zweig, G. (1964a) "An SU_3 model for strong interaction symmtry and its breaking: I and II", CERN preprint TH-401 (January 17 1964) and TH-412 (February 21, 1964).
Zweig, G. (1964b) "Fractionally charged particles and SU(6)" in *Symmetries in Elementary Particle Physics* (ed. Zichichi, A.; Academic Press) 192–243.

Author index

Subject index

Made in the USA
Lexington, KY
03 July 2015